EARTHQUAKES

EARTHQUAKES

Fourth Edition

BRUCE A. BOLT

University of California, Berkeley

W. H. Freeman and Company

New York

Acquisitions Editor: Melissa Wallerstein
Senior Marketing Manager: John Britch
Project Editor: Jane O'Neill
Cover and Text Designer: Cambraia (Magalhaes) Fernandes
Illustration Coordinator: Cambraia (Magalhaes) Fernandes
Illustrations: Fine Line Illustrations
Production Coordinator: Susan Wein
Composition: Progressive Information Technologies
Manufacturing: The Maple-Vail Book Manufacturing Group

Library of Congress Cataloging-in-Publication Data
Bolt, Bruce A., 1930–
Earthquakes / Bruce A. Bolt. — 4th ed.
p. cm.
Includes bibliographical references and index.
ISBN 0-7167-3396-X
1. Earthquakes. I. Title.
QE534.2.B64 1999
551.22- -dc21 99-29955
CIP

Printed in the United States of America

First printing 1999

W. H. Freeman and Company
41 Madison Avenue, New York, NY 10010
Houndmills, Basingstoke RG21 6XS, England

Contents

Preface

Earthquakes has been well received in the 20 years since it was first published as *Earthquakes: A Primer.* It has been used as a textbook in both high schools and colleges and enjoys a wide general readership as well. Reviews by professionals and teachers have generally been favorable, and from their suggestions I have clarified the text and in a number of sections have replaced some explanations and illustrations with topical case histories and up-to-date discussion.

It is particularly pleasing that *Earthquakes* has been distributed internationally through the publication of adaptations in several languages: Russian, Spanish, French, Italian, German, Chinese, Greek, and Japanese. I would like to thank the editors and translators of these foreign-language editions for their attention to the spirit of the English text and for helpful special appendixes that have been added; some of the changes have been incorporated in this new edition of *Earthquakes*.

In this fourth edition, there are three significant additions. The most important is a completely new chapter called "Earthquakes and Plate Tectonics," written in response to the suggestion of numerous users of the book. It helps focus on the central place that plate tectonics now has in geological thinking as well as providing coherent explanations of so many earthquake features. It should be remembered, however, there is some circular-ity in such explanations because the pattern of earthquake occurrence and the mechanisms of their sources played a crucial role in the development of the plate tectonic model itself. It should be no surprise, therefore, that much of observational seismology can be inferred from the theory of plate tectonics.

The second important feature in this edition, along with the updating of references, is the inclusion of a comprehensive list of World Wide Web addresses. The reader can thereby use the Internet to access many of the imaginative Web pages that have been developed about the occurrence, study, and hazard of earthquakes, tsunamis, and volcanoes. Because of the rapid technical advances in scientific communication since the last edition of *Earthquakes* appeared a few years ago, now any interested person can, within a few hours, read considerable detail about a major earthquake anywhere in the world.

Overall, the reader will find a variety of fresh material in this revised and expanded *Earthquakes* with more emphasis on historical and geological background, the third significant added feature in this edition. Chapter 2 provides a more substantial description of earthquake seismicity. Up-to-date ways of recording earthquakes are explained in Chapter 3. A fuller account of the extraordinary deep-focus earthquakes is given in Chapter 4. The measurement of crustal strain using Global Positioning Satellites is included in Chapter 6. The new Chapter 7 recounts the history of the formulation of plate tectonics, beginning with Sir Charles Lyell, and it classifies earthquakes in terms of their plate geometry. Chapter 9 describes the severe tsunami that washed ashore on Hokkaido, Japan, in 1993. Chapter 11 has additional material explaining the difference between earthquake hazard and risk. Chapter 12 recounts the tragic 1995 Sakhalin and 1997 Assisi earthquakes, both of which involved critical questions on retrofit policy. The appendix on world earthquakes and the library reference list have been brought up to date. An improved projection net (the equal-angle Wulff net) for constructing fault properties from distant wave measurements is now provided in Appendix F. Finally, a comprehensive list of World Wide Web addresses is included.

I appreciate the help of a number of fellow teachers of seismology in the preparation of this new edition. In particular, special reviewers Charles J. Ammon, Gregory C. Beroza, A. C. Johnson, Robert C. Lieberman, George H. Myer, and R. L. Nowack provided very helpful critiques of the third edition, and Douglas Dreger kindly commented on some rewritten features.

Preface to *Earthquakes: A Primer*

Oddly enough, I felt my first earthquake in 1959 in the Sydney suburb of Coogee, Australia, while I was sitting quietly in my study. It was the place of occurrence that was so unlikely because Australia is not very active seismically, although earthquakes do occur in various areas in that great continent. On that day in September 1959, a sudden movement had occurred in the Earth's crust some 350 kilometers away from Sydney, near the Snowy Mountains in southern New South Wales.

In the years since, I have felt many earthquakes: quite a few at my home in Berkeley, California; aftershocks of the main earthquake in 1971 in San Fernando, California; and many earthquakes of the famous swarm of hundreds of thousands that occurred in Matsushiro, Japan, in 1965. I have felt earthquake jolts in Tokyo, in the Venezuelan Andes, in Seattle, in Romania, and elsewhere. So far, I have not experienced a major earthquake.

Seismology is truly an international science. As I was putting the finishing touches to this manuscript, a tragic earthquake struck Romania. At about 9:30 p.m. on March 4, 1977, a magnitude 7.2 earthquake spread out from a center under the Carpathian Mountains and heavily damaged part of Bucharest, Ploesti, and other Romanian towns. It was probably the strongest earthquake to occur in Central Europe in modern times: its shaking was felt as far away as Rome and Moscow, and damage occurred in Yugoslavia and Bulgaria. About 2000 persons were killed. In an effort to learn from the disaster, in order to reduce the hazards in future earthquakes, a group of seismologists and engineers, of which I was a member, left the United States for Romania within a few days. Some of the lessons we learned will be incorporated in seismological knowledge and applied as the years go by.

It is the task of the seismologist to study all aspects of earthquakes, their causes, their occurrence, and their properties. The seismologist also makes use of seismic waves to study the interior of the Earth, to assist in oil and mineral exploration, and to detect secret underground nuclear explosions in distant places. Of course, earthquakes are also of strong interest to other professional people, particularly engineers, architects, city and regional planners, and even politicians. There is an abiding fascination with earthquakes among the public, not only in earthquake country such as California, New Zealand, and Japan, but in areas where earthquakes are never experienced as well.

But in spite of the continuing demand from colleges, schools, professional people, and nonspecialists for popular books on earthquakes, very few are available that make easy reading for the person who is curious but has little or no background in earth science. The aim of this book is to provide a short, simple, and up-to-date account of our present knowledge of earthquakes that will be of general interest to people from various countries. I have used some of the more interesting studies of earthquakes as illustrations; they do not require any special mathematical or technical knowledge. Naturally, I have drawn on my 15 years' experience as Director of the Seismographic Stations at the University of California at Berkeley. Although this earthquake observatory is in great part concerned with advanced research on earthquakes, it also has the important responsibility of providing details on the seismicity of California. Numerous requests for earthquake information come in every week. Some are letters, and others are made in person by visitors. In particular, children write in for help with school projects. After finishing this book, readers will be able to answer questions about the causes of earthquakes—where and when to expect them—and to specify what protective measures can be taken against them. As a challenge, a list of questions is given at the back of the book. The answers to the questions can be obtained from the pages of the book. Important terms are defined in the glossary, and a list of suggested further readings is given in the bibliography. The appendixes provide useful but more specialized details in certain aspects of seismology.

A number of colleagues and friends have furnished me with photographs, references, examples, and criticism. I would like to extend to all of them my personal thanks. I am particularly indebted to Dr. P. Byerly, Dr. D. Boore, Dr. L. Weiss, Dr. L. Drake, S. J., Dr. F. Wu, Dr. D. Tocher, and Mr. T. R. Monteath either for material or for reading parts of the manuscript and making valuable suggestions for its improvement.

The basic draft was written while I was a guest of the Departamento de Fisica de la Tierra y del Cosmos, University of Barcelona. I am most grateful to Professor A. Udias, S. J., for his hospitality and comments. The task of preparing figures and tables was greatly lightened by the assistance of Mr. R. Miller and Mr. R. McKenzie. Mrs. A. McClure and Mrs. L. Martin ably typed and proofread the manuscript. My wife, Dr. Beverley Bolt, and my daughter Gillian helped construct the index.

Bruce A. Bolt
Berkeley, California
January 1978

EARTHQUAKES

1 What We Feel in an Earthquake

Damage in Charleston, South Carolina, from the earthquake of August 31, 1886. This earthquake is the largest ever reported in the eastern region of the United States. [Courtesy of J. K. Hillers, USGS.]

A bad earthquake at once destroys the oldest associations; the world, the very emblem of all that is solid, had moved beneath our feet like a crust over a fluid; one second of time has created in the mind a strong idea of insecurity, which hours of reflection would not have produced.

— *Charles Darwin (reflecting on the devastating February 20, 1835, earthquake in Concepción, Chile)*

There are many voices about earthquakes from the past (see Plate 1). The earthquake that has long held a place among the greatest in the modern world is the Lisbon earthquake of November 1, 1755. It owes this distinction to the great destruction in Portugal, the deaths of over 60,000 people, an affected area of more than a million square miles, and the catastrophic sea wave. Actually, reports on the shaking indicate that there were three substantial, separate earthquakes within 3 hours. In the first, Lisbon was shrouded in thick dust, and the screams of injured survivors added to the tragic scene. People of the city ran to any open space, particularly along the banks of the Tagus River. A British merchant ship was among the assembled shipping at the mouth of the Tagus. The captain described the first shock:

> I felt the ship have an uncommon motion, and could not help thinking she was aground, although sure of the depth of the water. As the motion increased, my amazement increased also, and as I was looking round to find out the meaning of the uncommon motion, I was immediately acquainted with the direful cause; when at the instant looking toward the city, I beheld the tall and stately buildings tumbling down, with great cracks and noise.

Terror-stricken Lisboans had congregated on the newly built marble quay, the Cais de Pedra. At the strongest motion, the quay disappeared into the river, taking with it hundreds of people who had there sought sanctuary, and drawing into the turbulent waters small boats moored alongside. The sea waves came and went three times, submerging the lower part of Lisbon, adding to the destruction wrought by the earthquakes. Many of the ships anchored in the river or tied to the wharves and quays were engulfed.

The shaking caused many candles on altars to topple over, setting fire to vestments and church decorations; roofs and timbers collapsed into kitchen and hearth fires. Fanned by a steady northeast wind, the fires burned and spread until a huge pall of smoke from the firestorm hung over the city. Spreading through residences, churches, and palaces, the fire destroyed an invaluable quantity of furniture, pictures, tapestries, books, and manuscripts.

Eighty years later, Charles Darwin, the most influential biologist to have ever lived, encountered a powerful earthquake during his 5-year circumnavigation on the *H.M.S. Beagle* under Captain Fitzroy. On February 20, 1835, they had reached Valdivia on the Pacific coast of Chile (see Figure 1.8). Here they experienced an earthquake, which did great damage to the town and reduced the bigger city of Concepción, 200 miles farther north, almost to a heap of rubble.

Darwin recounted his own impression:

> I was on shore and lying down in the wood to rest myself. It came on suddenly and lasted two minutes (but appeared much longer). . . . There was no difficulty in standing upright, but the motion made me giddy. I can compare it to skating on very thin ice. In the forest no damage was done, and even in the town, where all the houses were of wood, only a few people were injured. Back in the harbour, the tide had been low: the water simply rose to the high-tide level and then receded.

Leaving Valdivia 2 days after the earthquake, the *Beagle* spent 10 days surveying up the dangerous coast to Talcahuano, the port of Concepción. Darwin went ashore to inspect the coast, which "was strewed over with timber and furniture as if a thousand great ships had been wrecked," the effect of a "great wave"* that had inundated the shore. The following day he and Captain Fitzroy rode through Talcahuano and on to Concepción. According to Darwin, "the two towns presented the most awful yet interesting spectacle I ever beheld. . . . In Concepción each house or row of houses stood by itself in a heap or line of ruins: in Talcahuano, owing to the great wave, little more was left than one layer of bricks, tiles and timber with here and there part of a wall yet standing up."

Not many years later in the United States, Dr. Francis L. Parker, surgeon for The Citadel, describes his experience when a damaging earthquake struck Charleston, South Carolina, on August 31, 1886, at 9:51 p.m.:

**Now called a tsunami (see Chapter 9).*

> I had just reached a point on Tradd Street when I heard a roaring sound, apparently in the direction of James Island Cut, which was southwest of where I stood. I made up my mind that a cyclone was coming, and instinctively turned towards the direction indicated, confidently expecting to see the air filled with flying debris from James Island. I then began to feel the vibrations of the earth very distinctly, and realized that they were produced by an earthquake. From that instant, the vibrations increased rapidly, and the ground began to undulate like a sea. I could see perfectly and made careful observations, and I estimate that the waves were at least two feet in height.

A number of scholars involved in scientific earthquake studies, such as the author of this book, have personally felt and been impressed by the forces produced by substantial earthquakes. A famous case is John Milne, often called the "Father of Modern Seismology." Milne, an English mining geologist, accepted a position as Professor of Geology and Mining in Tokyo in 1876. The major turning point in Milne's life, his final conversion to a dedicated seismologist,* was the direct outcome of an earthquake which shook the Tokyo-Yokohama area at about 10 minutes to one o'clock on Sunday morning, February 22, 1880:

> I was asleep. It, however, quickly woke me up and by means of my watch and a lamp which was burning close to my head to enable me to observe the time of the earthquake shocks, I obtained very fairly the time of commencement and the duration of the shock. At the time the house was swaying violently from side to side, and when I thought the motion had finished the distance through which the building was oscillating was so great that I was unable to walk steadily across the floor. Immediately after this I visited two long pendulums which, with various other apparatus, I had been using for making experiments on earthquakes' motion.

Japan continues to suffer from heavy earthquakes (see Chapter 5).

These descriptions raise many questions: Why did the earthquakes occur where and when they did? Why was the intensity so large? What was the huge sea wave? Why does the felt shaking have a defined pattern? What can be the geological cause? Can they be predicted? In the remainder of the book, answers to such questions will emerge. Perhaps sometimes they may be a little more complicated than we might wish, but at least they are more complete and satisfying than possible at the beginning of the twentieth century.

**From the Greek* seismos: *an earthquake.*

The 1906 Eye-Opening San Francisco Earthquake

Let us compare the conditions of today in an "earthquake country" such as California with those at the turn of the century. California's population has grown and spread: in the San Francisco Bay Area alone it has increased from approximately 800,000 to more than 8 million. Structures of new and different architectural types have been introduced into the area. At the beginning of the century, there were no large facilities or essential "lifelines," such as the Golden Gate and other Bay bridges and the Bay Area Rapid Transit system, BART, with its tube under the Bay. While many deaths and injuries will surely occur in a large earthquake, fortunately, most Californians still live in the types of wood-frame houses that can withstand earthquakes. If an extreme earthquake should strike California today when families are at home, casualties may amount to 1000 persons. If it should occur at a busy time, when traffic on the freeways and activity in downtown areas are heavy, the toll would be substantially higher — perhaps more than 5000 people. (Some studies have projected an even higher figure, but such estimates are uncertain, and it is hard not to bias the values.)

In a highly industrialized society, the economic impact now of a major earthquake that damages structures and their contents over a wide area is serious indeed (see Appendix B). Even if there is little structural damage to buildings, often the interior walls and fittings, work areas, electrical and mechanical equipment, and plumbing are broken and out of use for many days, causing high loss of investment and production. In predicting such consequences, it is valuable to examine what actually happened in the great San Francisco earthquake of 1906. As we will see, it remains in many ways the prototype.

April 18, 1906, early morning in California. By the Golden Gate slumbered San Francisco, a city of 400,000 people. Built in a series of economic booms during the previous century, it was a mixture of old and new buildings, all constructed with little heed to natural hazards. Already the downtown area was dotted with steel-frame high-rises,* but it was still dominated by older buildings of wood and unreinforced brick that lined the narrow streets and unprotected openings. Around the wharves were more structures, erected on former marshland that had been used for so long as a construction and garbage dump that it was completely dry. Farther away from the Bay were two- and three-story wooden Victorian homes, more elegant but equally combustible.

**Such as the Spreckels building of 19 stories and the Chronicle building of 16 stories. These high-rise buildings were not so heavily damaged as to be unsafe.*

At 5:12 a.m., a few kilometers from the Golden Gate, a section of rock snapped along the San Andreas fault. The break spread quickly along the fault southward and northward. As this rupture in the rocks grew, seismic waves radiated out through the Earth, shaking the ground surface across a wide area of California and Nevada.

Professor Alexander McAdie, head of the San Francisco Weather Bureau, wrote soon after:

> My custom is to sleep with my watch open, notebook open at the date, and pencil ready — also a hand torch. They are laid out in regular order, torch, watch, book, and pencil. I entered in the book, "Severe shaking lasting forty seconds." I remember getting the minute-hand position after waking, previous to the most violent portion of the shock.

Cool accounts like this one, by reliable eyewitnesses, are as important as they are rare. How many people living in earthquake country make the kind of preparation described by McAdie? But, as McAdie demonstrated, it is possible to think clearly even during strong earthquake shaking. He was calm enough to carry out a simple scientific experiment, which yielded a useful measure of the duration of strong shaking in this great earthquake.

The long rupture of the San Andreas fault (see Figures 1.1 and 1.2) occurring on that day was later mapped by field geologists, who concluded that it extended 430 kilometers, from near Cape Mendocino in Humboldt County to San Juan Bautista, near Hollister, in San Benito County.* The offset along the fault was mainly horizontal and reached 6 meters in Marin County, just north of San Francisco, with the west side moving northwest relative to the east side. The maximum vertical displacement across the fault was less than a meter. Near San Juan Bautista to the south, the fault displacement declined gradually to a few centimeters, then disappeared.

The total area significantly affected by an earthquake of this magnitude was surprisingly small. When the strength of the shaking was mapped, it was seen that the zone of intense ground motion was long, narrow, and parallel to the San Andreas fault. The places of most severe damage (called *meizoseismal* areas) were generally restricted to within a few tens of kilometers of the fault rupture. The earthquake was felt as far north as Oregon and south to Los Angeles, a total distance of 1170 kilometers. In general, the intensity died off markedly toward the east; Winnemucca in Nevada, 540 kilometers from the San Andreas fault, was the easternmost point at which

**Oceanographic evidence of offset submarine canyons and other features on the seafloor suggests the San Andreas runs under the ocean from Point Delgado to Point Arena (see Figure 1.2).*

Figure 1.1 Aerial panorama of part of Marin County, California, showing the majestic San Andreas rift from Bolinas Lagoon (foreground) to Bodega Head (see Figures 1.2 and 1.4). In the 1906 San Francisco earthquake, fault offset in this section ranged from 3 to 6 meters. (An interesting place for a field visit is along Highway 1 to Olema, where there is an "earthquake trail" near the Point Reyes National Seashore Park Headquarters.) [Sutherland photograph.]

the earthquake was reported. In sum, perceptible shaking occurred across an area of about 1 million square kilometers, which is much less, for example, than the felt area of over 5 million square kilometers in the December 16, 1811, New Madrid earthquake in Missouri (see Figure 8.1 and Plate 4) or the 4 million square kilometers reported for the great earthquake centered off the coast of Portugal on November 1, 1755. In the ensuing months of 1906, strong aftershocks were reported in California.*

**Those who would like to judge for themselves the intensity of the 1906 earthquake can do so by reading the* Report of the State Earthquake Investigation Commission. *This fascinating and very readable account has been reprinted at a bargain price by the Carnegie Institution of Washington.*

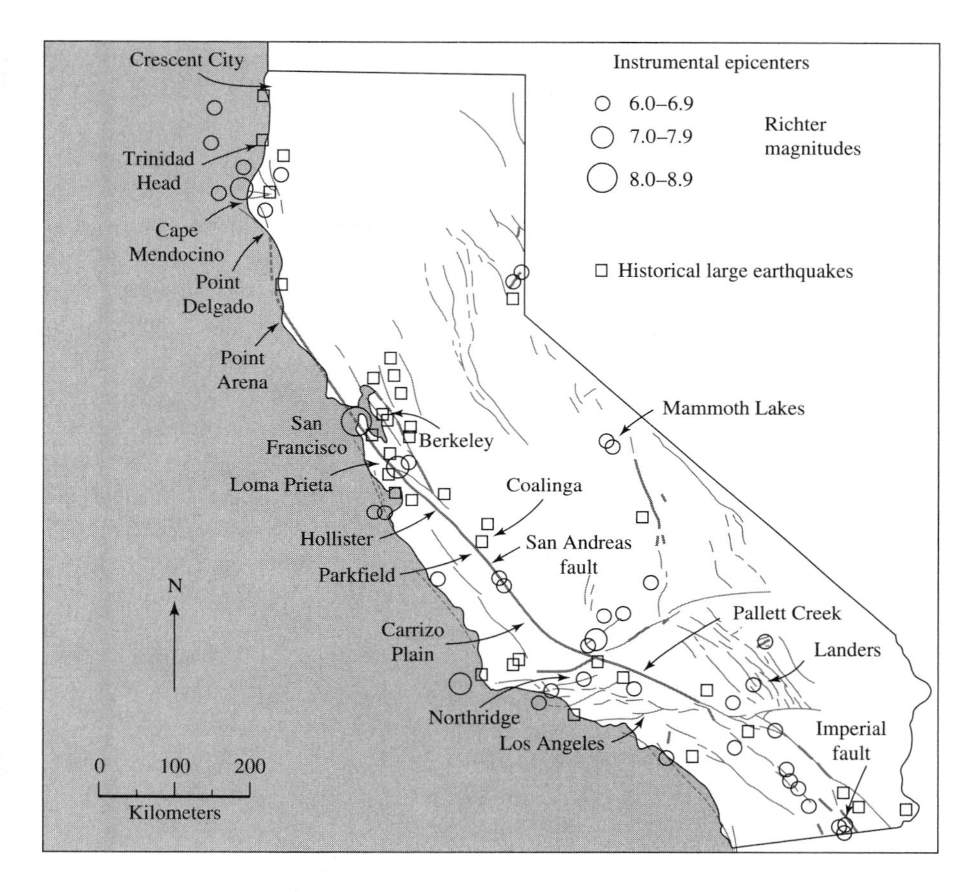

Figure 1.2 Epicenters of the major earthquakes of California in historical times from 1800 to 1998. The "large" earthquakes (open squares) were so designated on the basis of felt effects. Trends of major mapped faults are shown as continuous lines or (if the location is uncertain) as dashed lines on land and as dotted lines under the ocean. The thick black line is the San Andreas fault.

The restricted meizoseismal area resulting from such a large fault rupture is somewhat reassuring for those Californians who do not live in close proximity to the active San Andreas fault zone. It is probably due to a shallow rupture depth along the fault (believed to be about 10 to 15 kilometers) and the high rate of damping of seismic waves in California. There were, however, some exceptional pockets of high intensity that should be considered in the planning and engineering of any construction projects (as there were also in the smaller 1989 Loma Prieta earthquake produced by slip on the San Andreas fault south of San Francisco; see the next section). For example, there was heavy damage at Los Banos, 30 kilometers east of the San Andreas fault, and at the southern end of San Francisco Bay (including the present city of San Jose), where the surficial material is alluvium or recent fluvial deposits. But towns along the east side of San Francisco Bay, such as Berkeley, 25 kilometers east of the San Andreas fault, suffered little damage. Similarly, Sacramento, 120 kilometers east of the rupture, showed no notable destruction, even to the capital dome.* (A resident, Mr. J. A. Marshall, reminisced later that, "I was awakened by my wife's remark that she believed we were having an earthquake. We arose and observed and verified the phenomena.")

An often-quoted total estimate is that 700 lives were lost, with 315 known deaths in San Francisco. In recent years, additional study of city and hospital records and out-of-town newspapers suggests that up to 2500 people died in San Francisco alone. The correct number will no doubt remain controversial. The degree of destruction due to the earthquake itself, moreover, is difficult to estimate because of the fire that broke out almost immediately afterward and raged for 3 days (see Figure 1.3). The fire produced perhaps 10 times more damage than did earthquake shaking.

Sudden massive rupture of rocks like that of 1906 cannot be reliably predicted at present. Nevertheless, because the San Andreas fault is in a tectonically active region (at the boundary of the North America and Pacific plates, as described in Chapter 7) and because direct geodetic observations show that strain is now building up along it, we *can* predict with confidence that another great earthquake will occur along it someday (see Chapter 10). The question often asked by the public is, "What will happen when it does?"

Yet there is reason for optimism in California: although much remains to be done, building codes have been gradually strengthened over the years, originally in response to the 1933 Long Beach earthquake and the 1971 San Fernando earthquake, both in southern California. Building practices and earthquake-resistant design have improved, particularly for schools, hospitals, freeway overpasses, and most major construction projects. This progress — in stark contrast to conditions in many other parts of the world — is important, because studies of earthquake damage have clearly demonstrated that catastrophic numbers of deaths and injuries are caused, not by the trembling itself, but by the collapse of human-made structures that cannot withstand even small amounts of ground shaking. A striking test of current earthquake vulnerability came in October 1989.

**Engineering studies in the 1960s indicated that the capitol, built in 1904, constituted an earthquake risk; strengthening measures were completed in 1982.*

Figure 1.3 Damage in San Francisco in the 1906 earthquake. This is a view of O'Farrell Street before fire swept the area. [Copyright 1906 by W. E. Worden.]

The 1989 Santa Cruz Mountains (Loma Prieta) Earthquake, California

Nothing so focuses the mind on a seismic threat as witnessing a damaging earthquake. The closer the shock, the more intense is our reaction. The large earthquake and the smaller aftershocks that hit the San Francisco Bay Area on October 17, 1989, proved this point. Not only was the mainshock felt strongly by the population of central California, but because of television, its immediate effects were witnessed not only by Americans but also by people all around the world.

The earthquake has been called "the Loma Prieta earthquake," after the mountain peak of 1100 meters high, which lies in the Santa Cruz Mountains, adjacent to and east of the subterranean source of the shaking (see Figure 1.4). Because the eyes of the nation were focused by means of television on the baseball stadium near San Francisco, where a championship baseball game was about to begin, perhaps the "World Series earthquake" is a more incisive designation in two significant ways. Foremost, this

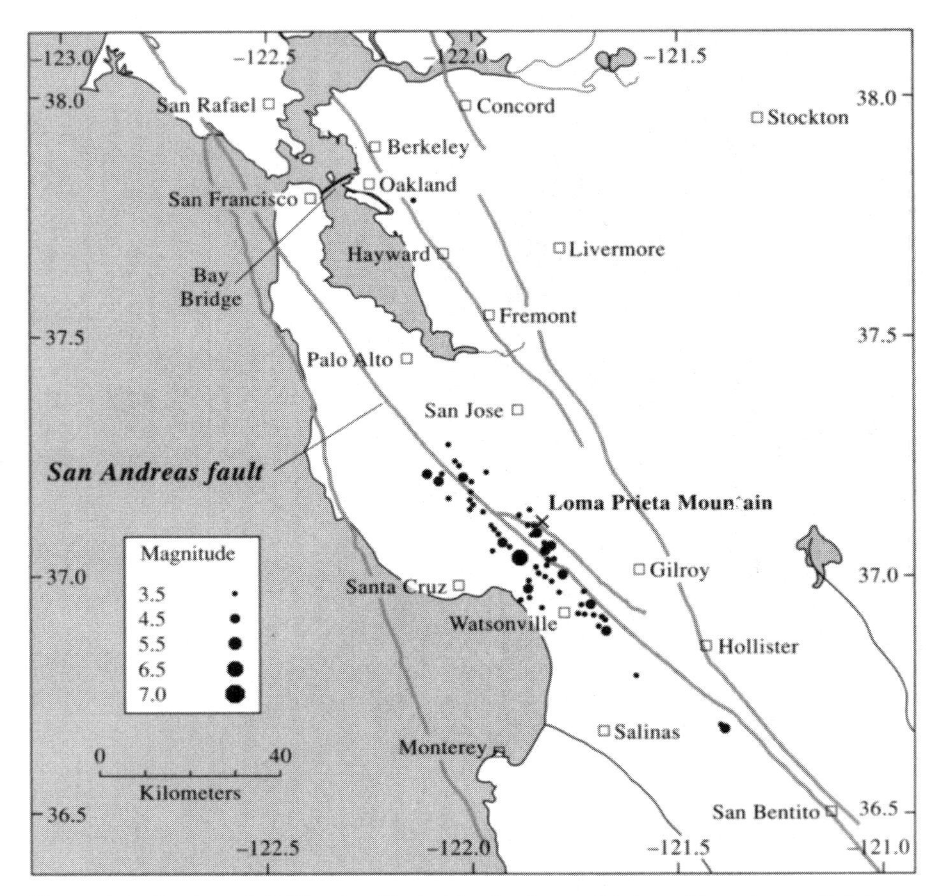

Figure 1.4 Region of California affected by the Loma Prieta earthquake on October 17, 1989. Heavy dots show the epicenters of the largest earthquakes in the sequence. The smooth lines are the main active faults in the region. The short dark line is the Bay Bridge joining San Francisco to Oakland across the Bay.

game provided the reason why many people were uncharacteristically, at 5:04 p.m., safely in their wood-frame homes rather than in the normal freeway commuter traffic or other congested areas of critical danger. Secondly, it was why the inevitability and reality of earthquake risk struck home across the whole nation. For me in my home in the Berkeley Hills across the Bay from San Francisco, there will always be the memory of that evening, and of the scene of a darkened city, with its great bridge to the East Bay severed, illuminated by flames arising from a waterside residential district.

Seismological recordings of ground shaking were widespread in this earthquake, both within 50 kilometers of Loma Prieta, "the dark rolling mountain," and further to the north in the cities of the San Francisco Bay

Area. These recordings allowed the cause of the earthquake to be determined quickly. From the sensitive seismographs it was calculated, for example, that the earthquake was produced by rupture of a fault segment related to the San Andreas fault system where it traverses the Santa Cruz Mountains.* Its magnitude was scaled from instrumental recordings at distant places to be more than one unit below that of the 1906 San Andreas fault earthquake. (Its surface-wave magnitude is rated as 7.1 compared with 8 1/4 for the 1906 San Andreas event. See Chapter 8.)

Because the more sensitive seismographs in northern California went off-scale immediately after the arrival of the first waves, the full pattern of strong shaking was captured by much less sensitive instruments called strong-motion accelerometers. These recordings were to prove a major scientific resource for subsequent studies of the earthquake.

In the main earthquake, the duration of the heavy shaking was recorded to be about 15 seconds near the earthquake center in the Santa Cruz Mountains. The strong-motion instruments also showed that there was great variability in the ground motion strength and duration in the San Francisco Bay Area due to significant changes in the type and thickness of soils. In the rocky parts of the hills around the Bay, the horizontal acceleration of the ground lasted for a second or two of intense but not excessive force. At the same distance from the epicenter, buildings on soft soils and areas of filled ground were subject to about three times those forces and five times the duration of shaking.

It was fortunate that the source of this earthquake in the Santa Cruz Mountains runs through a sparsely populated region; there, residents reported severe shaking took place. In homes, furniture was shifted by many feet, with one built-in oven ejected from its cabinet. Some residents who lived on sharp ridges where the topography may have amplified the shaking described being thrown to the ground by "explosion-like forces." Comparisons indicate the southern Santa Cruz Mountains and Monterey Bay regions suffered intensities comparable to the 1906 earthquake. The city of Watsonville was badly damaged in the earthquake, as were older buildings in downtown Santa Cruz. However, along the San Francisco peninsula segment of the San Andreas fault and around the margins of the San Francisco Bay, the shaking of the 1906 earthquake was significantly larger or more severe than in 1989.

Surprisingly, the highest intensity levels have been assigned to some isolated sites in distant San Francisco and Oakland. In San Francisco an

**On feeling the shaking begin at my home, I counted 7 seconds before the stronger shear waves arrived. I thus knew that the earthquake was centered 60 or 70 kilometers from Berkeley (see Appendix I).*

area along the waterfront called the Marina District suffered a great deal of ground failure and many collapses of wood-frame houses (see Figure 1.5). Some soils which moved were underlain partly by debris from the 1906 earthquake. Here also fires broke out early in the evening of October 17, but these were extinguished by firefighters using water from the Bay; such powerful pumps were not available in 1906. In the city of Oakland (population 350,000) there was considerable damage to a number of important buildings, such as the historic city hall and a large department store. Many old unreinforced masonry structures throughout this area also suffered structurally.

Most publicity centered on the collapse of a section of the freeway that connects downtown Oakland to the Bay Bridge between the East Bay, Yerba Buena Island, and San Francisco (see Figure 1.4). This bridge was completed in 1936. It consists of two sections, the West Bay crossing from San Francisco to Yerba Buena Island and the East Bay crossing from Yerba Buena Island to Oakland. The total distance from one end to the other is 7 kilometers. The bridge is a double-decked design with an upper deck carrying five lanes of traffic and the lower deck carrying five lanes of traffic in

Figure 1.5 Collapse of a four-story wood-frame apartment in the Marina District of San Francisco on October 17, 1989. [Courtesy of S. Mahin.]

the opposite direction. The West Bay section is a twin suspension bridge with its supporting piers founded on rock; it was not damaged in the earthquake. In the East Bay section, there are four simple spans and cantilevers. The bridge was designed for moderate earthquake intensities. Such designs were normal in 1930 when knowledge of damaging earthquake motions was limited. (The first measurements of strong ground motions were made at the time of the 1933 Long Beach earthquake in California.)

The main damage was the failure of the upper span between two of the piers. This caused the bridge to be closed for a month for repairs. The span fell from the supports when bolts that connected the east end were severed (see the front figure of Chapter 12). The span moved to the east, pulling it off the western support.

The second major collapse was the double-decked freeway structure called the Cypress Viaduct in Oakland, constructed between 1954 and 1957. The severe earthquake shaking in this area, due to the thick soil deposits under the section, caused a long portion of the viaduct to collapse. Even though traffic conditions were light, this structural failure was the most tragic consequence of the earthquake, with 41 people dying. Search-and-rescue operations continued for a week. The viaduct was California's first continuous double-decked freeway, and the design was adopted for freeway viaducts in San Francisco which were also damaged but did not collapse in this earthquake. There are no double-decked freeways of this design anywhere else in California.

The impact on the community affected by the earthquake was grave. There were 63 deaths and 3757 injuries. Over 8000 people were made homeless by the earthquake, with property damage of $5.6 billion. Estimates indicate that over 1300 buildings were destroyed and 20,000 buildings damaged. There was economic hardship caused by widespread disruption of transportation, utilities, and communications, with more than 3500 businesses damaged and about 400 destroyed. Thirteen state-owned and five locally owned bridges were closed to traffic following the earthquake — a small number considering there are over 4000 bridges in the area affected by the shaking. The cost of earthquake damage to the transportation system was about $1.8 billion. Beyond these known financial losses must be added the substantial hardship to individuals and businesses.

The 1964 Good Friday Earthquake in Alaska

The two California earthquakes just described were generated by rupture of the San Andreas fault. Let us consider now quite another type of earthquake source.

The Aleutian Islands and trench stretch in a sweeping arc across the northernmost Pacific Ocean between Kamchatka in Siberia and south-central Alaska. Into this trench the Pacific crustal rocks plunge downward and northward. Abundant volcanic and seismic activity occurs along the entire arc and extends eastward into the active and dormant volcanoes of the Rango Mountains (see Figure 7.6). Intermittent thrusting of the Pacific floor under Alaska occurs frequently, producing earthquakes over a wide region. The underthrusting rock slab may stick at any one place for centuries, while adjacent parts of it continue to progress onward (see Plate 13). Finally, a break occurs (see Figure 7.6).

Such an event occurred on Good Friday, March 27, 1964, at 5:36 p.m. The first slip occurred at a depth of about 30 kilometers under northern Prince William Sound, and the rupture in the rocks extended horizontally for 800 kilometers, roughly parallel to the Aleutian trench.

Hundreds of measurements along the shoreline later showed that beds of barnacles and other sea life had been raised above sea level about 10 meters (see Figure 1.6). From such observations and the uplift of tidal bench marks relative to sea level and from geodetic level lines surveyed carefully from the coast into Alaska, it was estimated that about 200,000 square kilometers of the crust were deformed in the Good Friday earthquake. It was the greatest area of vertical displacement ever measured in earthquake history. The underthrusting slip (see Chapter 7) occurred mainly beneath the ocean; only in a few places, such as on Montague Island in Prince William Sound, were fresh fault scarps visible. The vertical fault displacements on Montague Island amounted to 6 meters in places. Such great slips are not the record for Alaska, however; 14.3 meters of uplift occurred in the Yakutat Bay earthquake of 1899, centered about 320 kilometers to the east.

The sudden upward movement of the Alaskan seafloor along the rupturing fault generated large water waves, acting on the water of the ocean like a gigantic paddle. Such gigantic "tidal" waves, produced in some earthquakes, are called *tsunami* (see Chapter 9). The crests of the first waves struck the shores of Kenai Peninsula within 19 minutes and Kodiak Island within 34 minutes after the start of the earthquake. As the tsunami surged onshore, it devastated waterfront developments along the Alaskan coast, particularly at Valdez and Seward. About 120 persons drowned.

In Anchorage, 100 kilometers from the fault slip in Prince William Sound, strong ground shaking commenced about 15 seconds after the rocks first broke. The heavy shaking continued for more than half a minute, and strongly felt shaking continued for several minutes. After shaking began, the announcer at radio station KHAR, R. Pate, recorded his thoughts on a tape recorder:

Figure 1.6 Former seafloor at Cape Cleare, Montague Island, Prince William Sound, exposed by tectonic uplift in 1964. The surf-cut surface, which slopes gently from the base of the sea cliffs to the water, is about a quarter of a mile wide. [Photo by George Plafker, USGS.]

Hey, boy — Oh-wee, that's a good one! Hey — boy oh boy oh boy! Man, that's an earthquake! Hey, that's an earthquake for sure! — Wheeeee! Boy oh boy — this is something you'd read — doesn't come up very often up here, but I'm going through it right now! Man — everything's moving — you know, all that stuff in all the cabinets have come up loose. . . . Whooeee! Scared the hell out of me, man! Oh boy, I wish this house would quit shaking! That damn bird cage — oooo — oh man! I've never lived through anything like this before in my life! And it hasn't even shown signs of stopping yet, either — ooooeeee — the whole place is shaking — like someone was holding — Hold it, I'd better put the television on the floor. Just a minute — Boy! Let me tell you that sure scared the hell out of me, and it's still shaking. I'm telling you! I wonder if I should get outside? Oh boy! Man, I'm telling you that's the worst thing I've ever lived through! I wonder if that's the last one of 'em? Oh man! Oh — Oh boy, I'm telling you that's something I hope I don't go through very often. Maa-uhn! — I'm not fakin' a bit of this — I'm telling you, the whole place just moved like somebody had taken it

> by the nape of the neck and was shaking it. Everything's moving around here! — I wonder if the HAR radio tower is still standing up. Man! You sure can't hear it, but I wonder what they have to say on the air about it? The radio fell back here — but I don't think it killed it — Oh! I'm shaking like a leaf — I don't think it hurt it. Man, that could very easily have knocked the tower down — I don't get anything on the air — from any of the stations — I can't even think! I wonder what it did to the tower. We may have lost the tower. I'll see if any of the stations come on — No, none of them do. I assume the radio is okay — Boy! The place is still moving! You couldn't even stand up when that thing was going like that — I was falling all over the place here. I turned this thing on and started talking just after the thing started, and man! I'm telling you, this house was shaking like a leaf! The picture frames — all the doors were opened — the dishes were falling out of the cabinets — and it's still swaying back and forth — I've got to go through and make a check to make sure that none of the water lines are ruptured or anything. Man, I hope I don't live through one of those things again. . . .

Building damage in southern Alaska from the 1964 earthquake varied considerably, depending on the foundation conditions and the type of structure. In Anchorage, higher buildings suffered most, whereas wood-frame homes were reasonably unscathed, although their occupants were disturbed and furniture was thrown down. Because of the distance from the rupturing fault, ground shaking consisted mainly of long waves that do not affect small buildings (see the description of the Mexico earthquake in Chapter 12).

In all, Alaska sustained $300 million in property damage from the earthquake; about 130 persons died, only 9 from the effects of shaking. One serious secondary result of the shaking was the temporary change of soil and sand in many areas from a solid to a liquid state. The most spectacular example of such *liquefaction* was at Turnagain Heights in Anchorage, where soft clay bluffs about 22 meters high collapsed during the strong ground motion, carrying away many modern frame homes in a slide that regressed inland 300 meters along 2800 meters of coastline (see Figure 1.7). Throughout southern Alaska, rock slides, landslides, and snow avalanches were common, damaging roads, bridges, railroad tracks, power facilities, and harbor and dock structures.

The Great 1960 and 1985 Chile Earthquakes

During the last 130 years, Chile has experienced 25 major earthquakes. We have already mentioned the damaging earthquake witnessed by Charles Darwin in 1835. At present, Chile holds the seismological record as the site

Figure 1.7 Aerial view of the coastline of the Turnagain Heights area after the 1964 Alaska earthquake. Approximately 2.5 kilometers of the bluffs slid toward the ocean after liquefaction of the sand and silt in the clay formations. [Courtesy of G. Housner.]

of the largest earthquake in the world this century. The earthquake, which occurred on May 22, 1960, was so large* that it drove many seismographic stations records off-scale around the whole world, and it set our planet vibrating like a bell struck by a gong. It was generated by a rupture about 1000 kilometers long and 200 kilometers wide, along the Andean deep seismic zone between Concepción (37° south) and Taitao Peninsula (46° south) (see Figure 1.8). There was one aftershock of magnitude about as large as the 1906 San Francisco earthquake. It caused over 2000 deaths in Chile.

Coastline water height records in Japan show that the 1960 Chilean earthquake produced the largest tsunami in the Pacific region for at least

With a Richter magnitude of about 8.5 and a moment magnitude of about 9.9 (see Chapter 8).

500 years. The powerful ocean wave spread across the Pacific Ocean, hitting the Japan coast about 22 hours after it began. In Japan, about 120 people lost their lives, thousands of houses were washed away or flooded, and many ships sank.

The tsunami also attacked the Chilean coast, about 15 minutes after the earthquake, with three large waves. The inundation caused severe damage to coastal towns, particularly at Porto Saavedra. During a visit there in May 1998 I saw the ruined foundations of the houses in the former fishing town that flooded to a depth of 4 meters near the mouth of the Imperial River. Trees and palms planted as wind shelters, stripped bare by the surging water, can still be seen.

Ground deformations in Chile were widespread, with up to 2 meters of subsidence in the coastal mountains and uplifts of over half a meter along the foothills of the Andes, while to the west, offshore islands were uplifted as much as 6 meters. This pattern of ground elevation and depression is similar to that observed in the 1964 Alaskan earthquake described earlier (also situated along a Pacific plate subduction zone; see Chapter 7).

At 7:47 p.m., March 3, 1985, another in the tragic series of Chilean earthquakes occurred. There were 176 persons killed, 2483 injured, and 372,532 left homeless. The relatively light casualties were partly due to the timing: it was a summer Sunday evening when commercial buildings were closed and many people were outdoors. The epicenter was located off the coast, west of the capital city of Santiago (see Figure 1.8). The source of the earthquake was a sudden slip along the underthrusting slab that dips from the ocean trench along the Chilean coast under the country toward the Andes Mountains. (This thrust uplifted the continental shelf about a meter, producing a small ocean wave; just along the shoreline, the land subsided.)

The earthquake shook the most densely populated region of the country, and it was felt as far away as Buenos Aires, Argentina. Because of close acquaintance with several Chilean seismologists and engineers, I was able to spend several days immediately following the disaster examining the seismic effects. This intensity varied widely. Reassuringly, most of the modern reinforced structures were not seriously damaged. In Santiago (population of 2.5 million), there was only slight damage except to unreinforced masonry buildings, particularly in the old part of the city. Several serious fires broke out but did not spread.

In the modern cities of Valparaiso and Viña del Mar on the coast, near the northern portion of the energy-release zone (see Figure 1.8), the intensity was about the same as or a little greater than in Santiago. Generally, reinforced concrete high-rise buildings and other specially designed structures suffered little or no damage, and heavy damage to ordinary structures was not widespread. However, throughout these coastal cities, examples

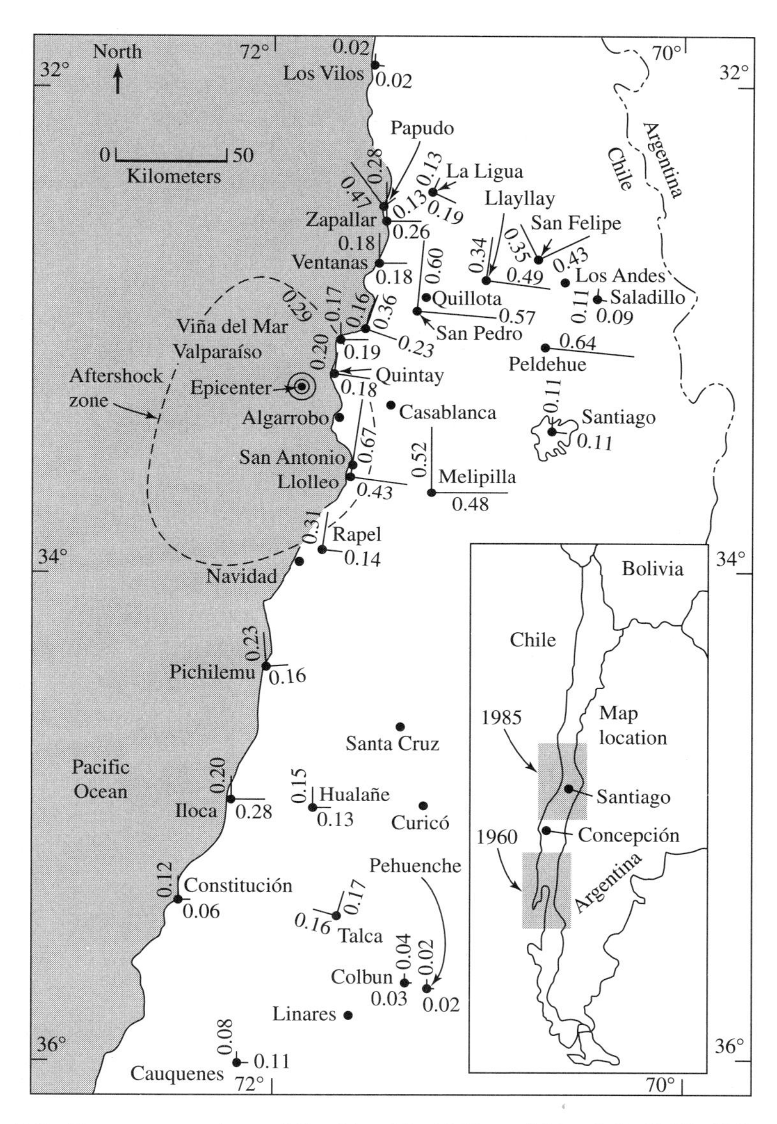

Figure 1.8 The source regions of the May 22, 1960, and March 3, 1985, Chile earthquakes and the 1985 aftershock zone during the first 10 days of the sequence. The recorded peak accelerations (as fractions of gravity) of the two horizontal components of ground motion in 1985 are also shown. [Courtesy of R. Saragoni and M. Pardo.]

could be found of adobe and panel walls thrown out of frame structures, slipping of tiles on roofs, and cracking of chimneys, columns, and walls. In high-rise buildings, brick or plaster partition walls were often seriously cracked and furniture toppled.

South of Valparaiso toward the old port city of San Antonio through Algarrobo, another coastal town, little general damage was observed even though the locations were closer to the earthquake center than Valparaiso. These observations reinforced the idea that the location of the slip center is not of great importance when considering the release of energy from an extended seismic source.

Damage to schools was widespread (in contrast, say, with the success of the Field Act in fostering safer schools in California; see Chapter 12). Although over 20 hospitals and clinics of the 80 in the most affected area were damaged, the health care system treated over 1600 injuries attributed to the earthquake. However, several hospitals were forced to relocate services after the earthquake. The lesson is that essential facilities must remain operational.

Generally, although in many places the intensity of shaking was very high and of long duration, most modern buildings were able to withstand the vibrations. This favorable result was partly due to building codes based on experience with past earthquakes in Chile. Of special note was the successful retrofitting of a church damaged in an earlier earthquake.

The 1985 Chilean earthquake made a lasting mark in earthquake science: not only were there visual observations of the effect of the shaking on structures of various designs, but also many instruments recorded the shaking of the ground (see Chapter 3 for a description of such instruments). Before the earthquake, a modern network of stations with sensitive seismographs had recorded many foreshocks. These records and those obtained with portable seismographs provided detailed locations of thousands of aftershocks and enabled seismologists to precisely map the extent of the seismic source of the principal earthquake along the underthrust slab.

Types of Seismic Waves in Earthquake Shaking

A handclap in the air sends sound waves outward to distant places as the air compresses and rarifies; the mechanical energy originally in the moving hands is transformed into air vibrations. A stone thrown into water sends waves spreading across its surface in the form of ripples. In a similar way, a sudden blow to gelatin and other elastic materials produces quivering as waves spread from the impulse throughout the elastic body. So too, the rocks of the Earth have elastic properties that cause them to deform and vibrate when pushed and pulled by forces applied to them.

Earthquake shaking and damage is the result of three basic types of elastic waves. Two of the three propagate *within* a body of rock. The faster of these *body waves* is called the *primary* or *P wave*. Its motion is the same as that of a sound wave in that, as it spreads out, it alternately pushes (compresses) and pulls (dilates) the rock (see Figure 1.9a). These P waves, just like sound waves, are able to travel through both solid rock, such as granite mountains, and liquid material, such as volcanic magma or the water of the oceans. It is worth mentioning also that, because of their soundlike nature, when P waves emerge at the surface from deep in the Earth, a fraction of them may be transmitted into the atmosphere as sound waves, audible to animals and humans at certain frequencies.*

The slower wave through the body of rock is called the *secondary* or S *wave*. As an S wave propagates, it *shears* the rock sideways at right angles to the direction of travel (see Figure 1.9b). Observation readily confirms that if an attempt is made to shear sideways or twist a liquid, it will not spring back. It thus follows that S waves cannot propagate in the liquid parts of the Earth, such as the oceans.

The actual speed of P and S seismic waves depends on the density and elastic properties of the rocks and soil through which they pass. (Some typical velocity values and a list of physical relations are given in Appendix I.) In most earthquakes, the P waves are felt first. The effect is similar to a sonic boom that bumps and rattles windows. Some seconds later, the S waves arrive with their up-and-down and side-to-side motion, shaking the ground surface vertically and horizontally. This is the wave motion that is so damaging to structures.

The third general type of earthquake wave is called a *surface wave*, because its motion is restricted to near the ground surface. Such waves correspond to ripples of water that travel across a lake. Most of the seismic wave motion is located near the outside ground surface, and as the depth below this surface increases, wave displacements decrease.

Such surface waves in earthquakes can be divided into two types, named after two English mathematicians who first described them. The first is called a *Love wave*. Its motion is essentially the same as that of S waves that have no vertical displacement; it moves the ground from side to side in a horizontal plane but at right angles to the direction of propagation (see Figure 1.9c). The horizontal shaking of Love waves is particularly damaging to the foundations of structures. The second type of surface wave is known as a *Rayleigh wave*. Like rolling ocean waves, the pieces of material

**Greater than about 15 cycles per second (Hertz).*

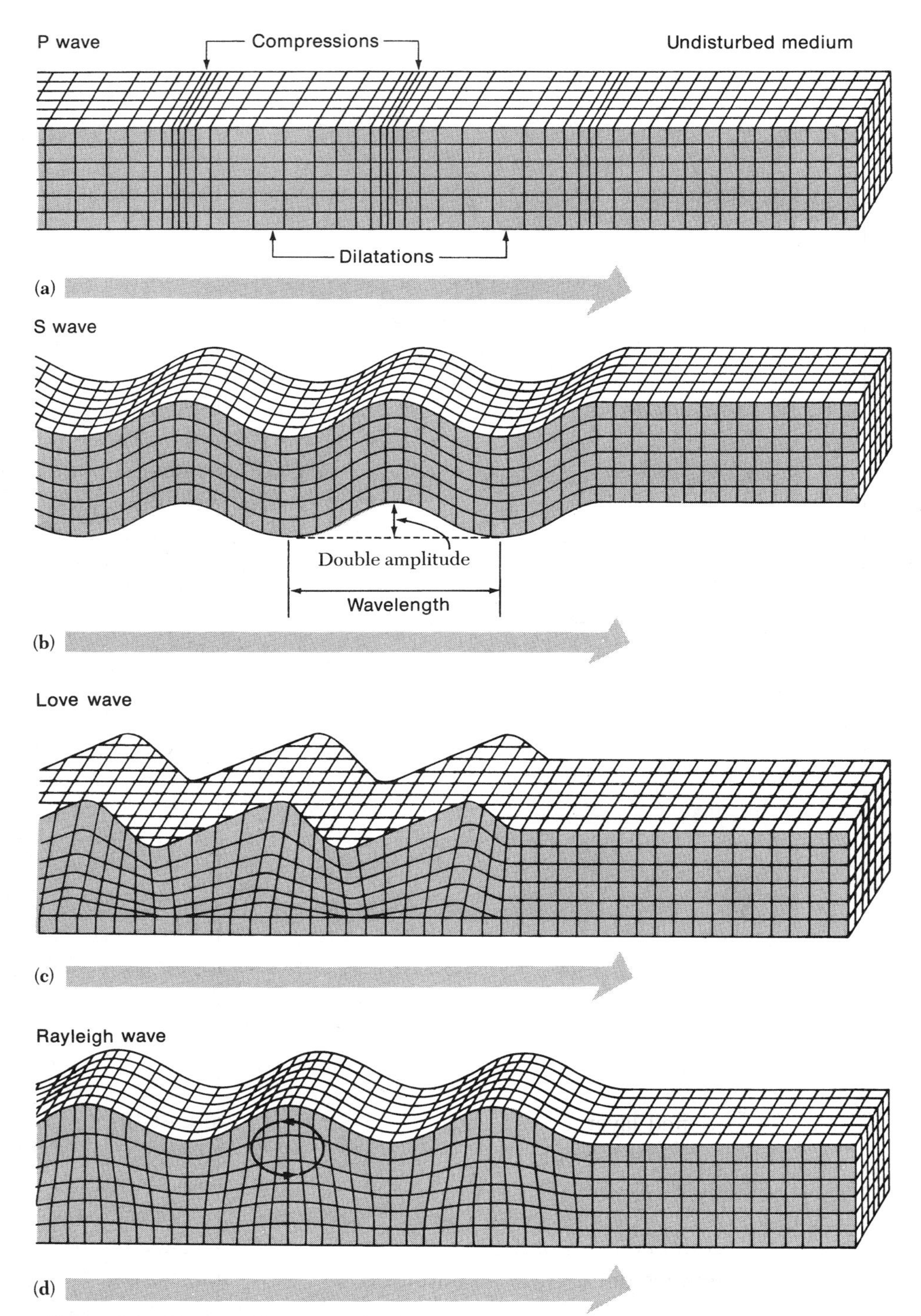

Figure 1.9 Diagram illustrating the forms of ground motion near the ground surface in four types of earthquake waves. [From Bruce A. Bolt, *Nuclear Explosions and Earthquakes: The Parted Veil* (San Francisco: W. H. Freeman and Company. Copyright 1976).]

disturbed by a Rayleigh wave move both vertically and horizontally in a vertical plane pointed in the direction in which the waves are traveling. The arrows in Figure 1.9d illustrate the elliptical movement of a piece of rock as the wave passes.

Surface waves travel more slowly than body waves; and of the two surface waves, Love waves generally travel faster than Rayleigh waves (see Appendix I). Thus, as the waves radiate outward from the earthquake source through the rocks of the Earth, the different types of waves separate out from one another in a predictable pattern. (An illustration of the pattern occurring at a distant place is shown in Appendix G. In this example the seismograph recorded only the vertical motion of the ground, and so the seismogram contains only P, S, and Rayleigh waves, because Love waves do not possess vertical motion.) Rayleigh waves, because of the vertical component of their motion, can affect bodies of water such as lakes, whereas Love waves (which do not propagate through liquids) can affect surface water only insofar as the sides of lakes and ocean bays move backward and forward, pushing the water sideways like the sides of a vibrating tank.

The body waves (the P and S waves) have another characteristic that affects shaking: when they move through layers of rock, they are reflected or refracted at the interfaces between rock types, as illustrated in Figure 1.10a. Also, whenever either one is reflected or refracted, some of the energy of one type is converted to waves of the other type (Figure 1.10b). To take a common example, as a P wave travels upward and strikes the bottom of a layer of alluvium, part of its energy will pass upward through the alluvium as a P wave and part will pass upward as the converted S-wave motion. (Part of the energy will also be reflected back downward as P and S waves.)

Thus, in strong ground shaking on land, after the first few shakes, a combination of the two kinds of waves is usually felt. But if you are at sea during an earthquake, the only motion felt on ship is from the P waves, because the S waves cannot travel through the water beneath the vessel. A similar effect occurs as sand layers liquefy in earthquake shaking. There is a progressive decrease in the amount of S-wave energy that is able to propagate in the liquefied layers, and ultimately only P waves can pass through.

When P and S waves reach the surface of the ground, most of their energy is reflected back into the underlying rocks, so that the surface is affected almost simultaneously by upward- and downward-moving waves. For this reason, considerable amplification of shaking typically occurs near the surface — sometimes double the amplitude of the upcoming waves. This surface amplification enhances the shaking damage produced at the surface of the Earth. By contrast, in many earthquakes, mineworkers below ground report less shaking than people on the surface.

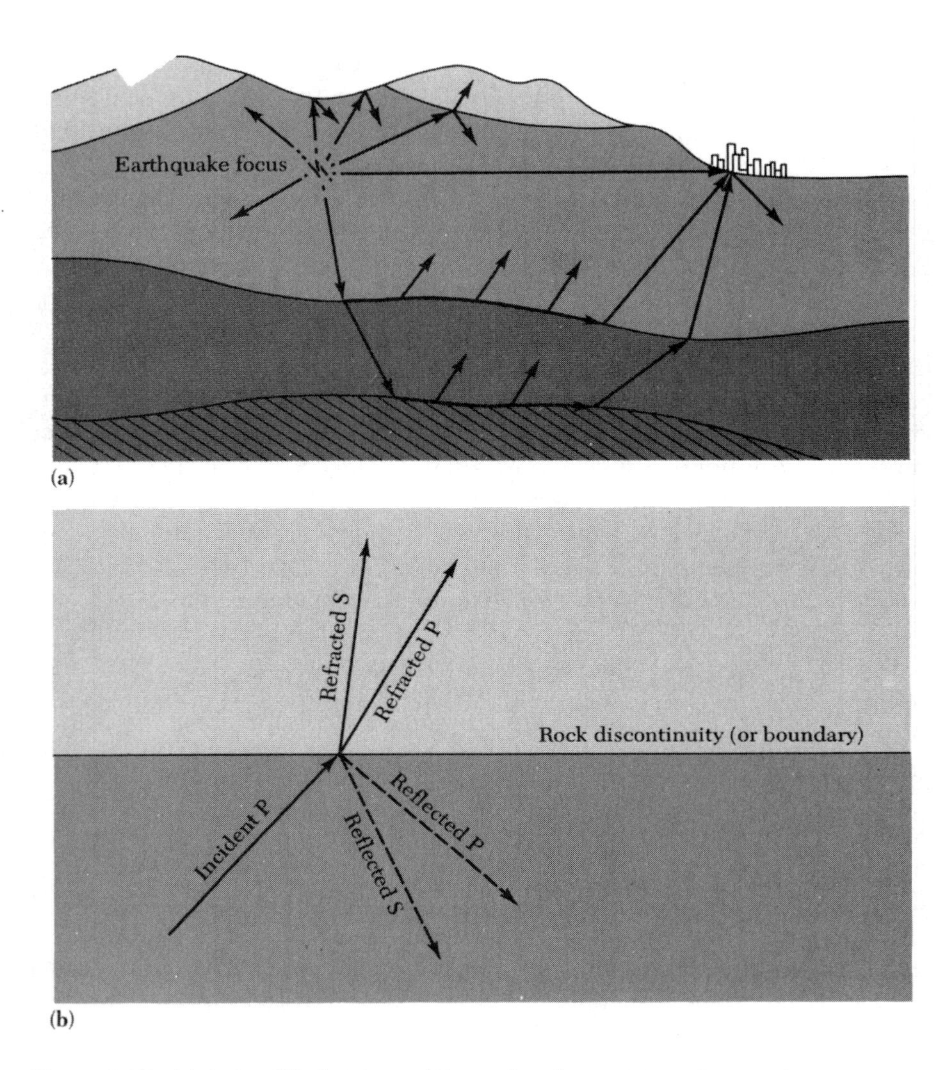

Figure 1.10 (a) A simplified picture of the paths of seismic P or S waves being reflected and refracted in rock structures of the Earth's crust. (b) The reflection and refraction of a longitudinal (P) wave in an earthquake after it hits a boundary between two types of rock. [From Bruce A. Bolt, *Nuclear Explosions and Earthquakes: The Parted Veil* (San Francisco: W. H. Freeman and Company. Copyright 1976).]

The above description, however, does not adequately explain the heavy shaking near the center of a large earthquake. Near a fault that is suddenly rupturing, like the San Andreas in 1906, the strong ground shaking in the associated earthquake consists of various kinds of seismic waves that are not distinctly separate. To complicate the matter, because the source of radiat-

Figure 1.11 Paramedics, police, and volunteers struggle to remove a victim from the upper level of the I-880 viaduct collapsed in the 1989 Loma Prieta earthquake, amplified in this area by soft soil. [Photo by Michael Macor, *The Oakland Tribune*.]

ing seismic energy is itself spread out across an area, the types of ground motion may be further muddled together. (In Chapters 6, 8, and 12, instrumental records of strong ground motion obtained near the source of an earthquake are analyzed to try to unravel these complicated motions.)

A final point about seismic waves is worth noting. Observational and theoretical evidence shows that earthquake waves are affected by both soil conditions and topography, sometimes dramatically (see Figure 1.11). For example, in weathered surface rocks, in alluvium and in water-saturated soil, the size of seismic waves may be either increased or decreased many times as they pass to the surface from the more rigid basement rock. Also at the top or bottom of a ridge, shaking may intensify, depending on the direction from which the waves are coming and whether the wavelengths are long or short relative to the size of the ridge. These waves properties may be critical for earthquake safety and will be revisited more fully in Chapter 8.

2 Where Earthquakes Occur

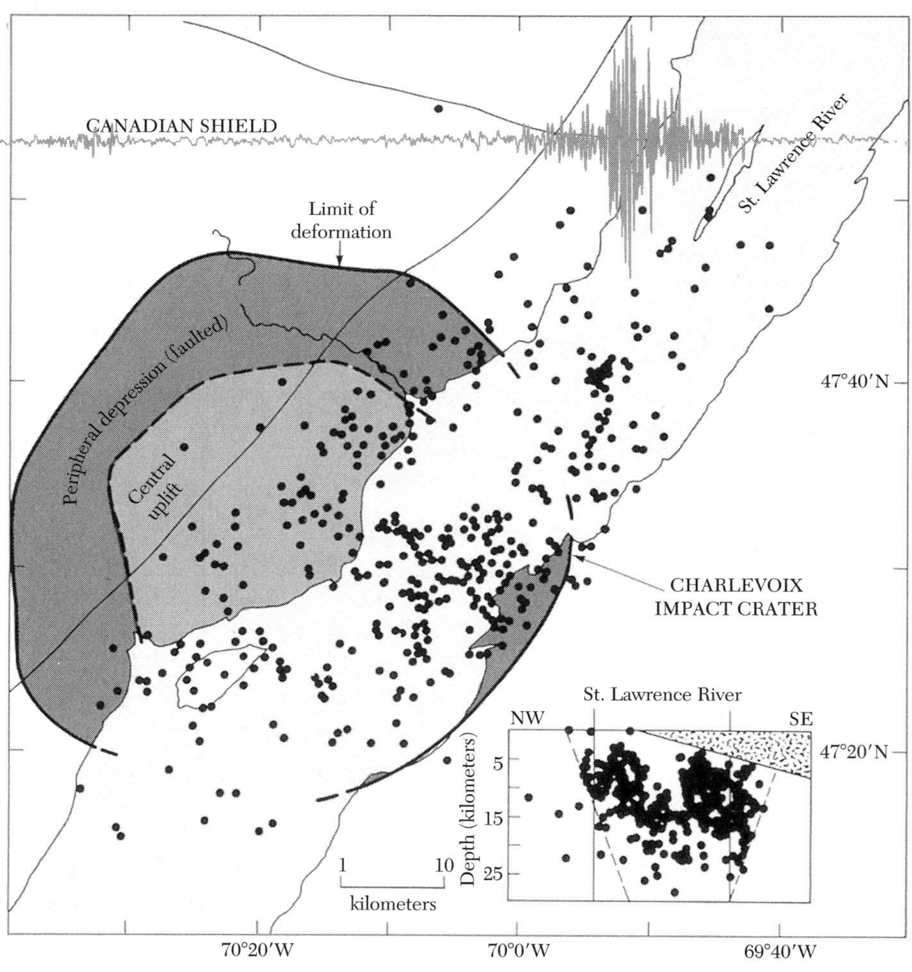

Epicenters and foci (insert) of recent earthquakes plotted on a map showing the site of the Charlevoix impact crater, along the St. Lawrence River, Canada. The meteor, radius 1 kilometer, fell 360 million years ago. The energy of impact reactivated (restrained) preexisting faults. [Courtesy of H. S. Hasegawa, Geological Survey of Canada.]

Some say the Earth was fevrous and did shake.

— *Shakespeare*. Macbeth, *II*, 3

Earthquakes can be violent, and they have been unpredictable. Their convulsions have so often produced helplessness and injury that people have always feared them. In many countries, popular legend attributed earthquakes to grotesque monsters that supported the Earth. In ancient Japanese folklore, for example, a great catfish *(namazu)* lay beneath the ground and caused earthquakes by thrashing its body (see the print at the beginning of Chapter 6). The namazu's activity was restrained by a god *(daimyojin)* who wielded a large stone mallet. But when the attention of the daimyojin wandered, the namazu moved and the ground shook.

Earthquakes in Past Times

The first systematic and nonmystical treatment of earthquakes occurred in Greece, where its people experienced Aegean volcanoes and earthquakes along the Mediterranean Sea, sometimes accompanied by "tidal" waves *(tsunamis)*. A number of the Greek philosophers offered mechanical explanations for these natural events. Strabo, for example, noted that earthquakes occurred more frequently along the coast than inland. He, like Aristotle, suggested that earthquakes were caused by rushing subterranean winds, igniting combustible materials underground.

As the practice of writing spread, descriptions of severe earthquakes around the world were recorded. The oldest of these are the Chinese records dating back 3000 years. This amazing catalog is thought to document every moderate to large earthquake in central China from 780 B.C. to the present. In Japan, the catalog of damaging earthquakes is not as long but is essentially without gaps from about 1600 A.D.; less reliable lists are available back to about 416 A.D. Such historical catalogs are crucial to our understanding of the relation of earthquakes to the geological features of our planet and to our assessment of seismic hazards for large engineering structures such as dams and nuclear reactors.

In the western hemisphere, there is a well-documented history for the eastern part of the Alpine belt, from Greece to Afghanistan, for about 17 centuries. Even earlier than this, sporadic allusions to large earthquakes in the Mediterranean region are found in classical Greek writings, in the Bible, and in Arabic chronicles. Some historians surmise that earthquakes helped end the Minoan civilization by destroying the city-state of Knossos in about 1700 B.C.

It has been claimed that the first biblical mention of an earthquake is the experience of Moses on Mount Sinai. More definite references are probably the accounts of the collapse of the walls of Jericho about 1100 B.C. and perhaps of the destruction of Sodom and Gomorrah. Palestinian earthquakes are associated with geological faults of the rift valley that runs north from the Gulf of Eilat through the Dead Sea. Although in this century the Jordan Valley rift has been the site of only a few small to moderate earthquakes, historical studies indicate that the whole area suffers, on the average, two or three damaging earthquakes each century.*

In the more recently settled parts of the world, such as the United States and Canada, the historical earthquake chronicles are, of course, quite short (see Plate 2). One of the first accounts describes an earthquake that struck Massachusetts in 1638, toppling stone chimneys to the ground. Somewhat more extensive reports describe a large Canadian earthquake in the Three Rivers area of the lower St. Lawrence River on February 5, 1663. For California, there are descriptions dating back to 1800 by the Franciscan fathers who documented the development of the Spanish missions. Thus we know that a series of earthquakes in 1800 damaged Mission San Juan Bautista, and that 1812 was called "The Year of Earthquakes" because of the great amount of seismic activity felt at that time.

Investigating earthquakes that happened long ago is frustrating work. There is a story of Professor George Louderback, a geologist at the University of California at Berkeley, who had a keen interest in disentangling the history of California earthquakes. The historical reports spoke of an earthquake on the morning of December 8, 1812 (a Tuesday), that destroyed Mission San Juan Capistrano, killing 40 American Indians attending Mass. Louderback asked: Why were they worshiping on Tuesday?

**Historical seismic activity has often occurred near the monastery of Saint John the Baptist, where Jesus was baptized. Damaging local shocks occurred in 1834 and 1927.*

He determined that that day was a holy day, so attendance in church was understandable. Further enquiry showed, however, that that particular holy day was currently not being celebrated in Rome. Why then were the American Indians in church on a Tuesday? Thus, historical enquiries sometimes lead to further puzzles.

By the mid-nineteenth century, documentation in California was fairly detailed. In the description of the great earthquake of January 9, 1857, for example, several independent references were made to extensive cracking of the Earth in central California near the settlement of Fort Tejon. This earthquake was one of the first indications of rupture on what is now called the San Andreas fault. The Fort Tejon shock is the most recent great earthquake to occur along the southern portion of the San Andreas fault.

Observatories to Study Earthquakes

Early in this century seismographic stations were established at many points throughout the world. At such stations, sensitive seismographs operate continuously and record earthquake waves that have been generated at distant places. For example, the 1906 San Francisco earthquake was well recorded at dozens of seismographic stations in a number of foreign countries, including Japan, Italy, and Germany.*

The significance of this worldwide network was that earthquake documentation no longer rested solely on subjective reports of felt and visual effects. There developed a cooperative international program in which earthquake readings could be exchanged to help pinpoint earthquake locations. For the first time, the temporal statistics of earthquake occurrence and the geological distribution of earthquakes, even in unpopulated regions, became known (see Figure 2.1). This information defines the *seismicity* of a region.

By 1960, about 700 earthquake observatories were operating in numerous countries, with a hodgepodge of types of seismographs. The ability to locate accurately earthquakes of moderate size at any place on the Earth's surface was subsequently greatly improved when the United States established the Worldwide Standardized Network of seismographs. By 1969, about 120 of these special stations were distributed in 60 countries. A comparable step forward was also made in the technology of earthquake

**In 1968, I used these records to compute the location of the point of initiation of the 1906 earthquake, using modern methods. It was only a few kilometers from the Golden Gate Bridge!*

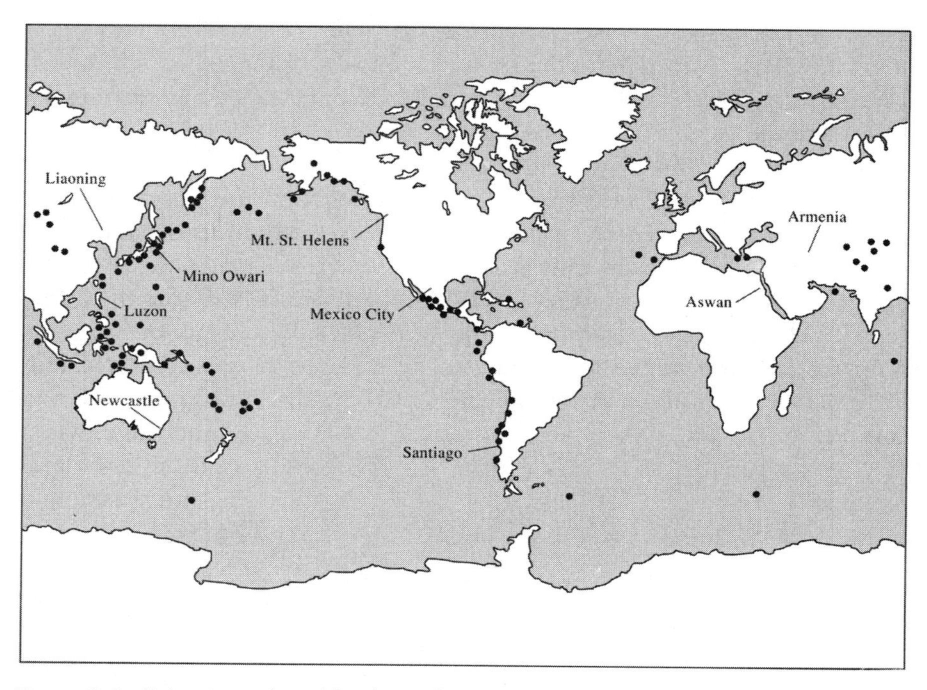

Figure 2.1 Epicenters of worldwide earthquakes having large magnitudes from 1897 through 1998. The positions of places discussed in the book are also marked.

observation in nonparticipating countries. Nowadays, there are probably over 1000 earthquake observatories that are more-or-less permanent and continuously operating worldwide (see Chapter 3).

In the last decade, even more useful global networks of latest-technology seismographs have operated. These seismographs record seismic ground motions on magnetic tapes or disks in a discrete ("digital") format that can be used directly in computers. By 1997 essentially all of the landmark Worldwide Standardized Network stations had been converted to the digital seismographic network standards that provide high-quality assurance, central data storage, and rapid individual access to personal computers anywhere in the world (see the list of Web addresses at the back of the book). In 1998 over 100 such stations came under the support and operational assistance of a consortium called the Incorporated Research Institutes for Seismology (IRIS).

The Global Mosaic of Earthquakes

The position of the center of an earthquake can be calculated from the earthquake wave readings at different seismographic observatories. In this way, a uniform picture of earthquake distribution around the globe has been obtained (see Figure 2.2 and current seismicity maps available from *http://quake.geo.berkeley.edu*). Clear belts of seismic activity separate large oceanic and continental regions, with interesting exceptions almost devoid of earthquake centers. Indeed some concentrations of earthquake sources can be seen in the oceanic areas; these are the sites of gigantic submarine mountain ranges called midoceanic ridges. The seismically active ridges of the Atlantic and Indian oceans meet south of Africa, and the mid-Indian ridge circles below Australia to connect with the East Pacific ridge, which extends eastward toward Central America and into the Gulf of California. The geological unrest that prevails throughout this global ridge system is evidenced by great mountain peaks and deep rift valleys. Volcanic eruptions are frequent, and earthquakes originating along these ridges often occur in "swarms," so that many hundreds of shocks are concentrated in a small area in a short time.

Dense concentrations of earthquake centers also coincide with beautifully symmetrical island arcs, all also volcano-rich, such as those of the Pacific and the eastern Caribbean. One of the finest examples of these island chains is the crescent-shaped Aleutian arc, swinging westward from Alaska toward Kamchatka. Southward, the islands of Japan form an arc that extends southward to the island chain of the Marianas. From Indonesia to the south Pacific, a number of seismically active arcs drape around Australia like a garland, with the Tonga-Kermadec trench as its eastern border.

There are a few isolated seismically active areas around the world that invite our curiosity and special study by seismologists (see Chapter 7). Some of these, such as the Hawaiian Islands and Emperor Seamount chain in the mid-Pacific, coincide with vigorous volcanic activity.

On the other side of the Pacific, the whole western coast of Central and South America is agitated by many earthquakes, great and small. High death tolls have ensued from the major ones, such as the 1976 Guatemala earthquake (see Appendix A). In marked contrast, the eastern part of South America is almost free from earthquakes and can be cited as a good example of aseismic country. Other seismically quiet continental areas can also be seen in Figure 2.2; earthquakes seldom occur in the large central and northern areas of Canada, much of Siberia, west Africa, or great parts of Australia. But note the long *trans-Asiatic* zone of high seismicity running

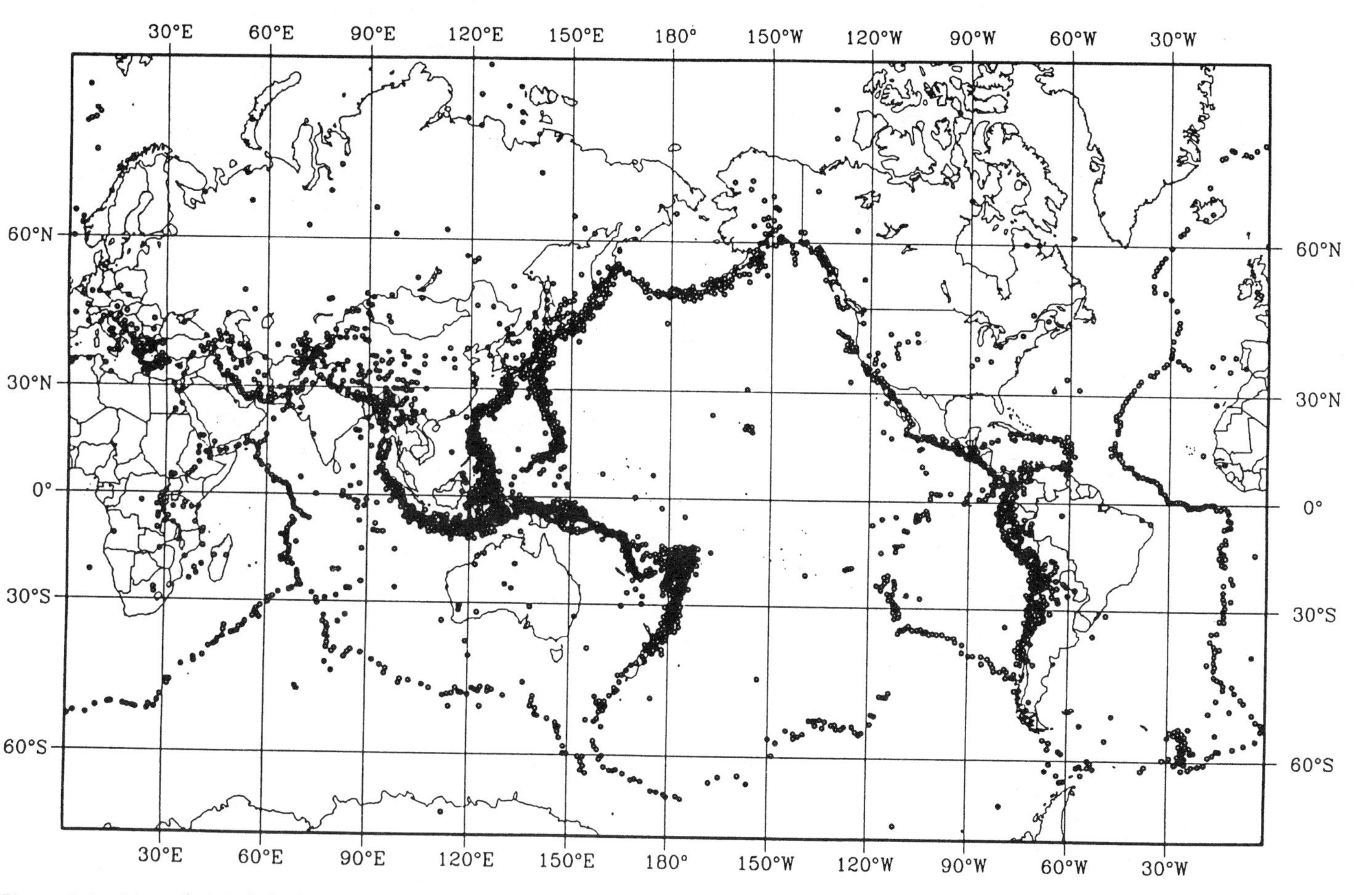

Figure 2.2 Map of global shallow seismicity, 1963–1988, magnitude greater than 5 (see Chapter 8) and depth less than 70 kilometers. [Courtesy of National Earthquake Information Center, USGS.]

approximately east-west from Burma* through the Himalaya Mountains and central Asia to the Caucasus Mountains and the Mediterranean Sea.

In Europe, earthquake activity is quite widespread (see Plate 3). To the south, in Turkey, Greece, Yugoslavia, Italy, Spain, and Portugal, large numbers of people have died in disasters throughout the years. An earthquake off southwest Iberia on November 1, 1755, produced a great ocean wave, which alone caused many of the 50,000 to 70,000 deaths occurring in Lisbon, Portugal, and surrounding areas; the ground shaking was felt in Germany and the Low Countries. In Alicante, Spain, on March 21, 1829, an earthquake killed about 840 persons and injured many hundreds more; total or partial destruction of more than 5000 houses was reported in and near Torrevieja and Murcia. A devastating earthquake hit Messina, Italy, on December 28, 1908, causing 120,000 deaths and widespread damage. On December 27, 1939, and on March 13, 1992, in Erzincan, Turkey, thousands of lives were lost from a major earthquake. Similar killer earthquakes have occurred in nearby Iran in recent years. Another earthquake struck Italy on May 6, 1976, in the Friuli region near Gemona; about 965 persons were killed and 2280 injured (see Figure 2.3). A similar tragedy occurred in southern Italy on November 23, 1980, when the Campania earthquake caused over 3000 deaths and left 250,000 homeless.

North of the Mediterranean margin, Europe is much more stable (Plate 3). However, destructive earthquakes do occur occasionally in Germany, Austria, and Switzerland, and even in the North Sea region and Scandinavia. For example, on October 8, 1927, an earthquake near Schwadorf in Austria caused damage in an area southeast of Vienna. This earthquake was felt in Hungary, Germany, and Czechoslovakia at distances of 250 kilometers from the center of the disturbance.

Damaging earthquakes have also occurred in Great Britain in historical times. On December 17, 1896, an earthquake series caused some damage in Hereford, a city of 4565 inhabited houses. About 200 chimneys had to be repaired or rebuilt, and the cathedral was slightly damaged. The area affected was just over 1000 square kilometers. A notable recent case was the earthquake in east Wales on April 2, 1990, which was felt over a wide area of England, Wales, and Ireland. Damage was minor but included some buildings in Manchester.

Seismicity maps such as that in Figure 2.2 are drawn up from data taken over a fairly short-term period. Consequently, if conclusions or pre-

**On July 8, 1975, a violent earthquake seriously damaged many temples and pagodas at Pagan in Burma. Ancient inscriptions there refer to restorations after former earthquakes.*

Figure 2.3 Damage to weak masonry structures in northern Italy in the Friuli earthquake of May 6, 1976. [Courtesy of I. Finetti.]

dictions about the likelihood of earthquake occurrence in a given area are founded only on such maps, they can be discredited by the abnormal occurrence of an earthquake in an area that is not usually regarded as seismically active. Of course, they can also be misleading by not highlighting a region that is quite seismically active but happens not to have had strong earthquakes recently (e.g. Plate 4).

An example of infrequent and dispersed seismicity is the occurrence of earthquakes in Australia (see Chapter 12). There are sound geological reasons why this is so. Much of the western part consists of ancient rocks of the Australian Precambrian shield, and the continent as a whole is remote from the active ocean ridges and island arcs that surround it. Nevertheless, this country does have some areas of significant present-day seismicity. Of particular interest is a damaging earthquake of moderate size that was centered near Meckering, western Australia, on October 14, 1968, and was associated with fresh surface faulting about 30 kilometers long.

Seismicity maps — carefully worked out by the cooperative efforts of hundreds of seismologists — have contributed in essential ways to our knowledge of the Earth. For example, global patterns of earthquake occurrence have helped us to understand the evolution of mountain ranges, continents, and oceans. As well, seismicity maps are consulted by planners, geologists, and engineers whenever the mitigation of earthquake hazard is a consideration in the construction of large structures.

Depths of Earthquake Foci

The literal meaning of earthquake is merely the shaking of the ground. The waves that make up the earthquake are called *seismic waves*. Like sound waves radiating through the air from a gong that has been struck, seismic waves radiate through the rocks from a source of energy somewhere in the outer part of the Earth. Although in natural earthquakes this source is spread out through a volume of rock, it is often convenient to simplify an earthquake source as a point from which the waves first emanate. This point is called the earthquake *focus* or, alternatively, the *hypocenter*. The foci of natural earthquakes are at some depth below the ground surface. For artificial earthquakes, such as underground nuclear explosions, the focus is essentially a point near the Earth's surface. The point on the ground surface directly above the focus is called the earthquake *epicenter*.

How far down in the Earth are the foci? One of the early intriguing discoveries by seismologists was that, although many foci are situated at shallow depths down to a few tens of kilometers, in some regions they are hundreds of kilometers deep. Such regions include the South American Andes, the Tonga Islands, Samoa, the New Hebrides chain, the Japan Sea, Indonesia, and the Caribbean Antilles (Figure 2.4); a striking feature is that each of these regions is associated with a deep ocean trench. On the average, the frequency of earthquake occurrence in these regions declines rapidly below a depth of 200 kilometers, but some foci are as deep as 680 kilometers. Earthquakes with foci from 70 to 300 kilometers deep are arbitrarily called *intermediate focus*, and those below this depth are termed *deep focus*. Some intermediate- and deep-focus earthquakes are located away from the Pacific region, in the Hindu Kush, in Romania, in the Aegean Sea, and under Spain.

When the foci of earthquakes near island arcs and ocean trenches are compared with their depths, an extraordinary pattern emerges. Consider the vertical section of the Earth at the top of Figure 2.4, which is drawn at right angles to the Tonga arc in the South Pacific. To the east of these

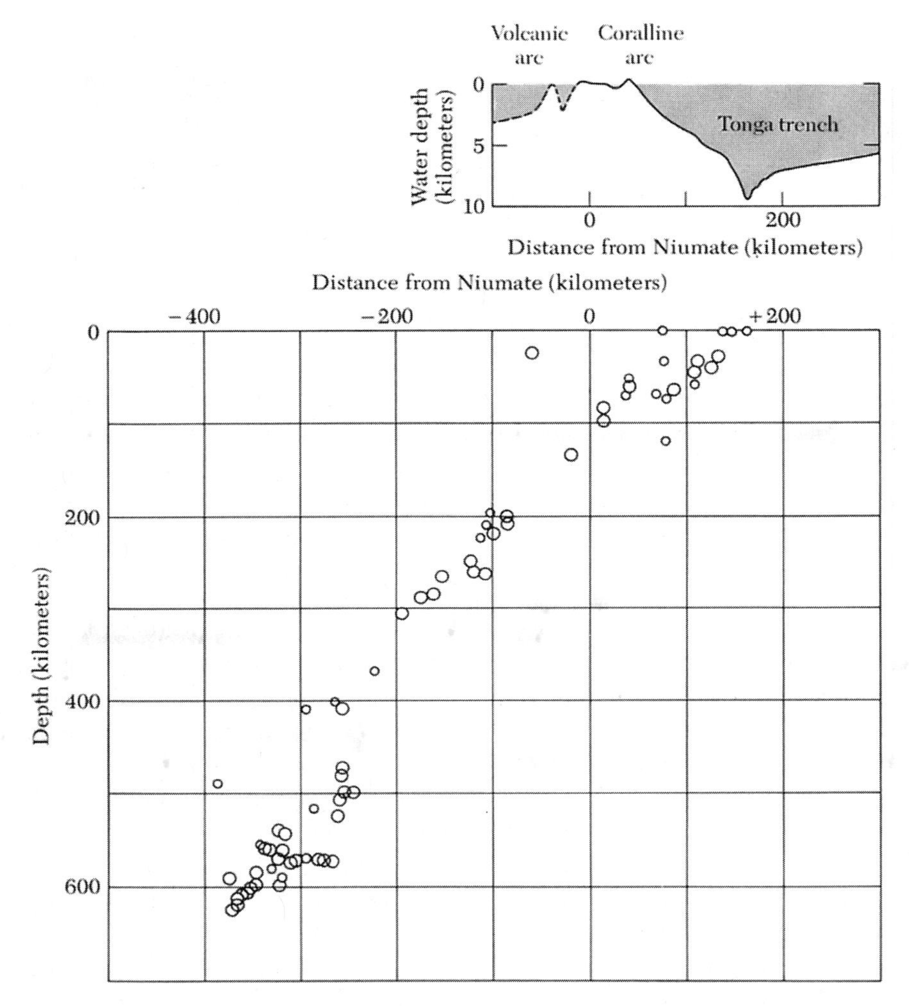

Figure 2.4 Foci of earthquakes in 1965 occurring under the Tonga arc in the southwest Pacific. The vertical section shows that most earthquake centers cluster along a narrow zone starting under the trench and dipping under it at an angle of about 45° to depths of more than 600 kilometers. [Courtesy of B. Isacks, J. Oliver, L. R. Sykes, and *J. Geophys. Res.*]

volcanic islands lies the Tonga trench, as deep as 10 kilometers in places. In the bottom part of the figure, the depths of the foci are plotted against their distance from Niumate, a point on Tonga Island. Notice that the foci lie in a narrow but well-defined zone, which dips from near the trench beneath the island arc at an angle of about 45°. In some dipping zones, the

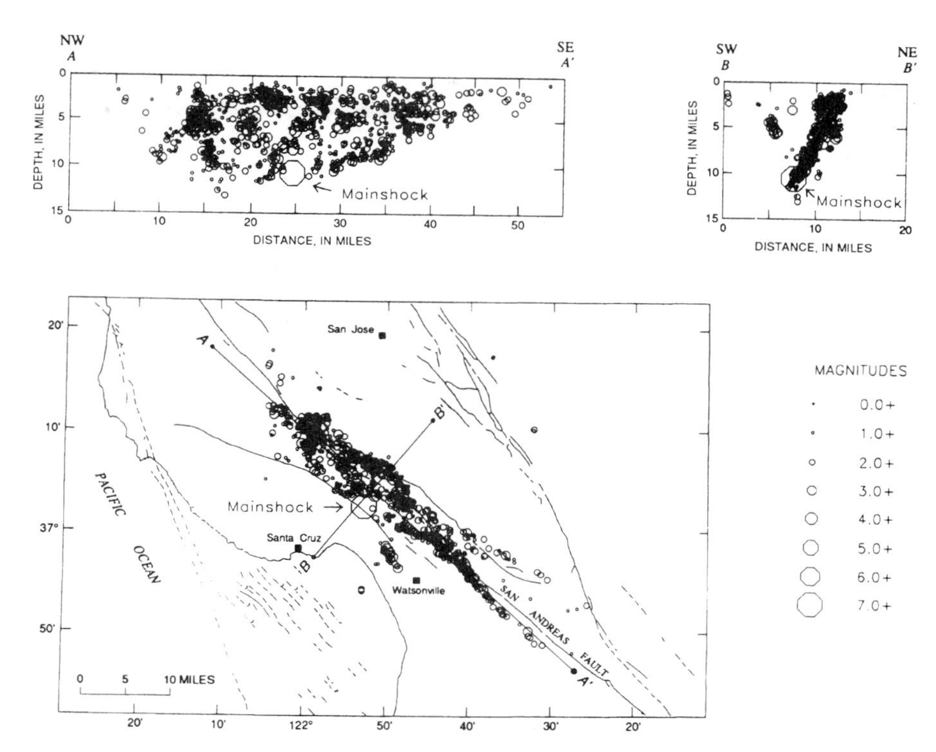

Figure 2.5 (a) Epicenters and foci of the main shock and aftershocks of the 1989 Santa Cruz (Loma Prieta) earthquake sequence. (b) Section AA′ along the San Andreas fault. (c) Section BB′ across the San Andreas fault indicating the dip of the main ruptured fault plane. [From Plafker and Galloway, 1989.]

foci actually lie along two parallel planes (as shown in Figure 4.5). Below depths of 400 kilometers, the active zone steepens, with some foci occurring below 600 kilometers. In other regions of deep earthquakes, some variation in the angle of dip and distribution of foci is found, but the general feature of a dipping seismic zone* is common to island arcs and deep ocean trenches. This universal but simple pattern is discussed later in Chapter 7.

This book concerns mainly *shallow-focus* events, but their foci are still many kilometers below the Earth's surface. Shallow earthquakes wreak the most devastation, and they contribute about three-quarters of the total energy released in earthquakes throughout the world. In California, for

**Called the* Wadati-Benioff zone *after famous Japanese and California seismologists, Professor K. Wadati (1902–1995) and Professor Hugo Benioff (1899–1968).*

example, all the known large earthquakes to date have been shallow-focus. In fact, in the coast ranges of central California the great majority of earthquakes originate from foci in the upper 10 kilometers of the Earth, and only a few are as deep as 15 kilometers. The 1989 Loma Prieta earthquake, discussed in Chapter 1, had an unusually deep focus of 15 to 18 kilometers, as shown by the plotted foci in Figure 2.5.

For various reasons, the determination of the depth of an earthquake focus is not as precise as the location of its epicenter on the surface. Yet depth can be of vital practical concern, because stronger ground shaking may affect a site when the focus is at a depth of 10 rather than 40 kilometers.

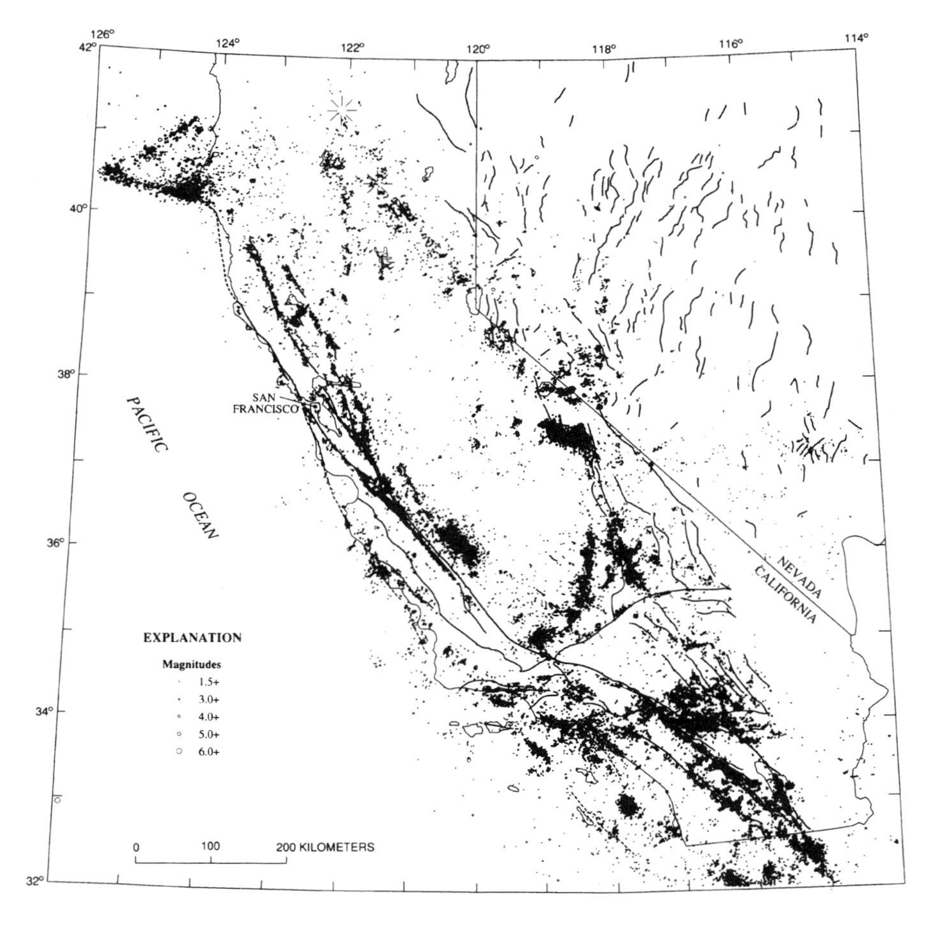

Figure 2.6 Earthquake epicenters in California for magnitudes greater than 1.5 from 1972 through 1983. [Courtesy of USGS.]

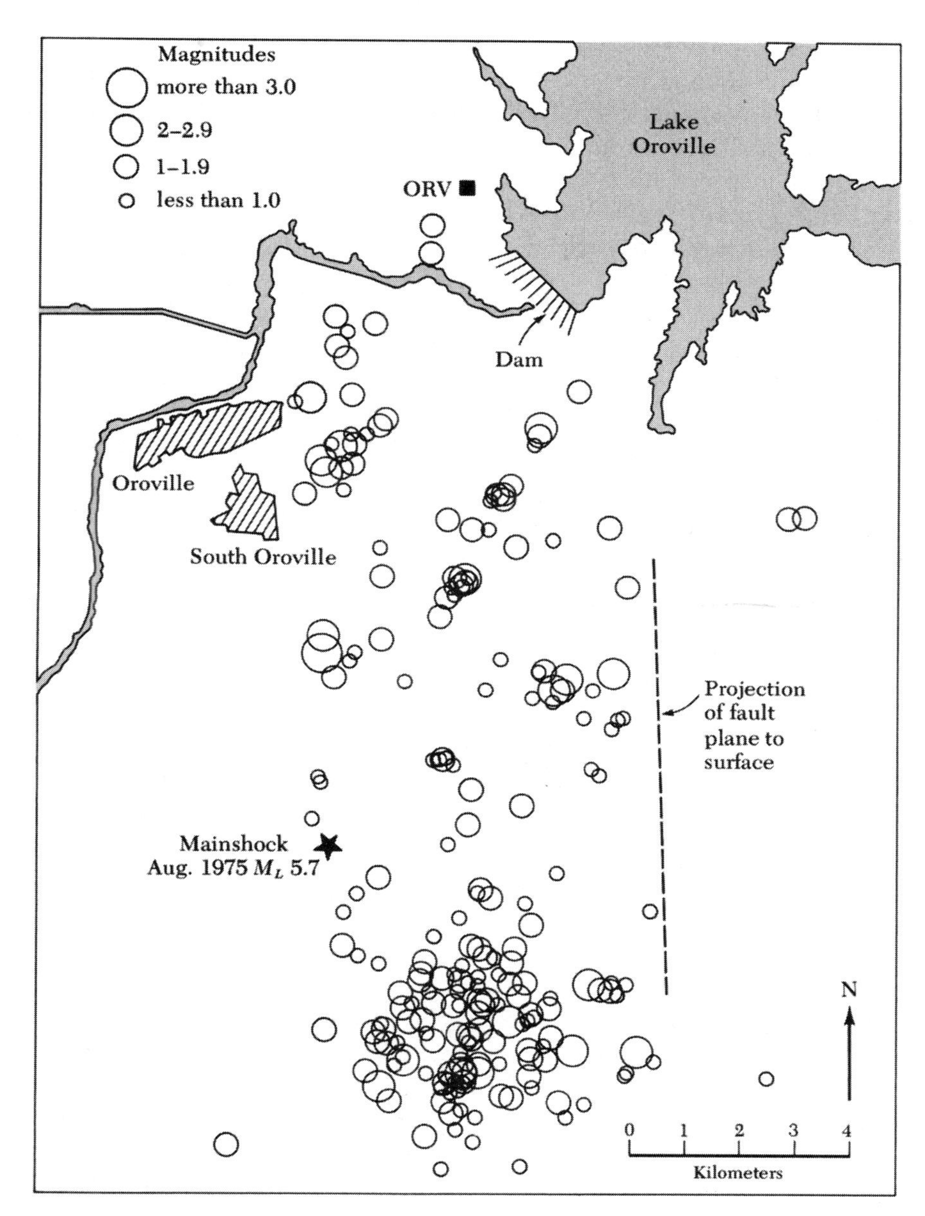

Figure 2.7 Epicenters plotted from the Oroville earthquake sequence, February through July 1977. The black square designates the location of the seismograph at the Oroville station. The dashed line represents the projection to the surface of the fault plane. [Courtesy of California Department of Water Resources.]

Most moderate-to-large shallow earthquakes are followed, in the ensuing hours and even during the next several months, by numerous smaller earthquakes in the same vicinity. These earthquakes are called *aftershocks.* The great Rat Island earthquake in the Aleutian Islands on February 4, 1965, was followed, within the next 24 days, by more than 750 aftershocks large enough to be recorded by distant seismographs. A few earthquakes are preceded by smaller *foreshocks* from the source area, and it has been suggested that these can be used to predict the mainshock (see Chapter 10).

Sometimes, if foci can be precisely located, as in the 1989 Loma Prieta sequence, they indicate the shape and size of the region beneath the ground surface that is the source of the main earthquake (see Figure 2.5). More generally they define the presently tectonically active regions and geological faults (Figure 2.6). This seismological mapping of the deeper rock structures is an extension of normal field methods used by geologists to map surface features. Another example demarcation of such a zone for shallow local earthquakes near Oroville, California, is given in Figure 2.7.

The 1975 Oroville, California, Earthquake

The case history of the Oroville sequence bears witness to the social relevance of earthquake occurrence around our planet. The mainshock of the sequence, which occurred in the afternoon of August 1, 1975, at 1:20 p.m., near the city of Oroville in northeastern California, did not cause major damage to Oroville and surrounding communities, but it attracted public interest because it was only 10 kilometers southwest of Oroville Dam (see Figure 2.7). This earthfill dam near the western foothills of the Sierra Nevada is the largest in North America; it is 236 meters high and has a capacity of 4365 million cubic meters. The filling of the reservoir began in November 1967 and was completed in September 1968.

The region is one of low seismicity, but generally a few minor earthquakes a year take place within 50 kilometers of the dam site. In 1963, before construction of the great dam, seismographs were installed nearby to monitor the background seismicity. These instruments detected no change in the low rate of earthquake occurrence within 30 kilometers of the reservoir either during construction, during filling, or after the water was raised to its highest elevation in 1969 through early 1975.

On June 28, 1975, a few small shocks were recorded to the southwest of the Oroville reservoir. It was not known whether these were foreshocks of a larger earthquake or merely an earthquake swarm of small-magnitude events, common in many parts of California. Nevertheless, some additional portable seismographic stations were installed to keep better track of the position of the earthquakes. About 20 small shocks were recorded through July in the same general area, the largest of magnitude 4.7. Then at the end of July the rate of occurrence fell.

However, in the early morning hours of August 1, the seismic alarm system at the University of California seismographic station was triggered. Seismogram readings indicated that an earthquake of magnitude 4.7 had occurred near Oroville. Later that morning at 6:30, another minor earthquake occurred nearby.

A personal anecdote may be in order here. Because I was at the time a member of the Consulting Board for Earthquake Analysis of the California Department of Water Resources, the owners of the dam, I had been watching the sequence of earthquakes closely. The reactivation of the sequence on the morning of August 1 led me to call Professor G. Housner at the California Institute of Technology, who was chairman of the Consulting Board. I mentioned to him that the earthquake sequence had reactivated, that there was a small but definite likelihood of a mainshock, and suggested that he might contact the Department. Subsequently he did so and engineers in charge of operating the dam facilities made a special service inspection of the dam, its facilities, and instrumentation.

While the dam inspection was taking place, the principal shock of the sequence, magnitude 5.7, occurred just after lunch on August 1. In a sense, then, this earthquake was forecast. The prediction was based upon a personal hunch that the unusual pattern of small local earthquakes indicated that they might be foreshocks of a larger shock. The incident illustrates that some success in practical prediction is feasible under very restricted conditions, such as when foreshocks occur (which is not always the case — see Chapter 10). A necessary condition for success is that, for whatever scientific, engineering, or social reason, informed persons are aware of changes in local conditions and able to think hard and continuously about the change.

The damage from the 1975 mainshock in the meizoseismal area was not severe. Some unreinforced brick chimneys toppled and some weak masonry parapets in the city of Oroville collapsed onto the street. The dam itself had not been affected by the shaking. Instruments on it that had been installed some years before for just such an occasion recorded the ground motion, which was quite moderate. Numerous aftershocks occurred. Subsequent careful locations of their foci with portable equipment defined a zone that dipped into the valley at about 60° to the west. These foci varied in depth from 12 kilometers to the west to nearly surface locations to the east.

If this zone of foci were to be projected to the surface (see Figure 2.7), it would intersect the surface to the south of the dam. Imagine the excitement, therefore, when after a day or two, field geologists found a line of surface cracks in open country at about the place indicated by this intersection. Subsequently, trenches were excavated in several places across the line of cracks, and these confirmed that the cracks in the soil were the surface expression of a preexisting fault zone. Fault gouge was present in most of the trenches, and the offsets in soil and rock layers mapped in the walls of the trenches indicated that vertical motions had probably occurred a number of times in the last 100,000 years in a normal dip-slip fashion (see the discussion of faults in Chapter 5), causing offsets of a few centimeters on each occasion. Field mapping later established that the total length of surface faulting across the grassy fields was about 5 kilometers.

After the Oroville earthquakes there was much speculation on whether they had been caused by the reservoir — an intriguing aspect of seismology that is discussed in Chapter 6. However, in this case there is only circumstantial evidence. The argument against the possibility of reservoir-triggering was that small-to-moderate earthquakes were not unknown in the nearby foothills of the Sierra Nevada, and the regional geological map shows that the line of cracking observed after the earthquake lined up with the extensive system of faults to the south of Oroville. Geomorphic expressions, such as small soaks and springs along the zone of cracking, indicated that spasmodic ground movement during the last few thousand years had occurred.

Further evidence against reservoir induction was that the focus of the mainshock lay at a distance down in the crust about 15 kilometers away from Oroville Reservoir. How could the reservoir affect such a distant point? Even the areal extent of the aftershocks defined a linear dipping zone that intersected the surface of the ground south of the reservoir itself.

On the other side of the argument, more quantitative but still circumstantial evidence comes from the measured rate of occurrence of foreshocks and aftershocks. This rate of occurrence of earthquakes above a given magnitude can be countered and measured by the factor b given in Appendix G. In most regions, b varies between 0.7 and 1.0; in northern California, b equals about 0.8, if averaged over a few decades. The b values for reservoir-induced sequences are usually found to be higher than the respective regional values. Yet for the Oroville aftershocks the rate of occurrence gives b a value equal to 0.6 — a value *less* than the regional value.

In the end, the strongest argument for earthquake triggering at Oroville is simply the presence of the nearby reservoir. Undoubtedly, it sent a pressure pulse through the water in the rocks of the crust (see Chapter 6).

Perhaps, as the pressure pulse spread out by percolation through the crustal rocks nearby, it eventually reached a weak place along an already existing fault zone (see Chapter 5). Although the pulse weakened as it spread outward, it may have been sufficient to open microcracks just enough to allow sudden fault slip — the straw that broke the camel's back.

Surprise Seismicity in Upper Egypt

In many lands, although earthquakes are extremely rare, the unexpected sometimes shakes entrenched views. One recent case was along the Nile River. Its rise and fall, as recorded by nilometers in Cairo since 622 A.D., has been so regular that Egypt became a food storehouse in time of drought and famine, punctuated by some very high floods and resulting disasters. The river is 6700 kilometers long, with the main stream formed by the confluence of the White Nile and the Blue Nile at Khartoum, 3080 kilometers above its mouth. Below Khartoum, the river reaches Aswan, located on its first cataract (see Figure 2.1). There the river is placid, with a strip of green vegetation between the water and the rocks of the desert.

A major change to the Nile system occurred in the 1970s when the High Dam was built of earthfill about 10 kilometers south of Aswan. Its maximum height is 111 meters and the length of its embankment is 3.6 kilometers. The dam impounded a huge artificial reservoir called Lake Nasser, approximately 300 kilometers long. Lake Nasser has a maximum capacity of $164{,}000 \times 10^6$ cubic meters, which is slightly larger than that of Lake Kariba in southern Africa. Filling of the reservoir started gradually in 1964 and reached a maximum water level of 177.5 meters in November 1978. Typically, the lake reaches a seasonal peak level in October-November; the annual level cycle is approximately sawtoothed in shape. There is a slow decrease from November to a low in July, followed by a more rapid filling. The usual annual variation in level is 4 to 5 meters.

On November 14, 1981, a moderate earthquake (local magnitude = 5.6; see the discussion of Earthquake Magnitude in Chapter 8) was felt strongly in Aswan. Although its exact hypocenter is unknown (the closest operating seismographic station at the time was at Helwan, about 690 kilometers from Aswan), the intensity data and aftershock locations (recorded by local seismographic stations placed after the mainshock) indicate a shallow hypocenter beneath an extensive bay of Lake Nasser, about 60 kilometers from the Aswan High Dam.

The earthquake was preceded by a few recorded foreshocks and followed by thousands of aftershocks in the same general vicinity (see

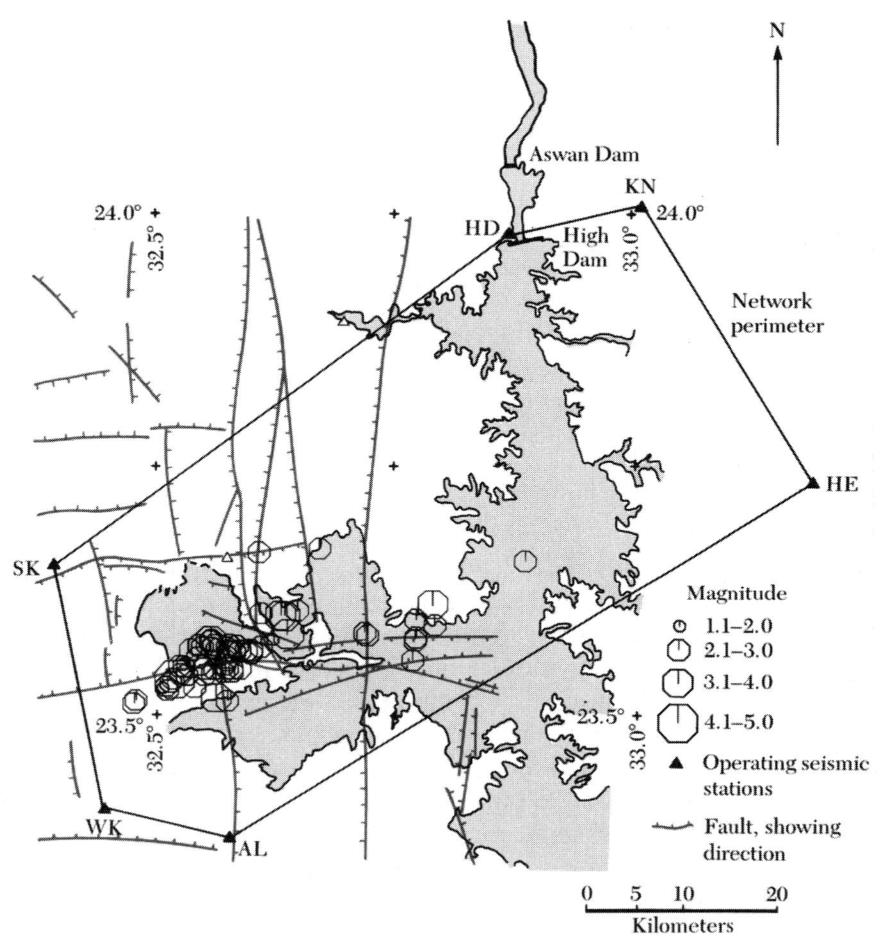

Figure 2.8 Map of aftershock epicenters of the Aswan sequence from May 7 to June 11, 1982, from the Egyptian Geological Survey and Mining Authority seismographic network. The Kalabsha fault trace passes east-west through the dense epicenter cluster. [From T. R. Toppozada and others, California Division of Mines and Geology, 1984.]

Figure 2.8). At Aswan, most people rushed out of doors and heard explosionlike sounds, often a sign of very shallow focal depth. Some minor damage was caused to some old buildings in Aswan, but no damage occurred to the dam or to its appurtenant structures.

The seismic episode was an unwelcome surprise. The Aswan area had been considered aseismic; no significant earthquake had been mentioned in historical records as far back as 3000 years anywhere in upper Egypt or located there since the introduction of global seismographic observatories at the turn of the century. It is therefore reasonable to assume that the Aswan area is not the source of significant earthquakes under natural geological circumstances (at least with return intervals of 5000 years). This conclusion makes it very likely that the November 14, 1981, earthquake and aftershocks are causally related to the water of Lake Nasser — just the situation explained in Chapter 6. The Nubian sandstones along the Nile River are very porous, so that the effective size of the reservoir is much larger than the surface of Lake Nasser itself, with a huge volume of water absorbed underground in the porous basement rock. The water load is therefore greater than might at first be calculated, with consequent large changes in water pressure in the pores of the rocks.

Were there definite forerunners of the main shock on November 14, 1981? A few seismographs had been installed in the Aswan area before the induced earthquake occurred, but their operation was irregular before 1981. Nevertheless, they did enable Egyptian seismologists to identify a few small local earthquakes, suggesting that low-magnitude seismic activity in the area may have begun after the reservoir began filling. The principal earthquake followed the 1981 seasonal maximum in water level. After that time, the water level began to decline because of the widespread drought of the watershed of the Nile River in that year.

Studies of the aftershock sequence (see Figure 2.8) mapped the dense zone of aftershocks close to the surface trace of the Kalabsha fault, which strikes almost east-west across the desert from Libya and terminates close to the west bank of the Nile about 45 kilometers south of the High Dam.

In subsequent years, because of the appalling risk downstream posed by dam failure, an extensive seismological and engineering study of the Aswan High Dam area was made in which I was involved. All indications were that the dam structures will adequately resist any future earthquake, reservoir-induced or otherwise.

Moonquakes and Marsquakes

Until 30 years ago we all wondered: Are there moonquakes, and if so, do they resemble quakes on Earth? Answers were provided by the lunar space program of the United States. Beginning in November 1969, seismographic stations were set up on the moon during the landings of Apollo 12, 14, 15, 16, and 17 (see Figure 2.9). Special seismographs — with power from solar energy and nuclear batteries — placed at the five sites operated continuously, sending back a steady stream of geophysical data on the

Figure 2.9 Photograph of a seismograph operating on the moon's surface. [Courtesy of NASA.]

moon's interior. The plan was to follow scientific successes on Earth where, since 1900, seismologists used the earthquake waves traveling through the Earth to obtain highly detailed information on the structure of the terrestrial interior (see Chapter 4). Also, study of the location and mechanisms of earthquakes have revealed a great deal about the way the Earth is deformed. It was expected that seismographs on the moon would provide similar information on its interior and deformation.

But earthquakes on Earth are a consequence of its massive dynamic processes (see Chapter 7). In stark comparison, the moon has, for many millions of years, been an unchanging planetary body with no visible active volcanoes and no fresh rift systems. It was therefore somewhat startling that each lunar seismographic station detected between 600 and 3000 moonquakes every year on the average. Most of the moonquakes were tiny, with magnitudes of less than about 2 on the Richter scale (reference Box 8.1). No atmosphere and oceans mean an extremely quiet ground noise background, so that the seismographs could be operated with very high magnifications, at least 100 to 1000 times that which is normally possible on Earth. This sensitivity raises the question: Is the large number of moonquakes

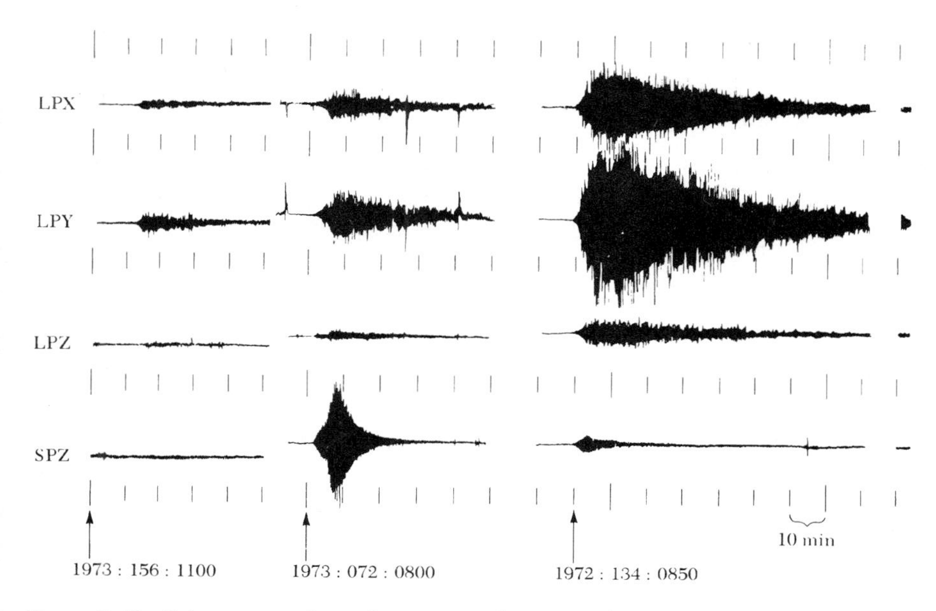

Figure 2.10 Seismograms from three types of moonquakes recorded at the Apollo 16 station. LPX, LPY, and LPZ are the three long-period components, and SPZ is the short-period vertical component. The first column shows a deep-focus moonquake; the center column, a shallow moonquake; the third column shows records of the impact of a meteoroid on the lunar surface. Vertical timelines are 10 minutes apart. [Courtesy of NASA.]

detected a consequence of the low level of background noise on the moon, or does it signify the presence of fairly active tectonic processes?

Sample seismograms of typical moonquakes, recorded at the Apollo 16 station, are shown in Figure 2.10). Three different kinds of events are indicated. First are the deep moonquakes, which have their foci at depths of 600 to 900 kilometers in the moon (radius 1740 kilometers). These deep events are surprising. They seem localized at a specific number of places in the moon's interior, of which more than 40 have been identified. At these active deep centers, moonquakes commonly occur within an interval of a few days during *perigee*, the point at which the moon's orbit is closest to the Earth. About equal numbers of deep moonquakes occur at these centers at opposite phases of this tidal pull, so that the most active periods are 14 days apart. These periodic properties at least suggest that the tidal pull of the Earth on the moon triggers the occurrence of the deep seismic-energy releases. On the Earth, similar conditions of the occurrence of earthquakes with perigee have been sought for many years, but generally without success (see Chapter 10).

The second type of moonquake shown in Figure 2.10 occurs in the shallower part of the moon. These moonquakes are not as common as the deeper events, and their locations do not exhibit any particularly regular pattern over the surface of the moon. It is thought that, like most earthquakes, they are due to the release of tectonic elastic strain in the rocks of the moon's crust. If so, then either groundwater is present in the moon or some dry fracturing is occurring in the unusual thermal conditions of the lunar surface, with its extreme range of cold and heat.

The third type of seismic event results from the impact of objects, both natural and human-made, on the lunar surface. An example of the seismograms from meteorite impact is also shown in Figure 2.10. The lunar seismographic stations were efficient detectors of meteorites hitting the surface even at a range of 1000 kilometers. In order to help with the determination of seismic-wave velocities of lunar rocks, parts of the lunar spacecraft were programmed to crash back on the lunar surface. These high-speed impacts generated seismic waves strong enough to produce clear recordings at lunar seismographic stations, and because the position of impact was known precisely, the travel times of the seismic waves could be easily calculated. This calculation was the first step to discovering the general architecture and properties of the moon's interior.

The waves recorded by lunar impacts and moonquakes indicated that most of the moon is solid rock. The layered crustal shell on the surface is about 60 kilometers thick, and below lies a denser solid mantle about 1000 kilometers thick. This overlies a central core that seems to be somewhat softer than the mantle but is most probably not liquid. The picture of the Earth's interior is markedly different (compare Chapter 4).

Moonquakes themselves sharply differ from earthquakes. Compare the seismograms in Figure 2.10 with that shown in Figure 3.5. A small earthquake may shake a remote seismograph for a minute or so, but on the moon, the recorded shaking of the lunar surface in a moonquake continues for as much as an hour. The wave patterns too are strikingly different: the secondary (S) waves and surface waves on lunar seismograms are not generally as clearly defined and distinct as are those of earthquakes. (In Figure 2.10, the small primary (P) wave onset for the moonquake can be seen on the SPZ record, and the S wave can be seen best on the LPY record.) After a rapid crescendo, the lunar seismogram of a moonquake shows reverberations with a slow decrease in amplitude for many tens of minutes. What is the explanation for this behavior? It is widely believed that both the lack of water saturation and the fractured nature of lunar rocks are contributory causes; the uppermost rocks are so dry the seismic waves attenuate very little, and at the same time the cracks in the lunar rock scatter the seismic waves in all directions.

Seismographs were also placed on Mars by the two Viking crafts sent by the United States in 1976. Unfortunately, the instrument in Viking 1 failed to return signals to Earth, but that on Viking 2 operated as planned, and signals of ground motion on Mars were recorded remotely on Earth. After about a year, scrutiny of the available Martian seismograms had found only one event that could reasonably be identified as a Marsquake. It is possible, however, that Marsquakes are simply not common, and it is too early to exclude the possibility of notable seismic activity there.

3 Measuring Earthquakes

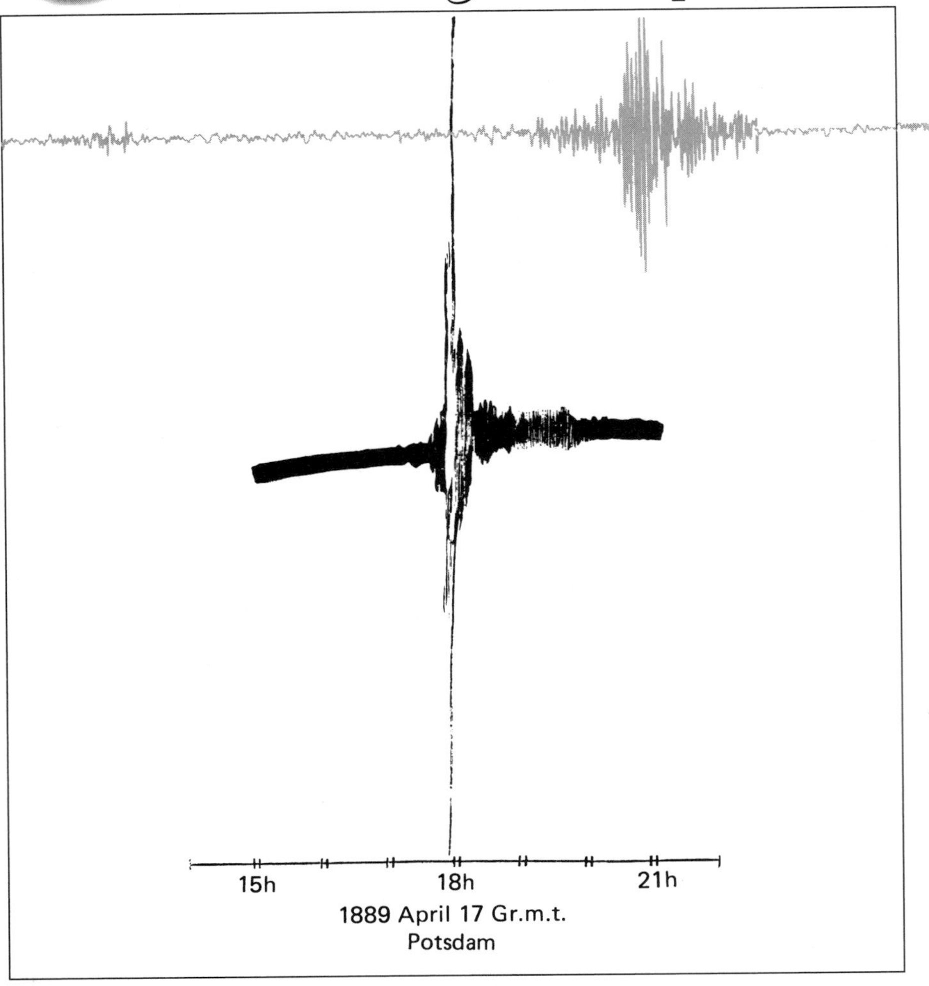

First identified seismogram of a distant earthquake. Response of a pendulum in Potsdam, Germany, to a large earthquake felt strongly in Japan on April 17, 1889.

It is not unlikely that every large earthquake might, with proper instrumental appliances, be recorded at any point on the land surface of our globe.

— *Professor John Milne, 1893*

What does the ground actually do during an earthquake? To answer this question, instruments have been designed to measure ground shaking in detail. We now know that earthquake waves of microscopic size can traverse the whole Earth. By contrast, in the severe shaking in the meizoseismal zone of a large earthquake the ground moves back and forward many centimeters.

Sensitive and Strong-Motion Seismographs

The first earthquake recorder described in any detail was an artistic device (see Figure 3.1) invented by the Chinese scholar Chang Heng about 132 A.D. Balls were held in dragons' mouths connected by linkages to a vertical pendulum. Shaking released the balls. The instrument was a *seismoscope* because, unlike a seismograph, it could not give the complete time history of the earthquake shaking but simply the direction of the principal impulse due to the earthquake. The ancient description states that,

> When an earthquake occurs, and the bottle is shaken, the dragon instantly drops the ball, and the frog which receives it vibrates vigorously; anyone watching this instrument can easily observe earthquakes.

The story continues:

> Once upon a time a dragon dropped a ball without an earthquake being observed; and the people therefore thought the instrument of no use, but after two or three days a notice came saying that an earthquake had taken place at Rosei. Hearing of this, those who doubted the use of the instrument began to believe in it again. After this ingenious instrument

Figure 3.1 The author with model of Chang Heng's seismoscope. Balls were held in the dragons' mouths by lever devices connected to an internal pendulum. The direction of the epicenter was reputed to be indicated by the first ball released. [Photo by National Geographic Magazine.]

> had been invented by Choko (Chang Heng), the Chinese Government wisely appointed a secretary to make observations on earthquakes.

Use of Chang Heng's instrument seems to have died out in a short time and it was not until just before the beginning of the twentieth century that the first effective seismographs were constructed. The first identified recording of a very distant earthquake was written by a pendulum instrument in Potsdam, Germany, from a large felt source in Japan on April 17, 1889. It was hardly recognized at the time that the photographic image reproduced opposite the opening page of this chapter was a gigantic step in understanding the structure and workings of planet Earth.

Although earthquake-recording instruments, called *seismographs,* are now more sophisticated, the basic principle employed is the same. A mass on a freely movable support can be used to detect both vertical and horizontal shaking of the ground. The vertical motion can be recorded, as illustrated in Figure 3.2, by attaching the mass to a spring hanging from an anchored instrument frame; the bobbing of the frame (as with a kitchen scale) will produce relative motion (Figure 3.3). When the supporting frame is shaken by earthquake waves, the inertia of the mass causes it to lag behind the motion of the frame, and this relative motion can be recorded

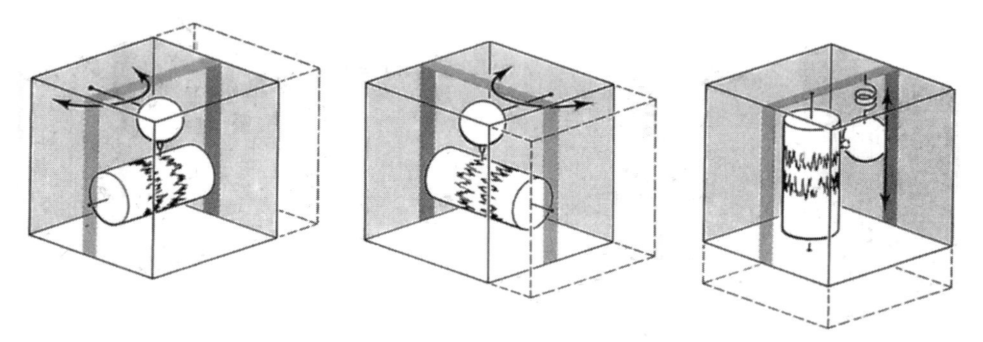

Figure 3.2 Simple models of pendulum seismographs recording the vertical and horizontal directions of ground motion. The pendulum must be damped in order to separate seismic pulses. [From Bruce A. Bolt, *Nuclear Explosions and Earthquakes: The Parted Veil* (San Francisco: W. H. Freeman and Company. Copyright 1976).]

as a wiggly line by pen and ink on paper wrapped around a rotating drum (alternatively the motion is recorded photographically or electromagnetically on magnetic tape or as discrete digital samples for direct computer input). For measurements of the sideways motion of the ground, the mass is usually attached to a horizontal pendulum, which swings like a door on its hinges (see Figure 3.4). Earthquake records are called *seismograms.* An example is given in Figure 3.5.

If you try to build a simple seismograph* by attaching a mass to the end of a spiral spring or rubber band and shake your hand to simulate the seismic waves, you will find that the mass continues to oscillate after the hand has been brought to rest. Such free motion of the pendulum tells nothing about the ground shaking and must therefore be damped by some mechanical or electrical means. In this way, the relative motion between the mass and the frame is a direct measure of ground motion. However, this relative movement is still not the true motion of the ground, so that most seismograms do not give an *exact* picture of what the ground did. The actual ground motion must be calculated by taking into account the mechanics of the pendulum — just as for a moving car, say, such motions can be stated as ground accelerations, velocities, or displacements.

In modern seismographs (see Figure 3.4) the relative motion between the pendulum and frame produces an electrical signal that is magnified electronically thousands and even hundreds of thousands of times before it is used to drive an electric stylus to produce the seismogram. In this way, very weak seismic waves from earthquake sources or large underground

**For instructions on how to build a working seismograph, see the World Wide Web address at the end of the book.*

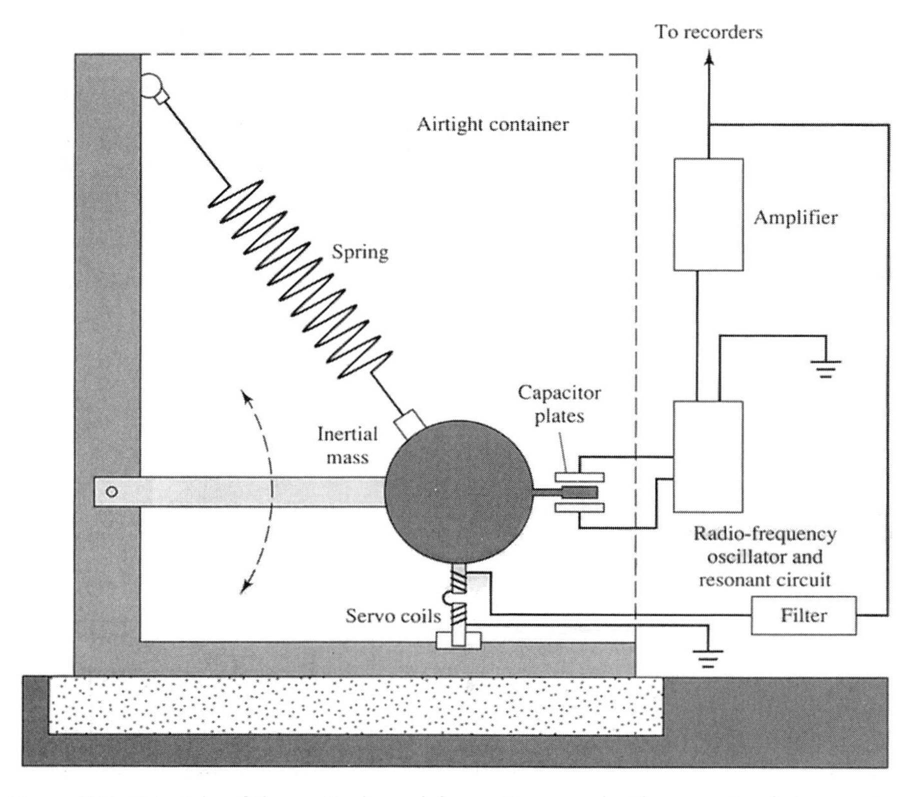

Figure 3.3 Principle of the vertical pendulum seismograph. The mass tends to remain stationary as the Earth moves. Relative motion at the capacitor plates generates an electrical signal that is fed to an analog or digital recorder. The filter feeds back spurious signals, representing undesirable ground motions, to coils that keep the mass centered. [From B. A. Bolt, *Inside the Earth*. (San Francisco: W. H. Freeman and Company, Copyright 1982).]

explosions at very remote sites can be detected. The electrical signals from a seismograph pendulum can also be recorded continuously onto magnetic tape (as sound waves are recorded by a microphone onto a tape recorder), or commonly nowadays, as a stream of numbers (e.g., the binary digits zero and one) representing the wave signal at many points each second. In this way, the ground motion can be preserved in a magnetic form and, when required, played out from computer memory through some visual recording device or screen or audibly to produce earthquake sounds.

In the zone near to an earthquake's source, no electrical amplification of the strong seismic waves is, of course, needed. For this reason *strong-motion seismographs* (Plate 5) have been specially designed to record the

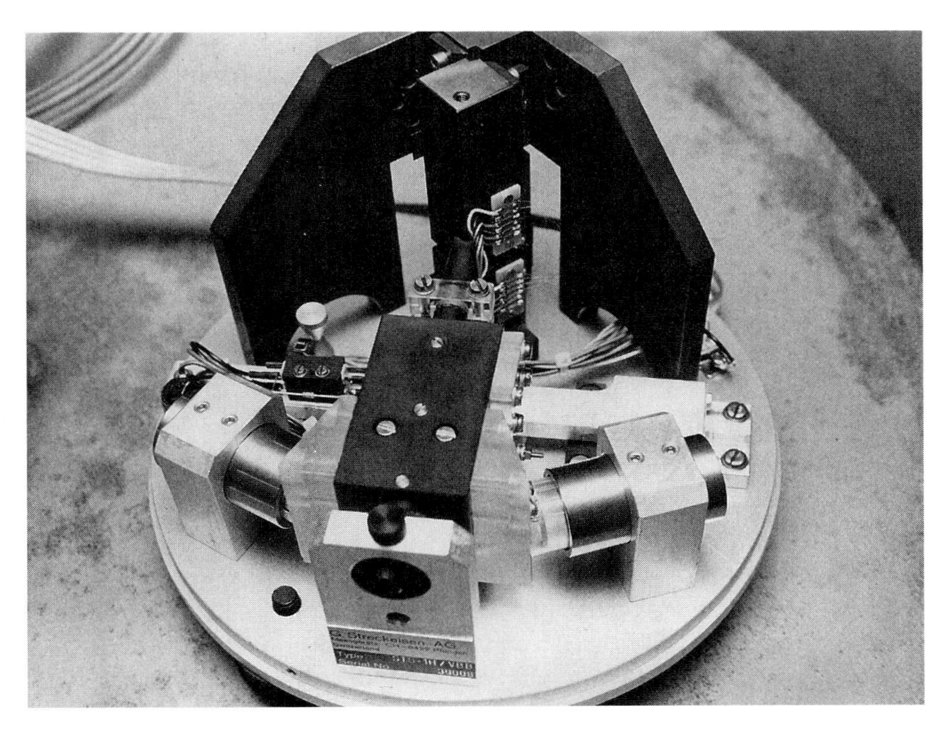

Figure 3.4 Photograph of a modern ground motion sensor of a digital seismograph. The horizontal pendulum is under the top rectangular plate, with the movable coils attached at right and left. Precise time can now be kept inexpensively using Global Positioning System (GPS) satellite clocks.

high-amplitude shaking of the ground. The records obtained can be directly read as acceleration, velocity, or displacement of the ground. The most common strong-motion seismographs record the ground acceleration directly and are called *accelerometers*. Usually, accelerometers do not record continuously but are triggered into motion by the first waves of the earthquake to arrive. This is because, even in earthquake country such as California and Japan, there may not be any strong ground motion from earthquakes to record for months, or even years, at a time. Consequently, continual recording of hundreds of such instruments would be a wasteful exercise. After triggering, the recording continues for some minutes or until the ground shaking falls again to imperceptible levels. These strong-motion instruments are capable of recording accelerations of the ground several times greater than that of gravity.

Strong ground acceleration appears as waves on the accelerogram (see Figure 8.4). When the recording is near the earthquake source, it is often difficult to distinguish on the accelerograms the regular types of seismic waves — such as P, S, and surface waves described in Chapter 1 (compare the seismograms in Figures 3.5 and 8.4). Seismologists are now making a major effort to understand more about these intriguing patterns of energetic waves because they contain information on the source process and effects of geological complexity.

Strong-motion seismographs are designed to record the strongest shaking, because these tell us about the felt and damaging motion near the source of the shaking. But most seismographs around the world — especially those at the more than 1000 continuously recording seismographic stations — are very sensitive "ears on the Earth." They can detect and record earthquakes of small size from very great distances, but they go off-scale (or "clip") at even moderate ground motions.

How to Understand Seismograms

A seismogram appears to be no more than a complicated series of wavy lines, but from these lines a seismologist can determine the hypocenter location, magnitude, and source properties of an earthquake. Although experience is essential in interpreting seismograms, the first step in understanding the lines is to remember the following principles. First, earthquake waves consist predominantly of three types — P waves and S waves, which travel *through* the Earth, and a third type, surface waves, which travel *around* the Earth (see Figure 1.9). If you look closely enough, you will find that almost always each kind of wave is present on a seismogram, particularly if it is recorded by a sensitive seismograph at a considerable distance from the earthquake source. Each wave type affects the pendulums in a predetermined way. Second, the arrival of a seismic wave produces certain telltale changes on the seismogram trace: the trace is written more slowly or rapidly than just before; there is an increase in amplitude; and the wave rhythm (frequency) changes. Third, from past experience with similar patterns, the reader of the seismogram can roughly identify the pattern of arrivals of the various phases.

Consider the seismogram shown in Figure 3.5, which is part of a longer record made by a sensitive digitally recording seismograph at Berkeley, California. The seismograph sensor responds to the vertical ground-wave motion being recorded continuously. The top trace indicates precisely the elapsed time. Accurate timing is a fundamental requirement of seismology, and most observatories today have crystal clocks that keep correct time relative to the worldwide standard, called Universal Time (UT), within a few thousandths of a second. Further time checks are made

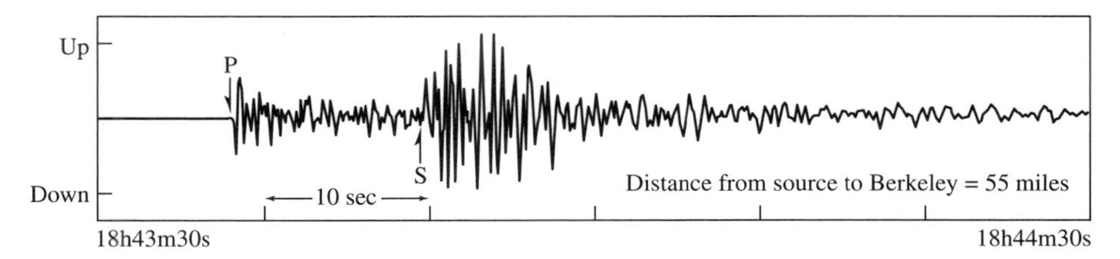

Figure 3.5 Seismogram at Berkeley, California, from a magnitude 5.3 earthquake located 90 kilometers away, northeast of Santa Cruz on June 27, 1988. This recording of the vertical component of ground motion clearly shows the separate onsets of the P and S waves. (Check the approximate epicentral distance from Box 8.1.) Time increases on the trace from left to right.

on the records from a radio time service (such as that provided by the Global Positioning Satellite System). The time offsets in Figure 3.5 are 10 seconds apart. The digital wave recordings are stored in a computer and the image is formed on the terminal screen.

The recorded ground trace is never without some small wiggles. These occur because this seismograph is so sensitive that it is able to detect the continuous, yet imperceptible, background noise of the Earth. These tiny shakings, called *microseisms*, arise from many local disturbances; traffic on streets, wind in trees, and other natural movements such as the breaking of the surf on the beach. Such microseisms have been numerically removed or filtered out in the image example in Figure 3.5.

Now examine the trace offset marked by the arrow labeled P in Figure 3.5. At this point in time, the seismograph detected a significant increase in ground shaking. The larger wave seen arriving at this point is the first P wave from an earthquake generated by the sudden rock slip at the focus of a magnitude 5.3 earthquake. Motion following this P wave continued for some 12 seconds. Next you can see a radical change on the seismogram. The height of the waves becomes suddenly larger at the arrow marked S. This point on the seismogram marks the arrival of the first S wave from the earthquake focus. These S waves continue for about another 10 seconds.

Now look a little farther to the right on the record; seismic waves running around the Earth's surface from the source to Berkeley arrive for over 30 seconds. These waves are mainly surface waves, which have taken longer to reach Berkeley than the body waves. Because the ground motion shown in Figure 3.5 is in the vertical direction, the later surface-wave train corresponds mainly to Rayleigh waves (see Figure 1.9).

It is important to note that seismic waves in the Earth gradually lose their energy as they spread over greater areas, just as the heights of ripples on a pond decrease as they spread out from the center of a disturbance.

P and S waves decrease because they spread more quickly than do seismic surface waves, and it is the surface waves that tend to persist at great distances. Even close to the earthquake center, however, a part of the shaking in an earthquake comes from surface waves, and they can cause damage to buildings.

A closer scrutiny of the seismogram in Figure 3.5 shows that although the wave motion is reasonably continuous, it is somewhat more complicated than can be explained in terms of the arrival of a single P wave, a single S wave, and a Rayleigh wave train. This complexity is due in large part to vicissitudes caused by variations in rock structure along the paths traveled by the waves. Just as sound waves echo back from high buildings or the walls of a canyon, seismic echoes reflect back from rock interfaces in the Earth, and these additional phases cause bumps on the record that, if observed carefully, can be interpreted (see Chapter 4).

A common time standard must be used to compare the arrival times of seismic waves between earthquake observatories around the world. Traditionally, seismograms are marked in terms of UT or Greenwich Mean Time (GMT), not local time. The time of occurrence of an earthquake in Universal Time can easily be converted to local time, but be sure to make allowance for Daylight Saving Time when this is in effect.*

Let us now briefly summarize the physics of the seismograph. Earthquakes produce swings of the pendulum relative to the frame, and these in turn record as a wavelike trace on a seismogram. On seismograms, we observe a series of peaks and troughs, resembling waves on the ocean or vibrations of a violin string. As illustrated in Appendix H, the height of a particular wave above its zero position is called the *wave amplitude*, and the time it takes to complete one cycle of motion (that is, from one peak to the next) is called the *wave period*. The frequency of a wave — measured in units called hertz — is the number of vibrations (cycles) per second, and the wave period is equal to the inverse of the frequency. Humans are able to hear sounds with frequencies of many thousands of hertz down to frequencies of about 15 hertz. In earthquakes the main shaking of the ground that is felt has frequencies of 20 hertz down to 1 cycle per second or even lower. The small overlap with human ear response is sufficient to allow earthquakes sometimes to be heard as well as felt.

As mentioned earlier, the amplitudes of the waves recorded on a seismogram do not constitute the real amplitude of the ground shaking that produced the record. This is because seismographs have amplifying devices that increase the ground motions by a desired factor (perhaps many

**California Standard Time is 8 hours behind UT.*

thousands of times). When the amplifying factor is taken into account, it turns out that the ground motion that produced the S waves marked on the seismogram in Figure 3.5 had an amplitude of only one thousandth of a centimeter.

How to Locate an Earthquake

The seismologist has the unique job of locating the center of an earthquake. At one time, this task was done solely by determining the strength of ground shaking from reports of human reaction and from damage; from these intensities, the position and extent of the source of the wave radiation could be roughly determined (see Chapter 8). In this way, the sources of large earthquakes were found to be not a point but spread out across a considerable area, some areas extending many tens of kilometers. These methods of locating roughly the *field epicenter* of damaging earthquakes remain valuable because in many areas historical earthquakes that occurred before seismographs were invented are important in the evaluation of seismic risk. Although these field methods give little indication of the depth of the focus of the earthquake, they sometimes define rather well (see Chapter 5) where geologists should search for surface fault rupture.*

Nowadays, for the great majority of earthquakes, the location is determined from the time taken by P seismic waves (and sometimes S waves) to travel from the focus to a seismograph. Many hundreds of modern seismographs are operated worldwide continuously and reliably for this and other purposes (see previous chapter). In some seismic areas, special local networks of seismographic stations have been installed to locate the foci of even very small earthquakes. For instance, around new large dams, sensitive seismographs are routinely operated to detect earthquakes that may have a bearing on dam safety. Sometimes the seismometers at each station of a network are connected over telephone lines to a central observatory where the signals are recorded side by side on film or magnetic tape. This procedure greatly helps seismologists to locate earthquakes accurately, not only because the pattern of P- and S-wave arrivals makes the rough

**I heard of a new twist to "field" location in California, when a geologist in Sacramento, speaking on the telephone to a colleague in San Francisco about 100 kilometers to the southwest, suddenly said, "Wow, I feel the building swaying from an earthquake." After a 35-second pause, his San Francisco colleague said, "Wait, so do I." Their conclusion was that the earthquake was to the northeast. (Would you agree?)*

location of the earthquake source immediately obvious but also because only one precise clock — at the central observatory — is needed.

Various modern methods of locating earthquake epicenters and foci differ little in principle. The reader will find details at many of the World Wide Web sites listed at the end of the book. All methods essentially depend on a single fact: the travel time of a seismic wave, such as a P wave, from the source to a given point on the Earth's surface is a direct measure of the distance between the two points. Seismologists have been able to determine by trial and error the average travel time of seismic P and S waves, for example, for any specified distance. The times have been printed in tables and graphs as a function of the distance. The appropriate distance between the observatory and focus can be read from such tables by comparing them with those that have been actually measured in an earthquake.

If arrival times at only one observatory are available, only the distance of the earthquake source from that observatory can be determined reasonably well, and not the geographical location. If arrival times at three observatories are available, then triangulation can be used to determine the latitude and longitude of the earthquake focus and the time of occurrence of the earthquake. Actually, it is common practice now to use the readings from many observatories. The International Seismological Centre in England, for example, might typically locate a moderate-size earthquake on the Mid-Atlantic ridge under the Atlantic Ocean, using the readings from 60 or more seismographic stations from around the world. The arithmetic is carried out with high-speed computers.

To demonstrate how to locate earthquake epicenters, one method that can be easily followed is displayed in Box 3.1. The problem is to determine the location of one of the aftershocks that followed the main 1975 Oroville, California, earthquake (see Chapter 2). Let us suppose that we have available only seismograms from three California stations — Berkeley (BKS), Jamestown (JAS), and Mineral (MIN). Suppose that scrutiny of the seismogram at Berkeley shows that the time interval between the onsets of the P and the S waves at Berkeley was 21.0 seconds. In the same way, time intervals between P and S can be obtained for the stations Jamestown and Mineral. Actual measurements of the arrival times of P and S waves at the three stations are listed at the top of Box 3.1.

From past experience, we know the average distance between an epicenter and a seismograph corresponding to each S minus P interval (see Box 8.1). Thus, the appropriate distances between the epicenter and BKS, JAS, and MIN have been figured to be 190 kilometers, 188 kilometers, and 105 kilometers.

Box 3-1

Sample Calculation of the Location of the Epicenter of an Earthquake (near Oroville, August 1975)

In this earthquake, P and S waves arrived at the stations Berkeley (BKS), Jamestown (JAS), and Mineral (MIN) at the following times (Universal Time):

	P			S		
	hr	min	s	hr	min	s
BKS	15	46	04.5	15	46	25.5
JAS	15	46	07.6	15	46	28.0
MIN	15	45	54.2	15	46	07.1

The following epicentral distances are estimated from the S minus P times above (from the left column of Box 7.1).

	S minus P (seconds)	Distance (kilometers)
BKS	21.0	190
JAS	20.4	188
MIN	12.9	105

With these distances as radii, one can draw three arcs of a circle, as shown in Figure 3.6. Note that these do not quite intersect at one point, but interpolation from the overlapping arcs yields an estimated epicenter of 39.5° N, 121.5° W, with an uncertainty of about 10 kilometers from these readings.

Then, applying the distance scale on a map of California, one can use a compass to draw three arcs of a circle, with the three observatories as centers (see Figure 3.6). The arcs will intersect, at least approximately, at some point. This point of intersection is the estimated location of the earthquake source. (The focal depth is still unknown, and more data are needed to calculate it.) The whole process is now easily performed on a personal computer with graphic screen.

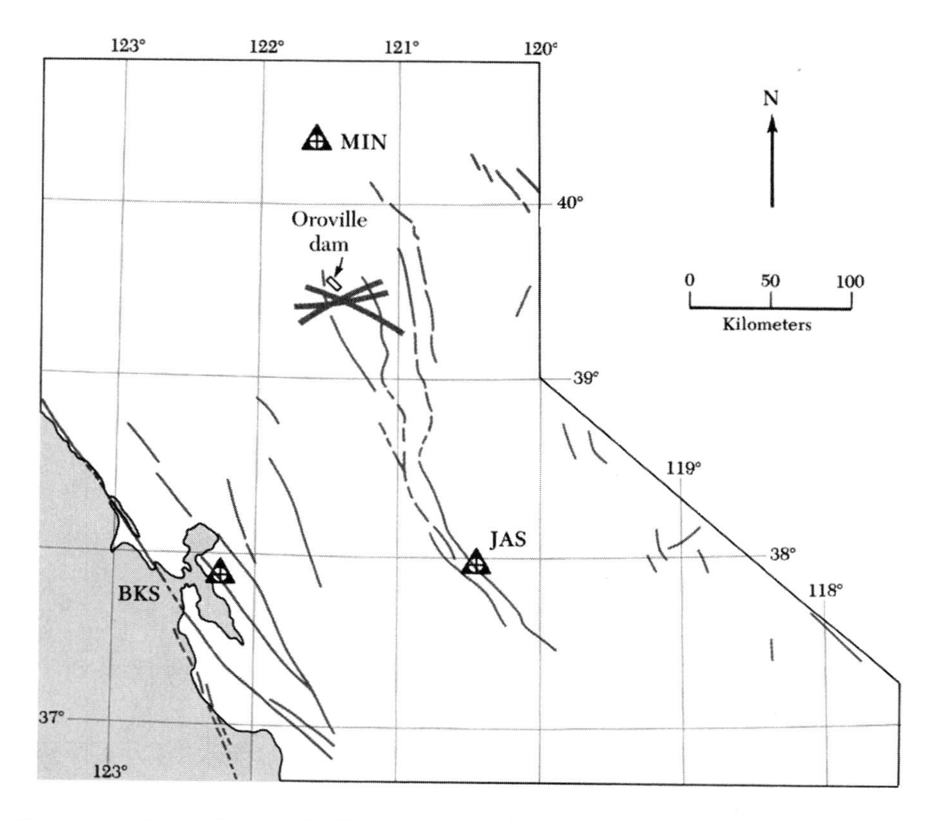

Figure 3.6 Map of central California showing these seismographic stations: BKS (Berkeley), JAS (Jamestown), and MIN (Mineral). The arcs near Oroville Dam are drawn as described in the text. The thin lines are surface traces of some prominent faults.

Sometimes, of course, even the use of basic methods such as those described above produces a false location because of a timing error from the clock, a mistake in identifying a seismic-wave type, a blunder in calculation, or a lack of data. My predecessor as Director, Professor Perry Byerly, told of his being called into the Berkeley Seismographic Station at 9:30 p.m., July 17, 1944, in response to a call from a news reporter who said that people had felt an "earthquake" somewhere in the San Francisco Bay Area. Byerly painstakingly made an epicenter location from the P and S waves he could see on the seismograms, then on photographic paper, and called back the newsroom to report his rough estimate (somewhere in the north*west* Bay). The janitor answered, saying, "All the reporters are away at Port Chicago (in the north*east* Bay) covering the great explosion at the dock there." "Well," replied Professor Byerly, "please tell them that the epicenter of the disturbance is at Port Chicago."

Seismograph Arrays—The Modern Earthquake Telescope

In some seismic regions, more powerful resolution of earthquake motions is now obtained by clusters of seismographs — called *seismograph arrays* — linked together with a common time-base. The scheme is similar to the use of groups of telescopes by astronomers to scan the sky. I was involved in the years after 1980 in designing a large seismograph array in Taiwan, a highly seismic area (see Figure 3.7). Recordings from it proved to be very helpful in understanding how much seismic shaking can vary over a short distance. As seen in the figure, this array consisted of 37 accelerometers arranged in three rings with radii of 100 meters, 1 kilometer, and 2 kilometers, and an additional central recorder. The seismic signals were recorded on ordinary magnetic tape cassettes in the discrete form *(digital)* that enabled them to

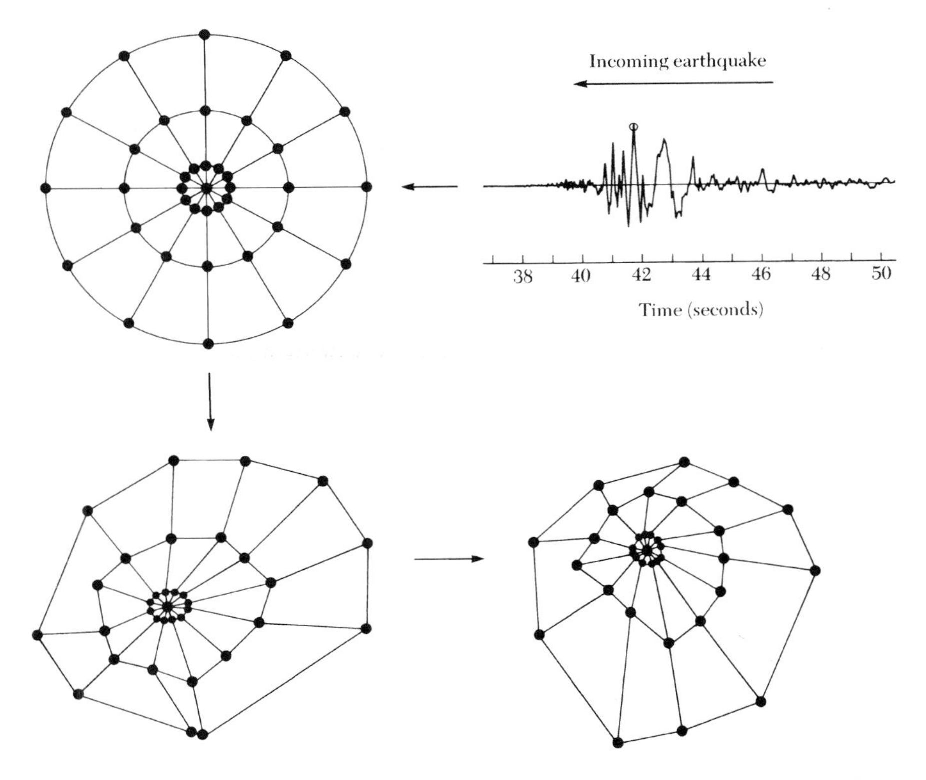

Figure 3.7 Dots denote sites of strong-motion seismographs operating at a circular array called SMART-1 in Taiwan. Incoming seismic waves produce a set of records like that at the top right. Each site is slightly displaced (a few centimeters for moderate-magnitude nearby earthquakes) by the wave motion, producing distorted rings (highly magnified) plotted by computer graphics at the bottom at two separate times.

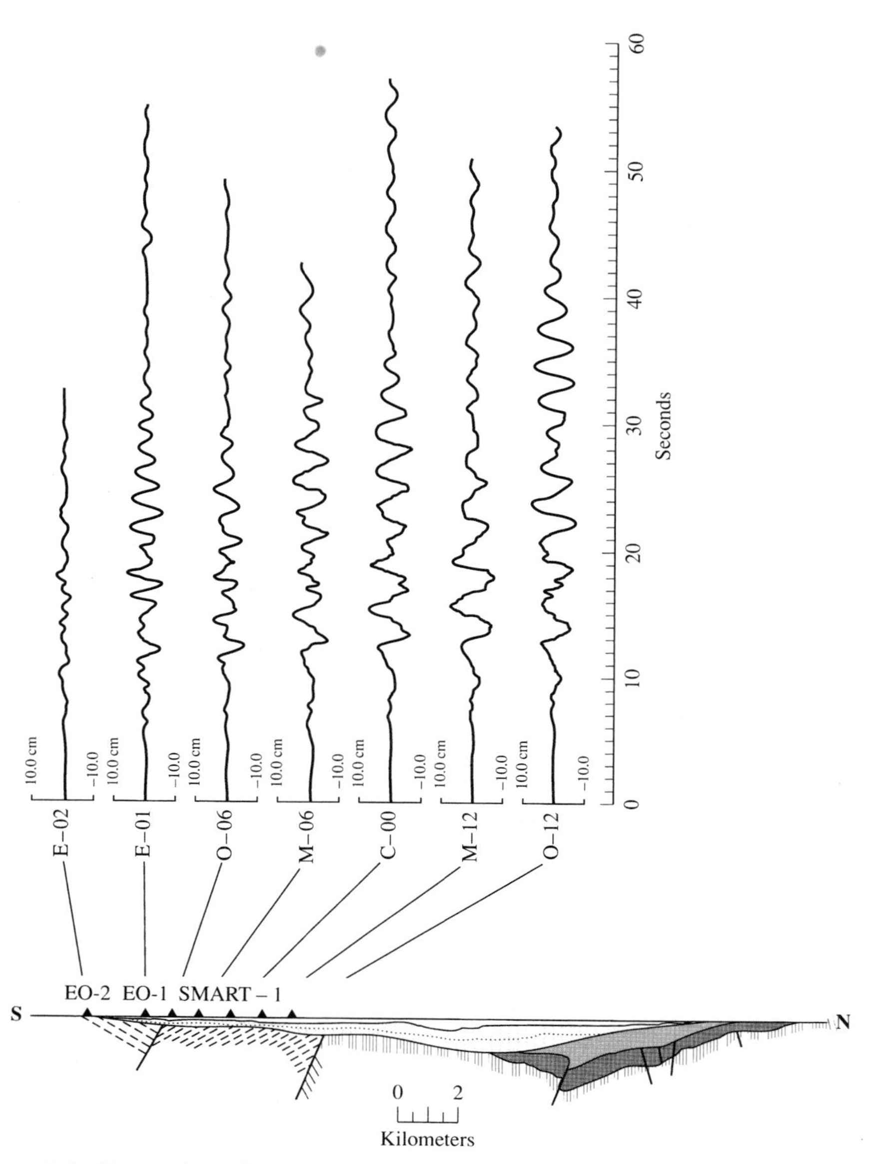

Figure 3.8 The geological cross section (bottom) shows the array called SMART-1 located over an alluvial basin (the instrument at EO2 is on rock) in Taiwan. At the top, the recorded ground displacements (in centimeters) from a November 14, 1986, earthquake 80 kilometers away to the south can be seen growing in amplitude and duration as it travels from south to north.

be read at once into the memory of a computer. Then the wave forms were played out in various ways for study.

Consider, in Figure 3.7, an incoming wave front (P or S waves) from a strong earthquake traveling across the array of seismographs, causing each to record the ground motion in turn. One such seismogram is shown in the top right of the figure in the form of the actual displacement of the ground surface. The direction of the seismic source relative to the array can be immediately determined by noting the sequences of arrival times of a wave front at each seismograph of the array. The differences in these arrival times from one side of the array to the other provide a measurement of the velocity of each wave front. In this way, the different types of seismic waves can be distinguished from each other (see Appendix I).

Using a computer, the set of 37 wave displacements in the earthquake depicted in Figure 3.7 has been combined to produce a series of snapshots of the actual time-variation of ground displacement. The two frames shown at the bottom of the figure highlight the manner in which strong ground shaking distorts the foundations of large structures in a continuously changing way. In this case, relative displacements of up to 5 centimeters occurred within distances of 200 meters.

Arrays of strong-motion seismographs also provide insights into the way that earthquake intensity varies with changes in soil and rock foundations. As pointed out in other sections of this book, in many earthquakes (e.g., San Francisco, 1906; Mexico City, 1985; and Loma Prieta, 1989) damage from ground shaking increased when the structures were built on less rigid geological materials. Such wave amplification is often associated, for example, with valleys containing deep sediments. In Figure 3.8, ground motion displacements from the Taiwan array clearly show the growth of the wave amplitudes and the duration of shaking as the waves travel from a rock site (at the left) across the sedimentary basin in which the softer soil and alluvium increases in depth to the north (right side).

4 Exploring Inside the Earth

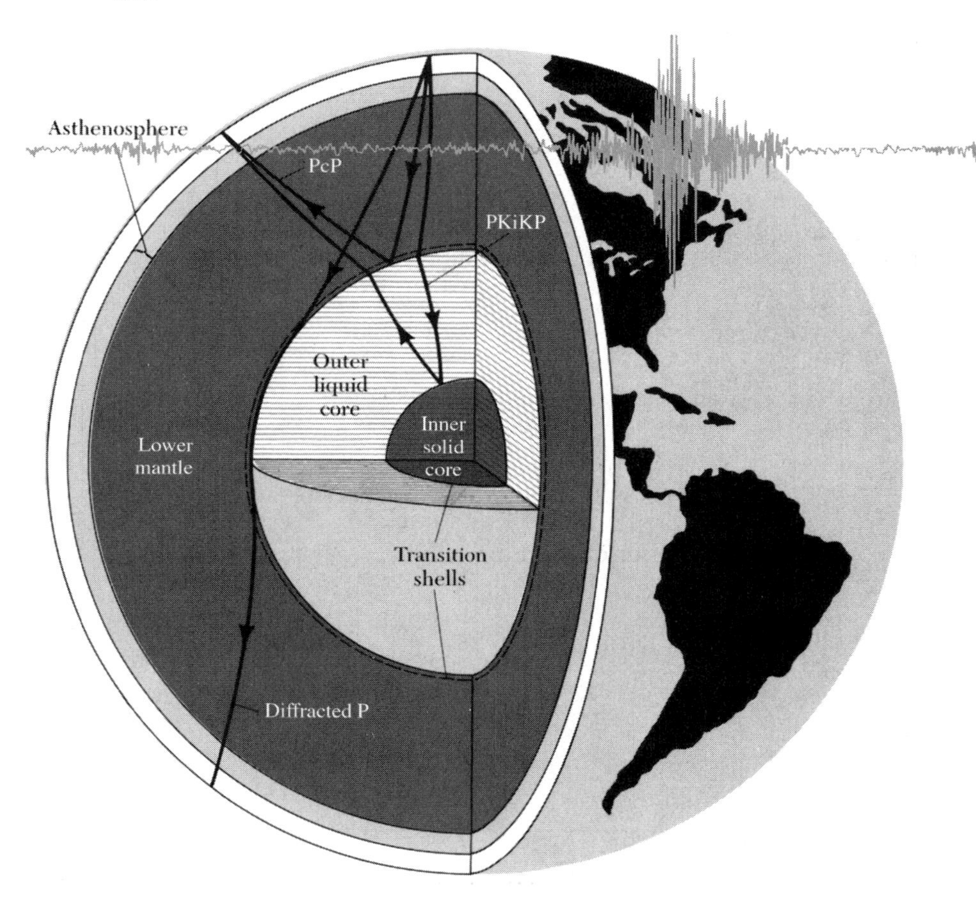

A cross section of the Earth based on seismological evidence. The paths taken by three major kinds of earthquake waves are shown. [From Bruce A. Bolt, *Inside the Earth* (San Francisco: W. H. Freeman and Company. Copyright 1982).]

The seismograph, recording the unfelt motion of distant earthquakes, enables us to see into the Earth and determine its nature with as great a certainty as if we could drive a tunnel through it.

— *R. D. Oldham, "The Constitution of the Interior of the Earth as Revealed by Earthquakes,"* Quarterly Journal, *Geological Society, 1906.*

Seismograms provide most of the detailed knowledge of the Earth's deep interior (Plate 6). Before the turn of the century and the development of the seismograph, the vast region that lies beneath our feet was the subject of much imaginative speculation based on very little information. Only the physical properties of surface rocks and the size and shape of the Earth were known with reasonable accuracy.

It was known that the Earth's mean density was about 5 1/2 times that of water and that the density increased toward the center because of the enormous pressures there. But was the material in the center solid, fluid, or gaseous? Was the interior of the Earth like a raisin pudding or was it composed of many uniform shells like the layers of an onion? These questions, as the British geologist R. D. Oldham pointed out in 1906, could only be answered by measuring earthquake waves that travel through the whole Earth. I much admire Oldham, who was one of the small group of scientists responsible for modern seismology. He was, for a time, Director of the Geological Survey of India. In this capacity he was responsible for a famous study of the great earthquake that devastated much of Assam, India, on June 12, 1897. For this report he made "every endeavor" to obtain as complete a set of instrumental records from the few seismographic observatories then operating in Europe. His study of these seismograms was a key step towards the remarkable discovery that I will now describe.

Earthquake Waves Through the Interior

The modern era of seismological recording of remote earthquakes began in 1887 when a German scientist, E. von Rebeur Paschwitz, noticed that ground motions in the form of waves had been registered by delicate horizontal pendulums operating in observatories in Potsdam and Wilhelmshaven in Germany. One of these very first identified seismograms is reproduced at the front of Chapter 3. Some time after a great earthquake was reported in Tokyo, von Rebeur Paschwitz realized that the seismic waves at the two German stations arrived about half an hour after the Japanese earthquake occurred. He then decided that the recordings in Germany were due to the great Japanese earthquake.

By 1900 it had been established, again, first by Oldham, that seismographs could detect both P and S waves (see Figure 1.9). The presence of both these kinds of waves generated by the sudden slip on a geological fault near the surface of the Earth was the essential tool long needed to unravel the deep Earth structure. If we think of the Earth as a giant glass spherical lens and the energy released by the fault slip as a lightbulb on its surface, the traveling seismic waves resemble the light *rays* refracting through the lens. The ray picture is shown in Figure 4.1. Because the Earth is a sphere, it is easiest to plot the distance between the earthquake source and the recording seismograph in terms of the angular distance subtended at the center of the Earth. Thus the distance of seismographic stations ranges from 0° to 180° at the antipodes of the earthquake source.

The first great triumph of the use of seismic rays to "x-ray" the Earth was in 1906 when Oldham suggested that the best way to explain the travel times of P and S waves from one side of the Earth to the other was to

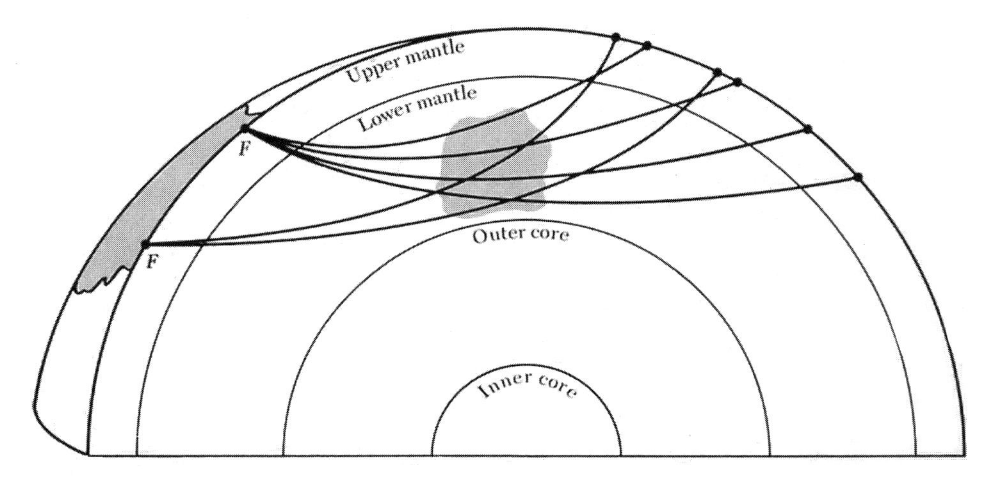

Figure 4.1 Paths of earthquake waves through the Earth's mantle from earthquake sources at F to seismographs at the surface. All the seismic paths pass through the shaded region and provide a tomographic scan of it.

invoke a large central core to the Earth. From suitable seismograms, Oldham plotted the travel times of P and S waves along their ray paths against the angular distance and found that the S waves could be followed as arrivals out to about 110° of arc, but not beyond it. He explained this change in behavior of the S waves by hypothesizing that they had penetrated a central core in which they traveled at a much lower rate. (It is now believed that S waves do not propagate at all through the outer part of the core because it is fluid.)

A few years after Oldham developed his theory, in 1914, Beno Gutenberg, a seismologist working in Gottingen, Germany, fixed the depth to the boundary of the separate core at about 2900 kilometers, a value that has required little correction since that time. Gutenberg found that the direct P waves through the Earth could be traced out to about 105°, after which they rapidly became very weak in energy. Beyond this distance, seismograms showed waves arriving that were delayed by up to 5 minutes after that expected for the direct P waves. Even more decisively as evidence for a separate core, he also observed waves that could be explained as reflected directly from its boundary.

Now think of the penetration of seismic waves to great depth in the same way that one observes rays of light spreading out from a bulb at the point *F* in Figure 4.1. Because the rocks of the Earth's interior are more compressed toward the center by the great mass of overlying material, the velocity of both P and S seismic waves generally increases with depth from the surface toward the center of the Earth. This increase bends paths of the seismic waves so that they dip down and then refract back to the surface along the paths illustrated in Figure 4.1. At this surface point they are reflected down again and produce another similar leg.

Each time a P or an S wave encounters the surface of the Earth or a boundary between two rock types in the interior, the reflected and refracted seismic waves are, in general, of two kinds (see Figure 1.10). In other words, in marked contrast to light waves, at such encounters an incident P wave produces both a reflected (and refracted) P and S wave; the same thing happens with incident S waves with vertical motion (called SV waves) but not for the purely horizontal SH waves.* This branching of wave types greatly complicates the groups of seismic waves that move through the Earth and, in due course, arrive at the surface to be felt as an earthquake or to be recorded on a seismogram.

**By considering the directions that the rocks are displaced in the P, SV, and SH waves, the reader should be able to demonstrate that this is so (see Figure 1.9).*

Seismic-Wave Probes

As was explained above, the P and S waves in earthquakes penetrate through the body of the Earth like x-rays. In modern medicine, a common technique used by doctors to obtain images of anomalous growths inside the human body is CAT-scanning (for computerized axial tomography): sensors on one side of the body show the way that variations in human tissues affect the intensity of x-rays or atomic particles applied to the other side. By analogy, probing the Earth's interior by P and S waves is called *geophysical tomography* (see Figure 4.1). In a complementary way, the Earth's outer structure can be resolved by comparing surface wave trains recorded at seismographic stations remote from the earthquake sources. The two types of surface waves, Love waves and Rayleigh waves (see Figure 1.9), are seismic companions with separate and distinguishable features. These waves do not penetrate through the whole Earth but are channeled by the Earth's outside surface; their properties depend on the structure and elasticity of the rocks through which they pass. The measurement of the speeds and wave forms of surface waves can be used as tomographic signals; these signals can be decoded to yield a picture of the tectonically complex regions in the upper part of the Earth.

Recordings of seismic Love (LQ) and Rayleigh waves (LR) are illustrated in Figure 4.2. The Love waves arrive first, followed by the Rayleigh waves; the Love-wave motion is restricted to the horizontal plane, while Rayleigh-wave motion is confined to a vertical plane. Surface waves do not occur as the concentrated pulses typical of P and S waves but are spread out into a train of many cycles of vibrations. This spreading is called *dispersion*. As the surface waves progress around the Earth's surface, they sort

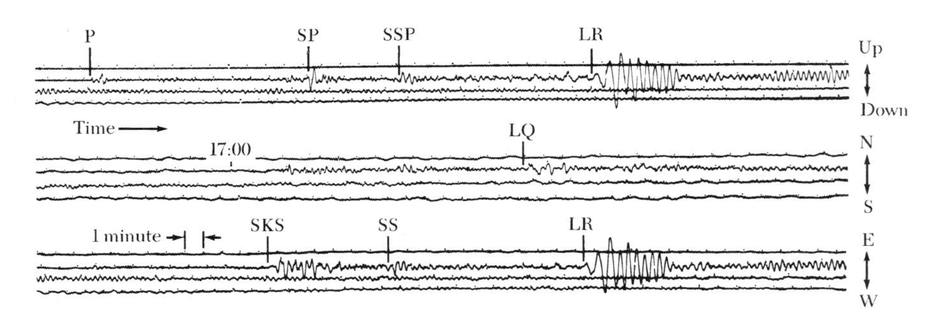

Figure 4.2 Three components of ground motion recorded by a seismograph in Berkeley at a distance of 88° from a south Pacific earthquake on May 7, 1972. The symbols LR and LQ denote Rayleigh and Love waves, respectively

themselves out into longer and longer trains of waves according to the period of the motion and the variation in the elastic properties with depth in the Earth. Waves with longer wavelengths travel more quickly than those with shorter wavelengths; because the longer waves penetrate deeper into the Earth, this observation implies that the deeper rocks transmit waves more quickly.

The time of arrival at the recording station of each of the separate wave components in a surface wave train can be measured and the wavelength speeds plotted as a function of the period of the wave. The resulting curve gives the crucial dispersion information on the deep structure and hence the elastic properties of the rocks through which the waves passed. This procedure involves what is known as an *inverse problem*. For its solution, the following direct (or forward) problem must first be solved. If the physical properties of the Earth under, say, the North American continent were known, then a computer could calculate the appropriate *theoretical* dispersion curve for the known structure. But in reality, it is the *observed* dispersion curve that is known, and from this cryptic information the elastic properties and layered structure of the rocks underground are inferred. A rough analogy is the figuring of the height and slope of a hill from measured speeds of a skateboarder at the bottom.

This way of looking at the exploration of remote depths of the Earth is recent. The inverse problem has been solved successfully for the broad average properties of the deep interior, and present research is aimed at refining the observations of earthquakes so that more tectonically complex regions (such as those between the continents and oceans and along the Benioff zones) can be explored.

Interpretations of Seismograms

Now that the paths of waves that constitute an earthquake have been explained, let us set as a challenge the interpretation of a seismogram recorded at a site remote from the earthquake source. The materials for the puzzle are contained in Figures 4.2 and 4.3.

Consider the three records shown in Figure 4.2. The top record is the vertical motion of the ground recorded by a seismograph at Berkeley located 88° away from an earthquake centered in New Ireland in the south Pacific on May 7, 1972. The lower two sets of traces are from the motions of the ground in the north-south and east-west directions, respectively. Several traces are shown in each of the three components because the

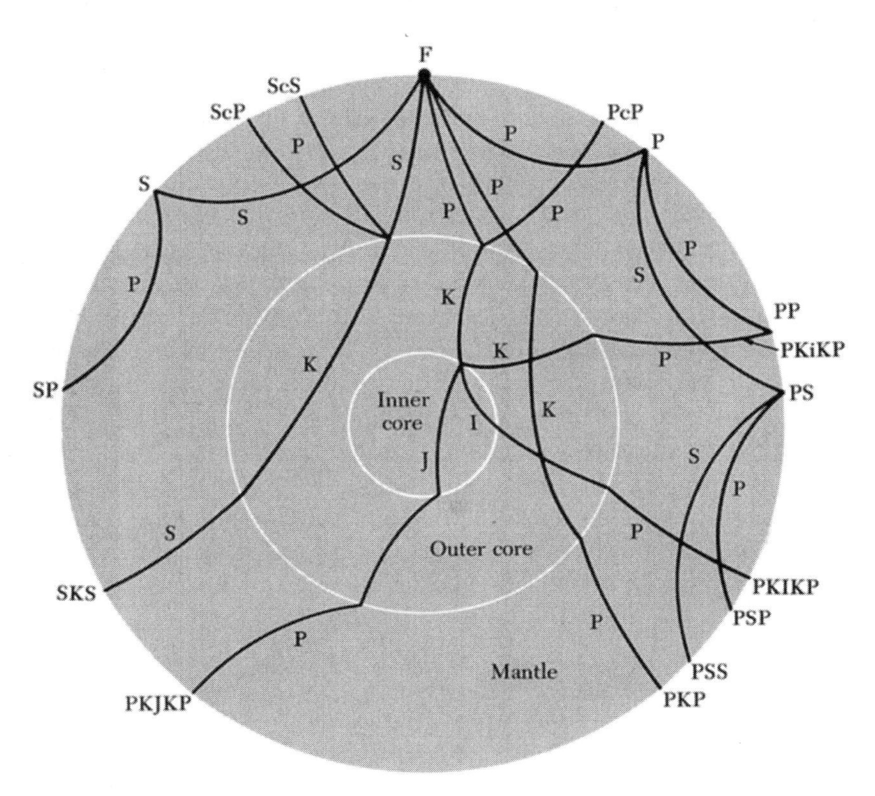

Figure 4.3 Sample seismic rays through the Earth. Begin at the focus of the earthquake F. The symbol c designates a wave reflected at the outer core's surface; thus PcP is a P wave through the mantle reflected at the core; ScP is an S wave reflected as P. The symbol i designates a wave reflected at the inner core's boundary: e.g., PKiKP. The symbols K and I refer, respectively, to P waves that have traveled through the outer and inner core. The symbol SP designates an S wave through the mantle reflected at the outer surface as P. The ray marked PKJKP, which travels as S through the inner core, has probably not yet been observed. [From Bruce A. Bolt, *Inside the Earth* (San Francisco: W. H. Freeman and Company. Copyright 1982).]

recorded waves overlap these lines, which were recorded continuously on a revolving drum.

The first-arriving P wave can be detected clearly in the recorded vertical ground motion. (Why is it more obvious on the vertical record?) From a separate calculation, it is known that it arrived 12 minutes and 53 seconds after traveling through the Earth from the earthquake focus. Along this curved path, the seismic wave penetrated as deep as 2000 kilometers into the Earth.

Interpretation of the various seismic wave onsets marked in Figure 4.2 can be understood by referring to the examples of types of ray paths drawn in Figure 4.3. For example, toward the beginning of the trace on the east-west component in Figure 4.2, the clear onset of waves is marked SKS. As Figure 4.3 shows, SKS denotes a wave that has traveled first as an S wave in the mantle and then as a P wave through the center of the Earth, along a path called K for the German word for "core." It then emerges on the other side of the Earth as an S wave in the solid mantle again.

Later along the record, onsets of waves are indicated as SP, SS, and SSP. These earthquake waves have traveled through the body of the Earth with at least one reflection at the outside surface producing echoes of various types. The reader might like to plot each of the ray paths of the New Ireland seismogram on a diagram like that in Figure 4.3.

Similar identification of thousands of seismograms like that analyzed here have led to a rather sharp image of the structure of the Earth's interior. The main results can now be described.

Interior Earth Structure

The Earth has four main shells: the crust, mantle, outer core, and inner core (Plate 6). This dominant concentric structure (shown in the figure opposite the chapter opening) has been established by many independent analyses of measurements of earthquake waves. Interpretations of the travel times, amplitudes, and other properties of recorded seismic waves of various types have provided the necessary tomographic images described in the first section above.

The *crust* is the worldwide outermost layer of rocks, ranging in thickness from 25 to 60 kilometers under continents and from 4 to 6 kilometers under deep oceans. It has complex internal structure in many places, such as in the tectonically active and transitional areas of the tectonic plates.

The underlying *mantle* extends from the base of the crust to a depth of 2885 kilometers. It consists of dense silicate rocks. P seismic waves as well as S waves (i.e., shear type) penetrate through almost all parts of the mantle; this demonstrates that the mantle is solid and strong, at least on short time scales. (There is separate evidence that over geological intervals of millions of years even mantle rocks flow slowly in giant convection cells because of their high temperatures and pressures. See Chapter 7.) The mantle has been subdivided further into shells and anomalous regions, the latter intimately connected with the dynamical geological processes that cause the topographic and tectonic variations of the surface. The behavior of earthquake waves indicates that the outermost 100 kilometers (including the crust) of the Earth, called the *lithosphere*, is relatively rigid, while below it there is a softer layer of about 400 kilometers thick called the *asthenosphere*. Refined tomographic seismic imaging is now being used to map these geologically important outer layers in more detail. All present indications are that within the whole mantle the rocks are not mixed homogeneously and that there are small variations in properties from place to place.

Further toward the Earth's center, below the solid mantle, is the *outer core* of the Earth, which R. D. Oldham discovered by using earthquake recordings explained earlier in the chapter. This enormous shell appears to be liquid and is composed mainly of iron, oxygen, and silicon. Among the arguments for the liquidity of the outer core (although not the initial crucial one*) is that, despite many attempts, no seismic waves that have traveled through it as S waves (shear type) have ever been detected on seismograms. Because S waves cannot propagate through materials with no rigidity (see Chapter 1), the strong inference is that the outer core is in a liquid state. Reflections of seismic P and S waves take place efficiently from the outer surface of the core, thus establishing a sharp boundary there.

By 1936, the reality of the crust, mantle, and liquid core was well established. But some major mystery remained when observers attempted to interpret seismograms of very distant earthquakes and found, disconcertingly, clear onsets that had no simple explanation. The problem was solved by a remarkable young Danish seismologist, Dr. Inge Lehmann.†

Illuminating the Innermost Core

The year 1986 was the fiftieth anniversary of the discovery of the *inner core* of the Earth. This achievement was made by Lehmann, who worked as the only seismologist in the Copenhagen Observatory. The observatory was well situated to detect the PKP waves (see Figure 4.3) that pass through the center of the Earth to Europe from earthquakes in the seismically active south Pacific region.

Lehmann was born on May 13, 1888, near Copenhagen, Denmark, where she grew up and spent much of her long, fruitful life. She had

**Given convincingly in 1926 by British astronomer, seismologist and statistician, Sir Harold Jeffreys, based on the Earth's tidal deformation and overlooked in many textbooks.*

†We became friends in her latter years. She went on to live to 104 years.

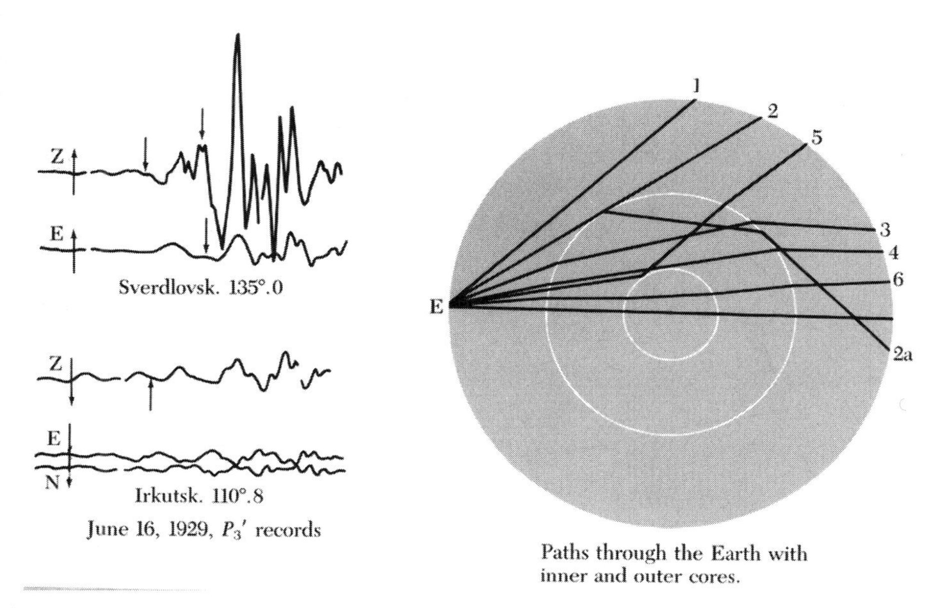

Figure 4.4 Figures reproduced from I. Lehmann, "P'," Bureau Central Seismologique International, Series A. Travaux Scientifique, 14, 88, 1936. This famous paper was the first to establish that there was an inner core at the Earth's center. [From Bruce A. Bolt, *Biographical Memoirs of Fellows of the Royal Society* (The Royal Society, London, Vol. 43, 1997).]

been sent by her parents to an enlightened coeducational school run by Hannah Adler, an aunt of the famous physicist Niels Bohr. She learned there that boys and girls could be treated alike for work and play: "No difference between the intellect of boys and girls was recognized, a fact that brought some disappointments later in life when I had to recognize that this was not the general attitude." In 1925 she was appointed assistant at the forerunner of the Geophysical Institute in Copenhagen and although she had no formal education in earthquakes, in this fortuitous way she entered the discipline of seismology, a rare profession in aseismic Denmark.

Figure 4.4 shows some of the seismograms from a 1929 New Zealand earthquake that Lehmann saw at the European stations of Sverdlovsk at a distance of 135° away from the focus (i.e., the angle subtended at the Earth's center) and Irkutsk 110.8° away. In 1929 there was no convincing explanation for the arrival of a seismic wave pulse about 19 minutes after the onset time of an earthquake at these distances.

The Earth was at that time conceived as having just three shells — a crust, a mantle, and a single core. The simplified Earth section in Figure

4.4 (taken from Lehmann's discussion) shows that for such an Earth model with constant wave speeds in each shell, rays like those marked 1, 2, 3, and 2a were to be expected. However, in order to account for the rays marked 5, 4, and 6 at the observed arrival times, an additional feature was needed. Lehmann stated "we take it that, as before, the earth consists of a core and a mantle but that inside the core there is an inner core in which the velocity is larger than the outer one."

If one assumes an inner core with a sharp outer boundary, a PKP wave (such as path 5 in Figure 4.4) would be reflected from this boundary back to shorter distances; this would explain the mystery waves from the New Zealand earthquakes.

Inge Lehmann's argument for the inner core was convincing enough that other seismologists working on the problem of the Earth's structure adopted her hypothesis. Within a few years, after more observational tests, the reality of the small spherical body at the center of the Earth, with a radius now estimated to be about 1216 kilometers, became accepted.

Of course, there is always uncertainty in such fundamental research, and Lehmann was careful to state in her 1936 paper* that her argument did not *prove* the existence of an inner core but merely established it with some degree of probability. Only in this way can earthquake probes "discover" deep structure; indeed, no Earth model is strictly unique. A great strength of the "model" of the Earth described by Lehmann was that it could be tested by independent means, such as the reflections from its surface. Later research in which I was involved in the 1960s and 1970s led me to identify many such sharp wave onsets on specially selected seismograms as P waves bounding off the inner core boundary. Specially clear were seismic echoes from the inner core after underground nuclear explosions in Nevada.

Additional seismological work over the last 50 years indicates that the inner core is a solid body in contrast to the liquid outer core (Lehmann did not discuss this question). Some of this more recent work involved measurements of the vibrations of the whole Earth that occur in the very largest earthquakes. The enormous release of energy in such cases causes the entire globe to ring like a bell does after being struck. The tones of vibration depend on the elastic properties throughout the Earth and hence give information about its structure.

**Lehmann's published paper, "P'," has one of the shortest titles in science. Her discovery perhaps would have rated a Nobel Prize in physics for the analogous detection of a new atomic particle!*

Very recently, earthquake waves passing through the inner core have provided evidence that it may be rotating slowly relative to the surrounding outer core. Inge Lehmann would have no doubt been delighted to see how others have built on her early inference as the seismographic recording of earthquakes has become more precise and widespread.

Fine Mapping of Deep Earth Structure

There is a very simple way in which earthquakes are used to map Earth structure near the surface. We have seen it already in a map of world seismicity shown in Figure 2.2. These plots of thousands of epicenters indicate the surface regions where there is tectonic activity. If we move to a three-dimensional picture by considering also focal depths of earthquakes, the plots of earthquake foci below the surface give some of the most basic — yet straightforward — clues on structure and geological forces known. Sometimes quite fine detail in the geological structure can be resolved.

An account was given in Chapter 2 (see Figure 2.4) of the way that deep earthquake foci were used to infer the presence and shape of Wadati-Benioff deep seismic zones. A particularly arresting illustration of such three-dimensional seismicity is reproduced in Figure 4.5. In this diagram we look sideways into a slice of the Earth running essentially east-west under the northeastern part of Honshu, Japan. Study of Japanese earthquakes has a peculiar advantage: because there is a network of modern seismographs across northern Honshu, seismic waves passing upward from the foci can be read accurately and, by tracing rays between stations and foci, the focal positions can be precisely located. To the east (right side) is the deep trench under the Pacific Ocean off Japan and to the west is the Japan Sea. The foci of many hundreds of recent earthquakes are plotted as circles. We see at once that many of the earthquakes have focal depths down to 250 kilometers, which places them far below the crust. The deeper foci mark out a zone that dips down beneath the ocean trench and underneath Honshu itself at an angle of about 30°. (The horizontal and vertical scales are not equal in Figure 4.5). It is evident that the dipping zone in Honshu has not a single, but two, seismically active planes.

The sandwich structure below Honshu has also been discovered in similar zones in New Zealand, the Aleutians, and the Kuriles and has caused considerable speculation about its cause among seismologists and geologists. Near the top of this slab in most dipping zones, earthquake foci lie in a narrow layer 20 kilometers thick and are often identified with mechanical thrusting along or near the top surface of the slab. For the

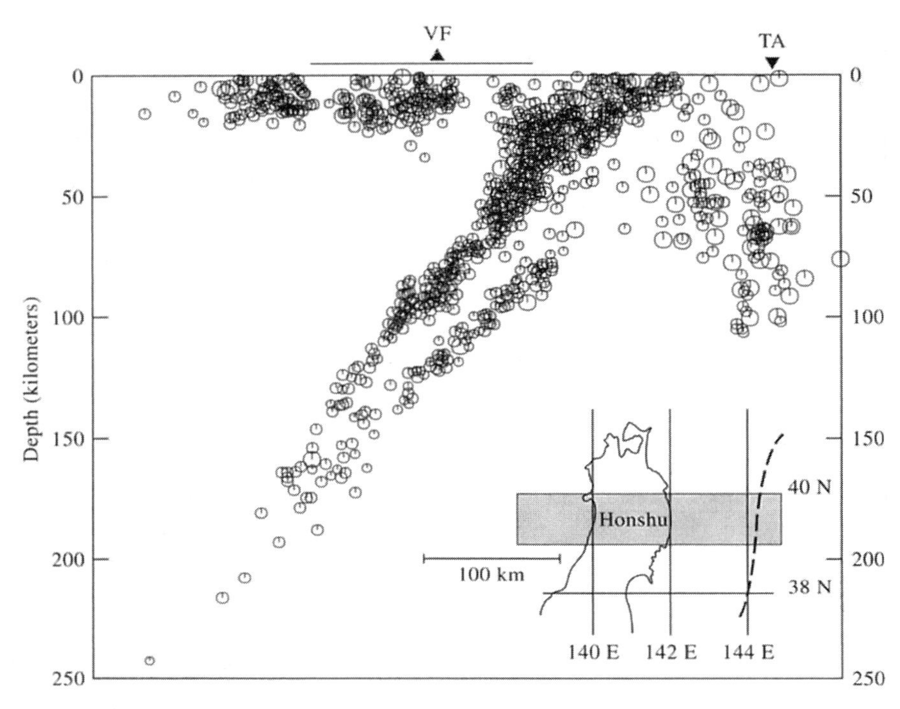

Figure 4.5 Foci of earthquakes recorded in 1975 and 1976 by a network of sensitive seismographs in Honshu, Japan. VF is the volcanic front; TA is the Japan trench axis. The foci are drawn as a side elevation with depth in kilometers. The Wadati-Benioff zone can be clearly seen dipping from the Japan trench just west of TA toward the west. The deepest focus is at about 240 kilometers. The striking feature of the Wadati-Benioff zone is the two distinct lines of foci about 20 kilometers apart. The seismicity plot also shows shallow earthquakes under Honshu within the crust as well as scattered earthquakes down to a depth of 100 kilometers to the east of the Japan trench under the marker TA. [Courtesy of A. Hasegawa, M. Umino, and A. Takagi, 1978.]

deeper earthquake foci, both within and along the top of the dipping slab, there is no universally accepted explanation, however, of their cause and mechanism, even though this puzzle has been made less cryptic by precise mapping like that shown on Figure 4.5.

One difference between shallow and deep earthquake sources is that deep earthquakes generally have very few aftershocks. For example, in 1970 probably the largest deep-focus earthquake in over a quarter of a century occurred under Colombia at a depth of 650 kilometers and a magnitude of 7.6. Seismographs detected no aftershocks at all following this event. After large shallow earthquakes, the foci of aftershocks usually follow close to the plane of the fault that has slipped. By contrast, even when aftershocks accompany deep earthquakes, they are more or less randomly distributed around the initial focus.

These differences suggested that the cause of deep earthquakes might be a sudden change in the volume of the rocks, resulting from a change in the phase state of the minerals, very much as water increases in volume when it changes to ice. The sudden expansion of the rock would produce seismic waves. This hypothesis would require that there be either an implosion or an explosion of wave energy; hence, seismographs around the world would detect only compressional or only dilational onsets, respectively, of the P waves. Such a consistent pattern of onsets is not found, however. The first motions of P waves on seismograms vary from upward to downward in various zones of the Earth, much as do the onsets of P waves from shallow earthquakes. In addition, as well as P waves, deep earthquakes produce significant S-wave motion, which would not be the case if an explosion or implosion were the only impulse, because such sources produce little rebound of the rocks.

There are at least two mechanisms recently proposed for deep earthquakes. The first suggests that water fundamentally influences the brittle and plastic properties of the rock at the high temperatures and pressures in the dipping slab. Many minerals in the Earth's rocks contain water of crystallization, which may become mobilized at high temperatures and pressures. Indeed, in laboratory experiments, rocks that contain the green, hydrous mineral serpentine are found to sustain brittle fracture under these conditions. What seems to be required is an intimate migration of fluid throughout the pores of the rock, which lubricates potential fractures and faults and allows slip to occur.

The second of these current hypotheses puts the unconfirmed hypothesis of a sudden change of mineral phase into a different form. The phase transitions are conceived to take place between the boundaries of rock lenses, perhaps where fluid conditions are particularly favorable for sudden transition. Along preexisting grain boundaries the crystal structure would change rapidly, thus weakening the bonds across the discontinuity.

To test these hypotheses, conditions in the deep Earth have been recreated in the laboratory with the squeezing of tiny samples of rock between two diamonds like a nutcracker. A laser beam shining through the diamonds heats the rock and also allows any sudden physical transition to be photographed. Acoustic sensors detect any sudden release of energy — the analogue of an earthquake. By these methods it is hoped that the old puzzle of the causes of deep subduction earthquakes can be resolved.

The role of subduction-zone earthquakes in the seismic hazard of the northwestern contiguous United States will be discussed further in Chapter 10.

5 Faults in the Earth

Fresh fault scarp on Johnson Valley below a damaged house after the June 28, 1992, Landers earthquake in the Mojave Desert, California. [Courtesy of E. Keller.]

The Mount of Olives shall cleave in the midst thereof toward the east and toward the west, and there shall be a great valley; and half of the mountain shall remove toward the north and half of it toward the south

— *Zechariah 14:4*

Most people — even in casual examination of rock quarries, road cuttings, and sea cliffs — have observed abrupt changes in the structure of the rocks. In some places one type of rock can be seen butting up against rock of quite another type along a narrow line of contact. In other places, displacements in strata of the same rock have clearly taken place, either vertically or horizontally. Such offsets of geological structure are called *faults*. Clear vertical offsets of layers of rock along an exposed fault in the wall of the Corinth Canal, Greece, can be seen in Figure 5.1.

Faults may range in length from less than a meter to many kilometers. In the field, geologists commonly find many discontinuities in rock structures which they interpret as faults, and these are drawn on a geological map as continuous or broken lines. The presence of such faults indicates that, at some time in the past, movement took place along them. We now know that such movement can be either slow slip, which produces no ground shaking, or sudden rupture (Plate 7), which results in perceptible vibrations — an earthquake. In the first chapter I discussed one of the most famous examples of sudden fault rupture — the San Andreas fault in April 1906. However, the observed surface faulting of most shallow-focus earthquakes is much shorter in length and shows much less offset in this case. In fact, in the majority of earthquakes, fault rupture does not reach the surface and is thus not directly visible.

Types of Geological Faults

The faults seen at the surface sometimes extend to considerable depths in the outermost shell of the Earth, called the *crust*. This rocky skin, from 5 to 60 kilometers thick (see Chapter 4), forms the outer part of the lithosphere.

Figure 5.1 Normal fault (see Figure 5.4) that has displaced the almost horizontal beds in young sedimentary rocks on the north side of the Corinth Canal, Greece. Height of the exposure is about 70 meters, and the total offset along the fault amounts to more than 10 meters. [Courtesy of J. Weiss.] Major earthquakes, often with tsunamis, occurred in the Corinth region in 227 B.C., 551 A.D., 1858, 1928, and on February 24, 1981. In the most recent earthquake, surface fault rupture was observed for a length of about 5 kilometers with vertical slip of up to 0.7 meters.

It must be emphasized that slip no longer occurs at most faults plotted on geological maps.* The last displacement to occur along a typical fault may have taken place tens of thousands or even millions of years ago. The local disruptive forces in the Earth nearby may have subsided long ago, and chemical processes involving water movement may have cemented the ruptures, particularly at shallow depth. Such an *inactive fault* is not now the site of earthquakes and may never be again.

Our primary interest is of course in *active faults*, along which crustal displacements can be expected to occur. Many of these faults are in rather well-defined tectonically active regions of the Earth, such as the mid-oceanic ridges and young mountain ranges. However, sudden fault displacements can also occur away from regions of clear present tectonic activity (see Figure 12.4).

It is possible to determine, by geological detective work, a number of properties of faults. For example, intermittent fault slip that has occurred in the past few thousand years usually leaves such clues in the topography as sag ponds, lines of springs, and fresh fault scarps. Many clues to movement along the San Andreas and similar fault zones can be seen in the landforms of Figure 5.2. But pinpointing the sequence and times of such displacements may be much more difficult. Such features as offsets of overlying soils and recent sedimentary deposits may provide this kind of chronological information. The digging of trenches a few meters deep across faults has also proved an effective means of studying displacements. Even subtle offsets in layers in the sides of the trenches can be mapped and the time intervals between fault offsets determined by fixing the ages of the various soil layers that have been displaced (see Figure 5.3 and Chapter 10). Sometimes the actual dates of movement can also be estimated from the known ages of buried organic material, such as leaves and twigs. Even along the seafloor, modern geophysical methods allow fairly accurate mapping of faults. From research vessels at sea it is possible to detect the passage of sound waves that have been reflected from the mud layers, and offsets in the layers indicated by these seismic records may be identified as faults.

Whether on land or beneath the oceans, fault displacements can be classified into three types (Figure 5.4). The plane of the fault cuts the horizontal surface of the ground along a line whose direction from the north is called the *strike* of the fault (see Figure 5.5). The fault plane itself is usually

**But sometimes faults not plotted on geological maps are discovered from fresh ground breakage during an earthquake. Thus, a fault was delineated by a line of cracks in open fields south of Oroville, California, after the Oroville earthquake of August 1, 1975 (see Chapter 2).*

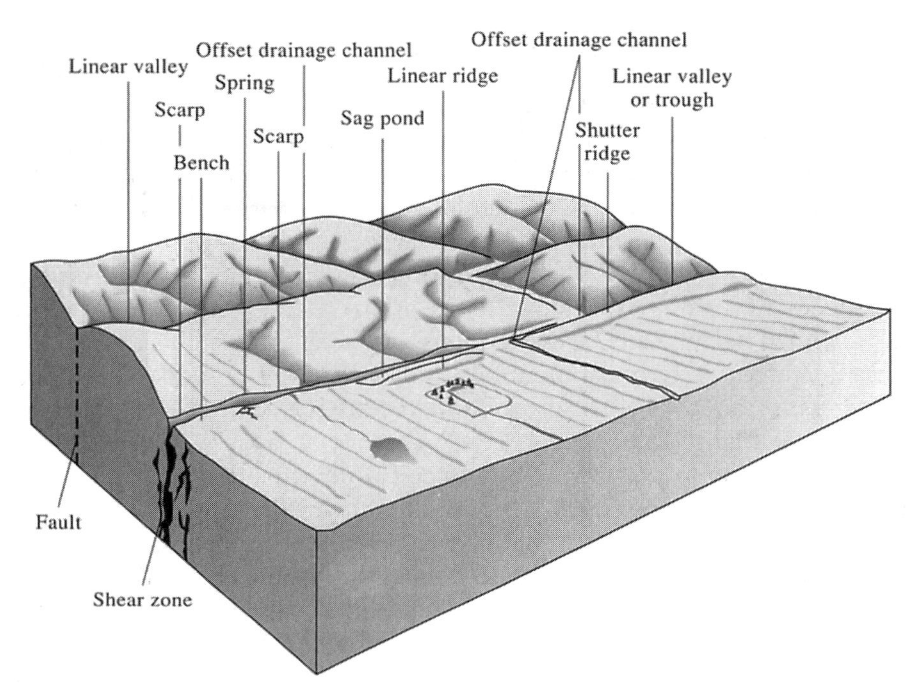

Figure 5.2 Common landforms along the San Andreas fault system. [From Vedder and Wallace, USGS, 1970.]

not vertical but dips at an angle down into the Earth. When the rock on that side of the fault hanging over the fracture slips downward, below the other side, we have a *normal fault*. The dip of a normal fault may vary from 0° to 90°. When, however, the hanging wall of the fault moves upward in relation to the bottom or footwall, the fault is called a *reverse* fault. A special type of reverse fault is a *thrust* fault in which the dip of the fault is small. The faulting in midoceanic ridge earthquakes is predominantly normal, whereas mountainous zones are the sites of mainly thrust-type earthquakes.

Both normal and reverse faults produce vertical displacements — seen at the surface as fault scarps — called *dip-slip* faults. By contrast, faulting that causes only horizontal displacements along the strike of the fault are called transcurrent, or *strike-slip*. It is useful in this type to have a simple term that tells the direction of slip. In Figure 5.4, for example, the arrows on the strike-slip fault show a motion that is called left-lateral faulting. It is easy to determine if the horizontal faulting is left-lateral or right-lateral Imagine that you are standing on one side of the fault and looking across it. If the offset of the other side is from right to left, the faulting is left-lateral,

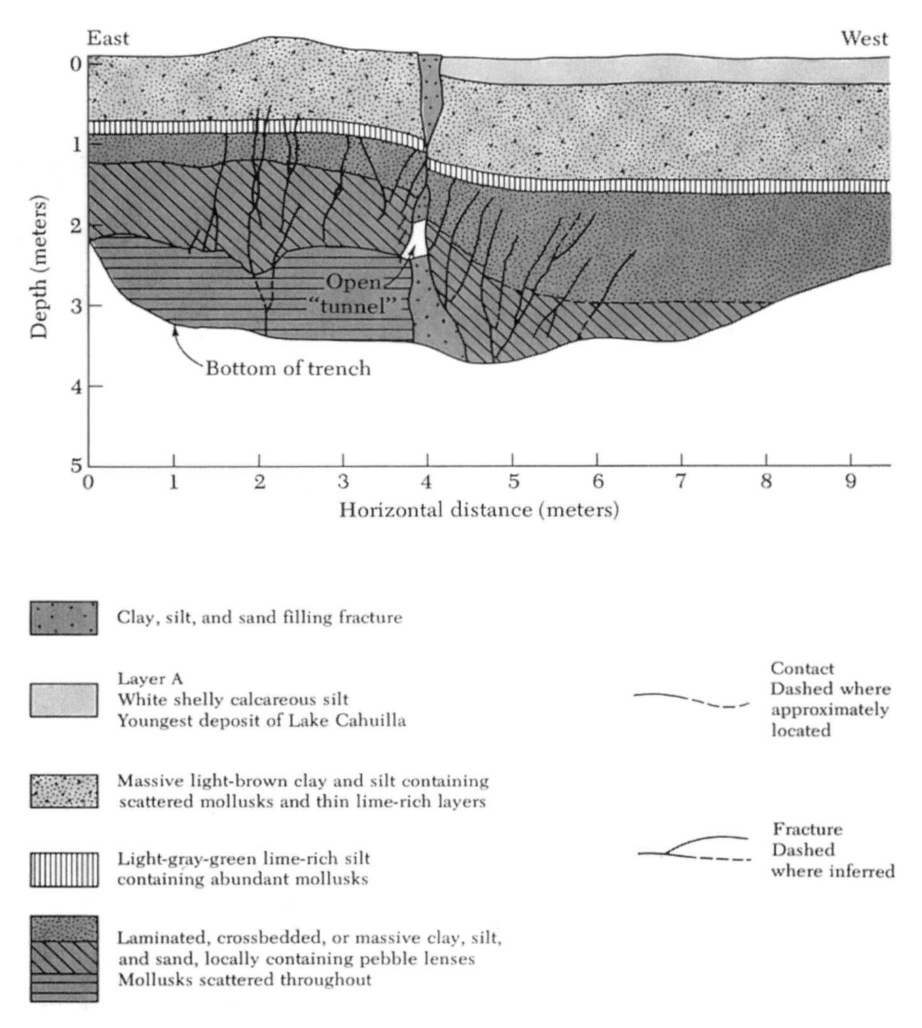

Figure 5.3 Map of a wall of a trench excavated across a fault that slipped in the 1968 Borrego Mountain earthquake in California. Each time a fracture occurs, there is more displacement between once-continuous soil layers. [After M. M. Clark, H. Grantz, and M. Robin, USGS Professional Paper 787, 1972.]

whereas if it is from left to right, the faulting is right-lateral. Of course, sometimes faulting can be a mixture of dip-slip and strike-slip motion.

In an earthquake, serious damage can arise not only from the ground shaking but also from the fault displacement itself, although this particular earthquake hazard is very limited in area. It can usually be avoided by the simple expedient of obtaining geological advice on the location of active

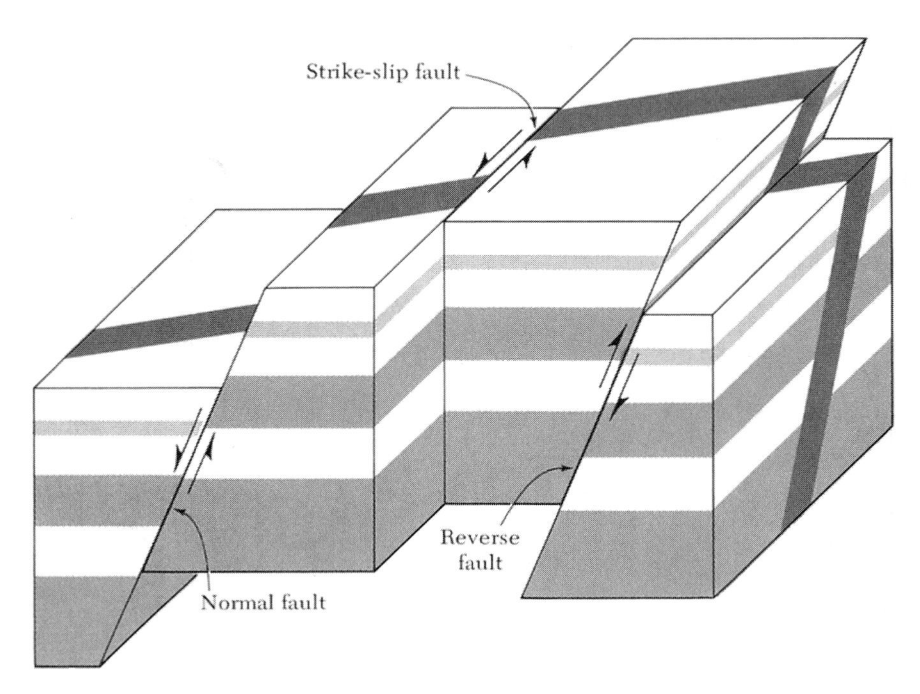

Figure 5.4 Diagram showing the three main types of fault motion.

faults before construction is undertaken. Areas astride active faults can often be set aside for open areas — for public recreation, parking lots, roads, and so on.

In land-use planning, one also needs to know that damage adjacent to fault ruptures — caused by the sliding and slumping of the ground — varies according to the fault type. In dip-slip faulting the scarp produced may spread the damage (by local ground sliding, cracking, and slumping) over a relatively wide zone along the fault itself. But in strike-slip faulting, the zone of ground disturbance is usually much less extended, and buildings just a few meters away from the rupture may not be damaged by it (compare the photo at the beginning of this chapter).

The 1891 Mino-Owari (Nobi) Earthquake, Japan

The Mino-Owari earthquake of October 28, 1891, was the greatest inland earthquake experienced in the Japanese islands in recorded history. Tremendous devastation occurred throughout central Honshu, particularly in the provinces of Mino and Owari. The number of deaths was 7270, and more than 17,000 persons were injured. More than 142,000 houses

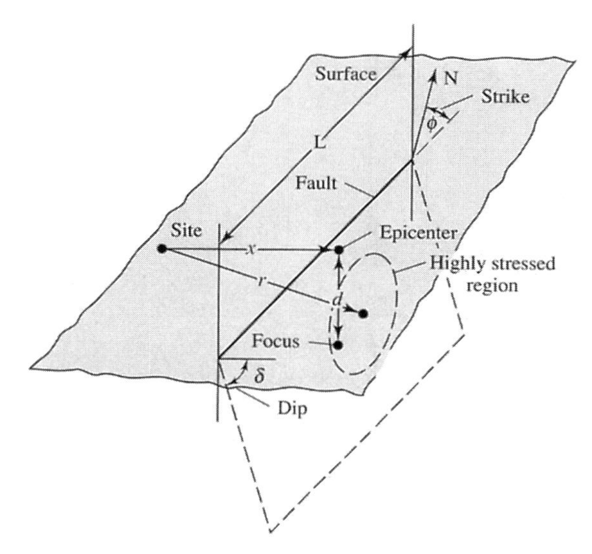

Figure 5.5 Diagram showing a section of a dipping fault length L, which strikes at an angle ϕ east of north and dips at an angle δ. An earthquake is generated by sudden fault slip in highly stressed rock, starting at the focus with depth *d* and distance *r* from the surface site of observation.

collapsed altogether, and many others were damaged. Some 10,000 landslides are known to have occurred throughout the area. Extraordinary surface breaks appeared that could be clearly traced for about 80 kilometers across the countryside. They exhibited maximum horizontal offsets of 8 meters and vertical offsets of 2 to 3 meters in numerous places.

At the time it was widely believed that great shallow earthquakes were caused by underground explosions or magma movements. But Professor B. Koto of the University of Tokyo was so impressed by the extent of faulting in the Mino-Owari earthquake that he departed from established opinion to assert that sudden fault slip had been the cause — a revolutionary idea then.

This earthquake has been restudied many times in an effort to assess confidently the likelihood of future earthquake occurrence in the heavily populated and generally highly seismic islands of Japan. The most recent evidence shows that the 1891 faulting was quite complicated, with left-lateral displacement on three major preexisting faults. The ruptures, however, were not visible at the surface over the entire length of each of them. Except in a few places, the fault dip was almost vertical, and a few minor fault breaks, shorter than a kilometer in length, occurred.

Among the minor breaks was the Midori fault near Neodani village (see Figure 5.6), which had a spectacular fault scarp more than 400 meters long with a vertical displacement of 6 meters (northeast side up) and a

Figure 5.6 Fault scarp near Neodani village, Honshu, Japan, as it appeared in August 1966. The faulting occurred in association with the Mino-Owari earthquake of October 28, 1891. Sliding and erosion have altered the original nature of the fault displacement, which was normal with the right (north) side up. [Photograph by Bruce A. Bolt.]

left-lateral horizontal displacement of 4 meters.* The direction of the vertical motion at Midori, however, is somewhat exceptional because most vertical motion is in the opposite direction at other places along the fault rupture.

During the 14 months following the Mino-Owari earthquake, more than 3000 aftershocks were felt at Gifu, the capital of Mino province. Recent seismographic recordings have indicated that small earthquakes are still common along the 1891 fault breaks. The focal depths of the recent earthquakes are confined mainly to the top 15 kilometers of the crust, suggesting that perhaps the 1891 faulting extended only to this depth.

Japanese surveyors made some detailed geodetic measurements around the Mino-Owari fault zone in the years 1894 to 1898 and compared them with similar measurements that had been made prior to the earthquake. Along certain level lines, there had been a widespread crustal uplift of

**In 1967 I visited Neodani and rephotographed the fault scarp, now somewhat smoothed by weathering. More recently, railroad construction through the valley threatened to obliterate this famous seismological landmark.*

about 70 centimeters, whereas elsewhere in the source region there had been appreciable subsidence, of 30 to 40 centimeters. Such crustal movements indicate the large extent of deformation of crustal rocks that is associated with the production of great earthquakes.

The 1979 Imperial Valley Earthquake, California

Some earthquakes have been instantly elevated to special status because they provide hitherto unavailable measurements. The moderate-size earthquake that occurred at 4:16 p.m. on October 15, 1979, in the southern Imperial Valley of California was one such earthquake. It was produced by right-lateral slip on the northwest-trending Imperial fault (see Figure 1.2). Fault slip was seen at the surface for approximately 30 kilometers south of the town of Brawley, and possibly for 37 kilometers to the international border. No one was killed.

The widespread instrumental measurements of this earthquake near its source shed light on four important questions regarding, first, the repetition of earthquakes along the same fault system and its implications for earthquake prediction; second, the vexing question of sympathetic slip on adjacent faults in earthquake sequences; third, the variation of strong ground shaking as measured by specially arranged groups of seismographs (called "strong-motion arrays"); and finally, the seismic resistance of structures built under modern seismic codes.

An almost twin earthquake occurred when the same fault ruptured 39 years earlier, on May 18, 1940. At that time, geologists realized that a previously unknown fault, designated the Imperial fault, had produced the earthquake through right-lateral strike-slip rupture. (This fault is actually a section of the San Andreas fault system; see Figure 2.6.) The fault slip was traced for at least 70 kilometers across the international boundary into northern Baja, Mexico.

There were both striking differences and similarities between the 1940 and the 1979 earthquakes. The two earthquakes were about the same size. The fault offsets in 1940 were up to 5.8 meters horizontally and 1.2 meters vertically. Nine people were killed, and structural and agricultural losses amount to $6 million. In 1979, the maximum right-lateral fault displacement was only about 55 centimeters, and the maximum vertical displacement was measured at 19 centimeters, down to the east. After both the 1940 and 1979 principal shocks, the locations of aftershocks shifted progressively northward along the Imperial fault, and each had damaging aftershocks near Brawley.

What are we to make of the repetition, after 39 years, of a similar-size earthquake caused by rupture of the same fault? It is not what might be expected from a simple application of the seismic-gap theory (see Chapter 7). According to the gap theory, because the 1940 earthquake would have relieved most of the strain on the Imperial fault from Brawley to south of the Mexican border, the next earthquake of comparable size might have been expected to occur either from fault rupture north or south of this section of the San Andreas fault system. Evidently potent strain energy persisted in the rocks.

A Recent Japanese Example, Kobe 1995

The fault sources of disastrous earthquakes in Japan have been revealed in many earthquakes in the last century. The 1891 Mino-Owari (Nobi) and 1943 Tottori earthquakes are examples. The 1995 earthquake that produced enormous grief and havoc in the port city of Kobe, located in the southern part of Hyogo Prefecture, Japan, is another (see Figure 10.1). The earthquake, called the Hyogo-ken Nanbu earthquake, occurred at 5:46 a.m. local time on January 16, 1995. The immediate death toll reached about 5100 dead, 30,000 injured, and 300,000 people homeless in a city of 1.4 million people. The hypocenter was at a depth of about 20 kilometers, near the northeast corner of Awaji Island (Figure 5.7), about 200 kilometers away from the tectonic plate boundary between the Philippine Sea plate and the Eurasia plate (see Figure 7.2). The right-lateral strike-slip on the Nojima fault was the culprit. Maximum horizontal offset on the fault was found to be about 1.5 meters, with some vertical displacement at places (Figure 6.5). The extent of the faulting is impressively defined by the mapping of the small aftershocks (see Figure 5.7). This map indicates that the fault tear probably extended horizontally for up to 50 kilometers; the position of the hypocenter indicates that the rupture spread in a bilateral sense (compare the 1989 Loma Prieta earthquake source, see page 9) with the northern rupture beneath the damaged city. This unhappy circumstance undoubtedly led to a sharp enhancement of the shaking as the northern fault rupture headed straight toward and underneath Kobe.

Some seismological circles in Japan were surprised by the earthquake. The region has many Quaternary faults, but the Nojima fault had not been explicitly marked as an imminent threat, perhaps because historically only two earthquakes (a magnitude about 7 in 868 and about 6 in 1916) have shaken the Kobe-Osaka Bay district hard. Although a large-scale earthquake prediction research program in Japan has extended over three decades (see

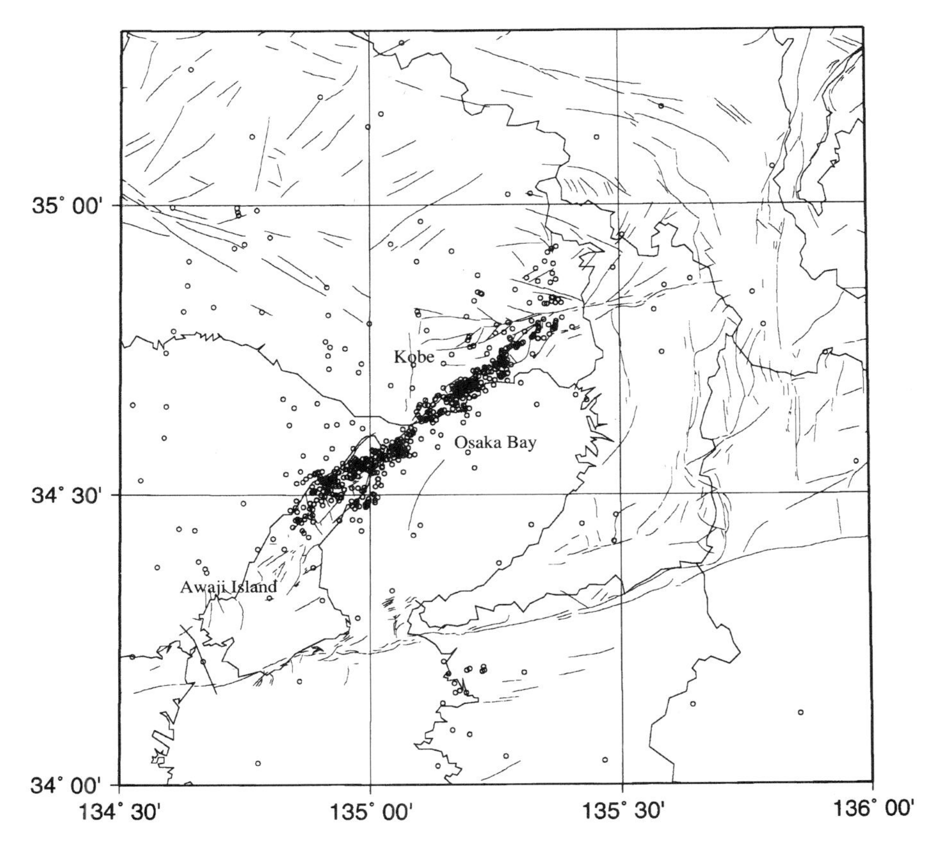

Figure 5.7 Fault map showing epicenters of aftershocks (January 16 to January 21, 1995) of the 1995 Kobe earthquake. The length (about 50 kilometers) of the ruptured Nojima fault is indicated by the dense lineation. The north end is under Kobe city. Fine lines are mapped faults. [Courtesy of Earthquake Research Institute, University of Tokyo.]

Chapter 10), this earthquake was not predicted. Nevertheless, there were many strong-motion seismographs in the area, both on the alluvium plain around the harbor and on the exposed rocks of the mountains that rise behind the city to the north. The peak horizontal acceleration was about 80 percent gravity.

The 1995 Hyogo-ken earthquake, with a local magnitude of 7.2, is one of the largest earthquakes to strike a dense urban area near its source anywhere in the world in modern times. The horror was revealed to viewers in many countries by means of contemporary television coverage available through satellite connections to cable television in private homes. The pictures of toppled expressways, crumpled train tracks, leaning buildings, and collapsed homes sent a stark message and led to the question of whether building codes in Japan are seriously flawed. Many factories, wharf facilities, bridges, and commercial buildings were damaged, including complete collapses of multistory structures; large portions of elevated highways failed (see the figure at the beginning of Chapter 10); and trains were derailed (see Plate 7). (Fortunately, regular train service did not begin until 6 a.m.). Because water, gas, and electric supplies were cut off for days, fires ignited throughout Kobe and spread over multiple city blocks; fire fighting was hampered by both lack of water and inaccessibility caused by debris-blocked streets. Emergency preparedness seemed quite inadequate and overwhelmed by the large number of homeless and injured citizens.

A detailed study of the distribution of damage in Kobe is important for many other urban areas subject to earthquakes. For example, the seismological situation at Kobe is mirrored in the eastern cities of San Francisco Bay, through which the active Hayward fault passes. The total number of buildings in Kobe was about 500,000, of which more than 100,000 were destroyed or seriously damaged with only about 20 percent of buildings in downtown usable immediately after the earthquake. The newly built ductile-frame high-rise buildings were generally not damaged. Many damaged commercial buildings, apartments, and lifeline structures had been built in the 1950s and 1960s after the bombing of Kobe in World War II. Most predated critical building code improvements enacted in 1971 and 1981 in parallel with similar upgrades in California and other earthquake-prone regions. Most collapsed freeways were built in the 1960s and 1970s; a notable feature was the lack of retrofit of most of these structures similar to that begun in California after the lessons of the 1971 San Fernando and the 1989 Loma Prieta earthquakes (see Chapter 12).

It was perhaps a surprise, given the long home-building tradition in earthquake-prone Japan, that residential houses proved to be highly vulnerable to the earthquake forces. Unlike the wood-frame houses typical in the western United States and New Zealand, for example, the homes in Kobe often had heavy clay tile roofs supported on vertical posts and horizontal beams, with little X-bracing or plywood sheathing to resist horizontal shaking.

The Tragedy of the 1988 Armenian Earthquake

In far too many parts of the Earth, modern industrialization has been at the root of tragedy. An indelible example is the earthquake sequence that struck northern Armenia on December 7, in the cold winter of 1988 (see

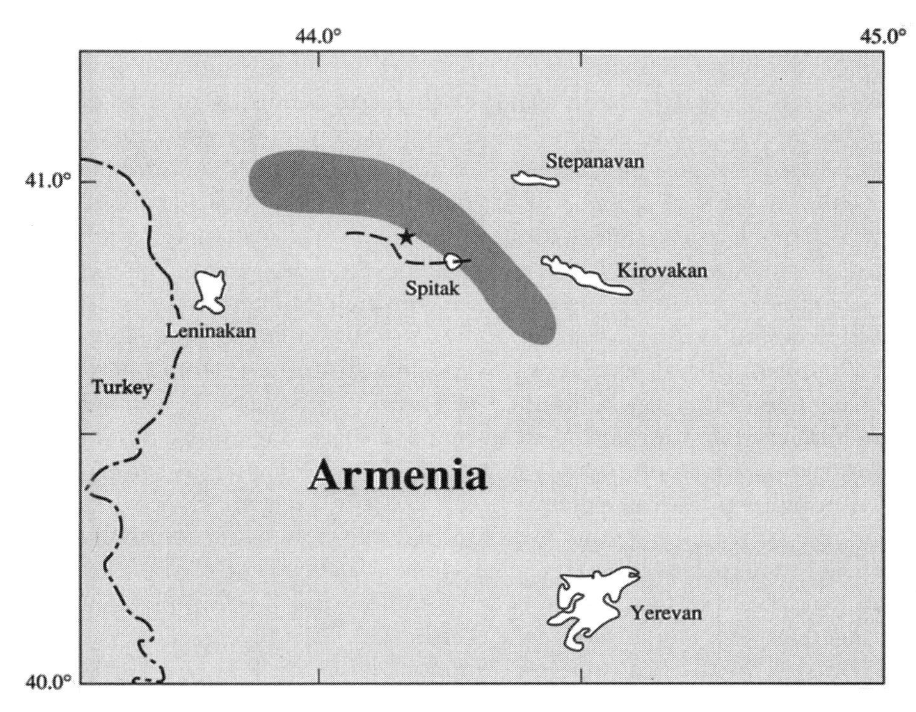

Figure 5.8 Map of highest intensity in the 1988 Armenian earthquake. The star is the epicenter, the dashed line is the surface faulting, and the darker gray the area of aftershock epicenters.

Figure 5.8). The first of these devastating earthquakes radiated out from a rupturing fault at 11:41 a.m. local time and affected a population of 700,000 people, destroying large sections of the cities of Spitak, Leninakan, and Kirovakan. In the surrounding countryside, 58 villages were leveled and 100 significantly damaged. One government estimate was that 25,000 people perished, based on the recovery of 24,944 bodies from the rubble. The earthquake left at least 4000 homeless and 30,000 injured.

What is today the Republic of Armenia, with its capital Yerevan, was part of a larger country which was an independent republic from 1918 to 1920. Historical earthquakes have been described there back to 550 B.C. In 1926, a notable earthquake occurred in Leninakan and destroyed many houses.

The geological source of the main 1988 earthquake was fault rupture about 40 kilometers south of the spine of the Caucasus Mountains. This magnificent mountain range was produced by pressure in the Earth's crust due to the convergence of the Arabian and Eurasian tectonic plates (see Chapter 7 and Figure 7.2). The Caucasus are thus a segment of the belt of high mountains ranging from the Alps across southern Europe to the Himalayas of Asia. Recall from Chapter 2, this belt is associated with continuous seismic activity, with major earthquakes occurring frequently from the Aegean Sea, across Turkey into Iran and western Afghanistan and Tazhikistan.* While the frequency of earthquake strikes in Armenia does not quite match that in some of the other segments of this active belt, geologically rapid crustal deformation in the vicinity of Armenia is marked by active thrust faults and volcanic activity. For example, Mount Ararat, the famous biblical mountain, is a 5165-meter-high Quaternary volcano 100 kilometers south of the epicenter of the 1988 Armenian earthquake.

On December 7, seismic waves radiated from the rebound of a previously strained but unnamed fault (see Figure 5.9) at least 60 kilometers in length. It extends west-northwest of Spitak close to the village of Nalband. The strike of the rupture fault was parallel to the Caucasus range and it dips toward the north-northeast. In 1992, I walked with some of my students along this seismogenic culprit still clearly etched on mountain slopes then covered with wildflowers. The vertical component of slip along the main part of the surface rupture was 1.6 meters near the southwest end and averaged 1 meter along most of the scarp.

Parts of the area of strongest ground shaking are highly industrialized with both light and heavy industry, including large chemical- and food-processing plants. There are a number of large electrical substations and thermal power plants, in this area; many were affected. A nuclear power plant near Yerevan, about 75 kilometers from the wave source, was subjected to only minor ground shaking and no damage to it occurred, but afterward the plant was closed for some years.

Many detailed accounts covering all aspects of the aftermath of the Armenian earthquake have now been published. A great deal of publicity was given to the heroic efforts of rescuers to save those trapped under fallen structures. Even in such tragic circumstances, there was time for local anecdotes. When a French rescue team pulled a man from the rubble after four days, he raised his hands and, thinking that World War III had started, said, "I surrender."

Everyone was interested to know why this earthquake caused so much destruction and death (Figure 5.10). Was the ground shaking exceptional? Can such terrible results be expected to be transferred to other seismic lands? The answers to these questions can perhaps best be given in the

**Recent highly destructive earthquakes along this belt include one on June 20, 1990, near the southern edge of the Caspian Sea. It destroyed many cities in northern Iran and killed upwards of 40,000 people.*

Figure 5.9 Scarp of causative fault in the 1988 Armenian earthquake. [Courtesy of A. Der Kiureghian.]

words of a frank assessment by Armenian and Soviet engineers. They summarized the problems as follows. First, the earthquake was more severe than the building code provisions allowed for. There were also design deficiencies in precast reinforced concrete structures, and the quality of construction was often poor. They wrote:

> The catastrophic earthquake that occurred on December 7, 1988, brought about heavy damage to most buildings and structures in many cities and villages. Initial results of our investigation revealed that in its manifestation most frame and nine-story frame panel buildings were completely destroyed. Stone buildings with no anti-seismic measures of construction collapsed. In extensive areas, there was large deformation of railroads and distortion of railtracks. Rock slides occurred in the mountains. Wide cracks appeared in the soil; there were massive slides along cliffs. Bridges were greatly damaged. In the city of Leninakan industrialized enterprises and trade centers collapsed and the chemical plant in the city of Kirovakan was wrecked. In Spitak, commercial

Figure 5.10 Collapsed concrete-frame apartment building in Leninakan (now Gumri), Armenia, 1988. [Courtesy of A. Der Kiureghian.]

enterprises were completely destroyed; educational institutions such as schools, nurseries, maternity wards and hospitals were lost in most cases.

This earthquake demonstrates the exceptional urgency of undertaking measures to build earthquake-resistant buildings and to protect the population and unique equipment in earthquakes. It is extremely important to conduct seismological research, to develop new approaches to ensure good quality construction in seismic regions, and to encourage a heightened sense of responsibility in participants.

We will see, as the description of earthquakes develops in the following pages, that this recommendation holds the key to ensuring maximum safety in all seismic areas.

The 1990 Philippine Fault Rupture

The Philippine Islands have long been harried by earthquakes. A conspicuous fault zone runs as a rift almost diagonally across the islands of Luzon, Leyte, and Mindanao. Slips within it and along its subsidiary faults have produced major earthquakes with great damage in past decades. The movement is left-lateral, opposite to the sense of displacement on the San Andreas fault in California. One of the foundation fathers of seismology, Professor John Milne,* wrote on the disastrous Luzon earthquakes that occurred in 1880, with buildings shattered in the city of Manila and elsewhere. He drew attention to the proclivity of the region to great earthquakes and pointed to the extreme seismic danger that exists there.

The latest violent earthquake to strike the Republic of the Philippines was of high magnitude. On Monday, July 16, 1990, at 4:26 p.m. local time, the region experienced severe shaking from the principal earthquake and its aftershocks. Figure 5.11 shows the main earthquake location in relation to the seismogenic tectonics of the region.

At least 1700 people were killed and 3500 seriously injured. Damage was extensive throughout the central region of Luzon and there was even some sporadic building damage in Manila about 240 kilometers away. The

**John Milne (1850–1913). This exceptional scientist was Professor of Mining and Geology at the Imperial College of Engineering in Tokyo, where he greatly advanced the study of earthquakes. Later he lived on the Isle of Wight, England, with his Japanese wife, and there developed the first global network of seismographs.*

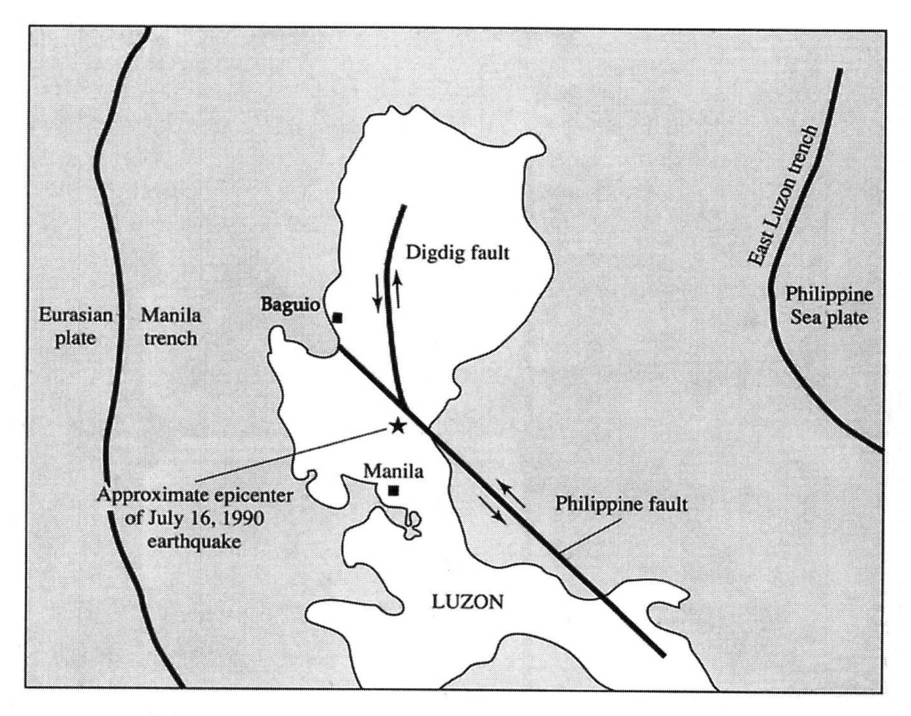

Figure 5.11 Schematic plot of the location of the Philippine and Digdig faults in Luzon with the epicenter of the main July 16, 1990, earthquake.

most serious effects were in areas, particularly in the resort city of Baguio, where structures were built on soft, alluvial material and along the river sediments. Massive landslides occurred in the rain-saturated tropical soils, burying houses and even entire villages, and blocking extensive lengths of the highways.

The cause of the mainshock and its aftershocks was slip along the Philippine and Digdig faults. The offsets could be measured over a distance of 110 kilometers. Because the region is mountainous and hard to access, even more extensive faulting may have reached the surface. The largest offsets were found on the Digdig fault with a maximum of about 6.1 meters near the city of Imugan, and to the south along the Philippine fault, on which the largest displacement was about 5.1 meters horizontally southeast of the town of Rizal. The surface rupture near the village of Digdig can be seen in Figure 5.12. In this photograph there are vertical as well as lateral offsets, but in general, vertical elevations were not consistently to the north or south of the fault trace — although they sometimes reached about 2 meters.

Figure 5.12 Surface rupture on the Digdig fault in pavement near the village of Digdig. Assuming a vertical fault plane, the left-lateral component (4.61 meters) and the east-side-up vertical component (1.49 meters) here combine to make a 4.84-meter slip vector, measured on the seventeenth post-earthquake day. [Courtesy of R. Sharp.]

As a footnote to this calamity, the U.S. Clark Air Force Base, located about 180 kilometers south of Baguio, suffered only minor damage so that personnel there were able to render assistance in the stricken area. Less than a year later, the base was so damaged by ash falls from the eruptions of the nearby Mount Pinatubo volcano that it had to be permanently closed. Until 1991, Mount Pinatubo was regarded as a dormant volcano with no eruptions for over 400 years. A series of small explosions from Mount Pinatubo beginning on April 2 led to the evacuation of at least 58,000 people prior to that volcano's climactic eruption on June 15, 1991. Although 320 persons died as a result of these eruptions, mostly due to the collapse of ash-covered roofs, the forewarning and subsequent precautions undoubtedly averted much greater loss of life and property. Earthquake warnings are harder to come by (see Chapter 10).

Fault Slippage and Fault Gouge

The huge concrete-lined water tunnel carrying water from the Sierra Nevada, California, to the cities on the east of San Francisco Bay seems protected and durable as it passes through the Berkeley Hills. The Memorial Football Stadium at the University of California, Berkeley, seems indestructible in its reinforced concrete frame. The modern Cienega Winery building along Cienega Road near Hollister in the beautiful coastal range of central California looks serenely permanent. Yet all these structures are unobtrusively and almost uneventfully being torn in two.

Earthquakes are not the cause. The tunnel and the stadium straddle the active Hayward fault (see Figure 6.2), along which slow right-lateral slip is steadily occurring. When the water tunnel was emptied in 1966, cracks several centimeters across encircled the concrete lining just where the tunnel and fault zone intersect. Under the Memorial Stadium, a concrete drainage culvert now shows considerable cracking where it crosses the Hayward fault trace, and instruments that I placed across the cracks in 1966 showed that the right-lateral slip inexorably continues about 2 to 5 millimeters a year.

The present Cienega Winery near Hollister is, surprisingly, the third to be built on its site — one located squarely across the San Andreas fault trace. The fault trace is detectable from the slight change in elevation running parallel to Cienega Road on the western side, through the rows of rapevines (see Figure 5.13). Small springs abound along this trace, and in places the rows of grapevines are offset in a right-lateral sense. From the road the walls of the winery buildings can be seen to be bent by the slow slip occurring under the building. Just to the south an open concrete culvert has been broken and offset by the slip along the fault (see Figure 5.14).* Measurements of the alignments of the culvert and floor slabs show a relative offset rate across the San Andreas fault at this point of 1.5 centimeters per year. Not far away, in the town of Hollister, the subsidiary Calaveras fault is also creeping, producing noticeable offsets and damage to curbs, sidewalks, fences, and even houses.

Horizontal fault slippage has now also been detected on other faults around the world, including the north Anatolian fault at Ismetpasa in Turkey and along the Jordan Valley rift in Israel. Usually, such episodes of fault slip are *aseismic* — that is, they are not accompanied by local earthquakes. When earthquakes do occur associated with the slipping faults, the rate of slip may increase for a short time after the earthquake.

**Here the visitor can stand astride the crack in the culvert and imagine one foot on the Pacific plate and one on the North America plate (see Figure 7.2).*

Figure 5.13 Aerial photograph of the Cienega Winery built across the San Andreas fault near Hollister, California. The fault (indicated by the horizontal arrows) can be seen extending from left to right, through the culvert on the left (south) side of the winery building and through the building itself. [Courtesy of D. Tocher.]

Typically, a single slip offset has an amplitude of a few millimeters, and the episodic slip lasts from a few minutes to a few days. Measurements in California indicate that after weeks of immobility, slip will commence and then progress along the fault for tens of kilometers at a speed of about 10 kilometers per day.

What is the nature of aseismic slip on faults? Let us look more closely at the types of rocks in a fault zone. Crushed and highly deformed rock occurs in the zone in a band many meters wide in some places (compare Figure 5.3). In the course of millions of years, intermittent yet frequent differential movement along a fault breaks and shears the rock into fine granular and powdery forms. These in turn are altered by percolating groundwater to produce clays and sandy silts. The resultant material is *fault gouge.* When a fault section is penetrated by a tunnel or trench, the gouge zone is often found to form a barrier that is fairly impervious to water; the water table sometimes stands at different levels on either side of the fault gouge, and this is why soaks and springs are found along faults.

Wet fault gouge feels like a soft deformable plastic and behaves more like a viscous solid than a brittle elastic one. It is thus hard to conceive that

Figure 5.14 Offset of the concrete drainage culvert by slow fault slip along the San Andreas fault at the Cienega Winery. This amount of offset had taken place in the course of 20 years. [Courtesy of W. Marion.]

it would resist slip strongly. The depth of the gouge zones varies considerably, but on major faults it may be several kilometers.

Because active faults do suddenly slip to produce an earthquake, there must be, below the weaker surface materials, stronger and more elastic rocks that are in bonded contact across the fault planes. Only in this way will the slow straining of the rock store enough elastic energy to produce earthquakes. It is therefore reasonable to surmise that major fault zones, such as the San Andreas, consist of a sequence of layers; at the surface, weak, plastic gouge would extend down for several kilometers but progressively give way to stronger crystalline rock in welded contact across the fault surfaces, down to a depth of 15 or 20 kilometers. Below that depth the increased temperature in the Earth again softens the rock so that elastic straining is not mechanically feasible. Strong support for this model comes from the discovery in the early 1960s that, in most of central California,

Figure 5.15 There is always a good reason for those "No Parking" signs. California earthquake, 1986. [Courtesy of D. B. McIntyre.]

earthquakes do not occur at depths below about 15 kilometers (see Figure 5.15). At these profound depths, the rock has become plastic again and is no longer capable of storing strain energy.

Sometimes aseismic creep is observed at the ground surface along a ruptured fault that has produced a substantial earthquake. For example, along the San Andreas fault break in the 1966 earthquake on June 27 near Parkfield, California, offsets of road pavement increased by a few centimeters in the days following the main earthquake. Such continued adjustment of the crustal rock after the initial major offset is probably caused partly by the minor slips that produce aftershocks and partly by the yielding of the weaker surface rocks and gouge in the fault zone as they accommodate to the new tectonic pressure in the region.

It is clear that fault creep, when it occurs in built-up areas, may have unfortunate consequences. This is another reason why certain types of structures should not be built across faults if at all possible. When utility lines, roads, and railroads must be laid across active faults, they should have jointed or flexible sections in the fault zone.

6 The Causes of Earthquakes

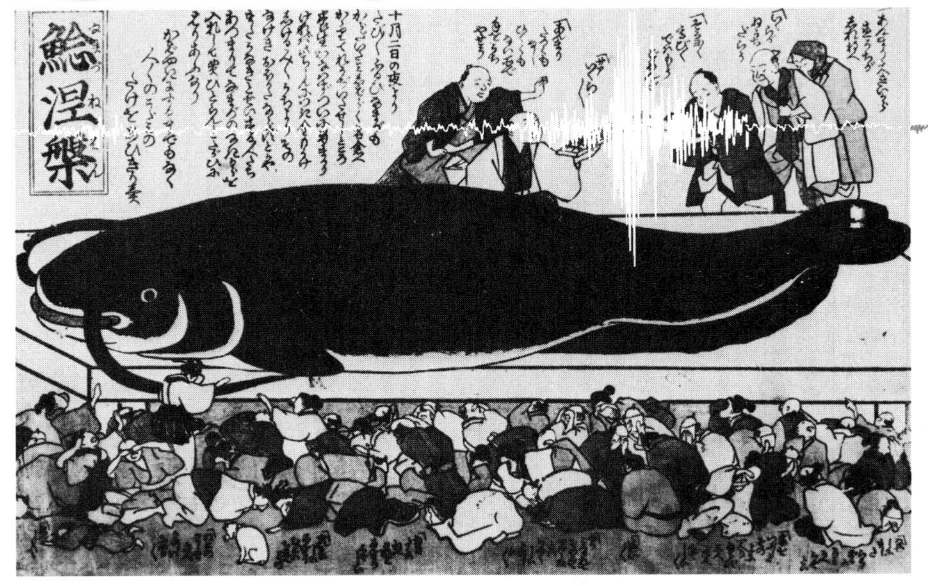

Fishmongers selling flesh from a *namazu* (catfish) whose excessive movement in a large Edo (Tokyo) earthquake has killed it. The people complain of the profiteering from their tragedy.

It is probable that the whole movement at any place (along the rupturing fault) did not take place at once, but that it proceeded by very irregular steps.

— *H. F. Reid, "Report of the State Earthquake Investigation Commission (The California Earthquake of April 18, 1906)*

Not so long ago people generally believed that the causes of earthquakes would always remain hidden in obscurity, because they originated at depths far below the realm of human observation. Moreover, for a long time the prevailing view was that earthquakes came as a punishment for human failings. "The Earthquake," a verse written in about 1750, summarizes this view:

What pow'rful hand with force unknown,
Can these repeated tremblings make?
Or do the imprison'd vapours groan?
Or do the shores with fabled Tridents shake?
Ah no! the tread of impious feet,
The conscious earth impatient bears;
And shudd'ring with the guilty weight,
One common grave for her bad race prepares.

— *Anonymous*

Today earthquakes and most of their observed properties are explained in terms of mechanical theory as developed under the name *plate tectonics*. Before describing this unifying geological scheme in Chapter 7, however, it is necessary to understand how the seismic waves in the ground are generated.

Types of Earthquakes

The first step toward grasping the modern view of seismic-wave generation is the appreciation of the close relation between those parts of the world that are most earthquake prone (shown in Figure 2.2) and the geologically new and active areas of the world (see Figure 7.2 and Plates 10 and 11). We have observed in Chapter 2 that most earthquakes occur in large-scale patterns associated with global-sized features such as mountains, rift valleys, midoceanic ridges, and ocean trenches. Fresh faulting at the surface is also often associated with earthquakes (Chapter 5). Indeed, many of the most widely damaging earthquakes — such as the 1906 San Francisco earthquake, the 1891 Mino-Owari earthquake in Japan, the 1976 Tangshan earthquake in China, the 1988 Armenian earthquake, and the 1995 Kobe earthquake (Plate 7) — were produced by extensive fault ruptures.

Before proceeding, it is helpful to classify earthquakes by their mode of generation. By far the most common are *tectonic earthquakes*. These are produced when rocks break suddenly in response to various geological forces. Tectonic earthquakes are scientifically important to the study of the Earth's interior and of tremendous social significance because they pose the greatest hazard. Consequently, in most of the book, we are concerned with this type of earthquake.

A second well-known type of earthquake accompanies volcanic eruptions. In fact, the idea that earthquakes are linked primarily to volcanic activity goes back to the Greek philosophers, who were impressed by the common occurrence of earthquakes and volcanoes in many parts of the Mediterranean. Today, a *volcanic earthquake* is still defined as one that occurs in conjunction with volcanic activity, but it is believed that while eruptions and earthquakes both result from tectonic forces in the rocks, they need not occur together. The actual mechanism of wave production in volcanic earthquakes is probably the same as that in tectonic earthquakes. (Volcanic earthquakes are discussed in more detail in Chapter 9.)

Collapse earthquakes are small earthquakes occurring in regions of underground caverns and mines. The immediate cause of ground shaking is the collapse of the roof of the mine or cave. An often-observed variation of this phenomenon is the so-called mine burst. This happens when the induced stress around the underground workings causes large masses of rock to fly off the mine face explosively, producing seismic waves. Mine bursts have been observed in Canadian workings, for example, and are especially common in deep South African mines.

Collapse earthquakes are also sometimes produced by massive landsliding. For example, a spectacular landslide on April 25, 1974, along the Mantaro River, Peru, produced seismic waves equivalent to a small-to-moderate earthquake.* The slide had a volume of 1.6×10^9 cubic meters

**A Richter magnitude of 4.5 was recorded.*

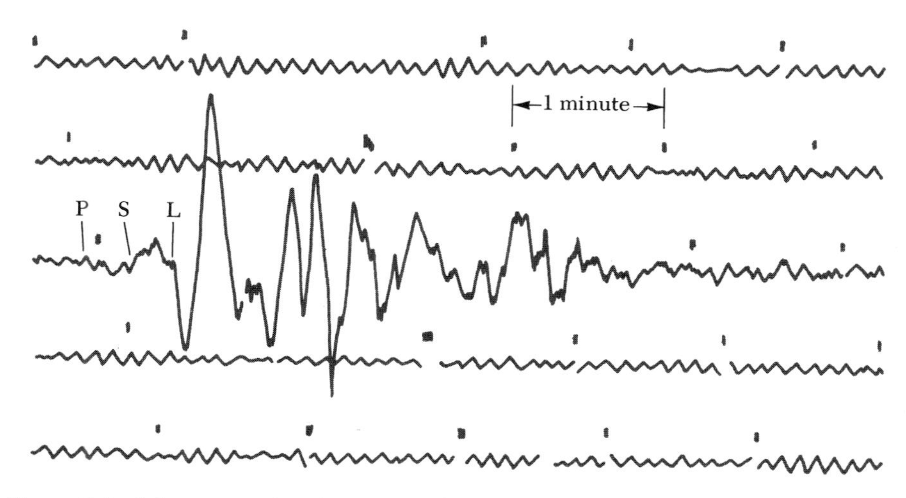

Figure 6.1 Seismogram showing east-west ground motion at Nana seismographic station, Peru, caused by the Rio Mantaro landslide, 240 kilometers away. The station reader has marked the arrival of the P, S, and surface (L) waves. Time increases from left to right on each trace.

and killed about 450 people. As far as we know, this landslide was not triggered by a nearby tectonic earthquake, as often happens. Part of the gravitational energy lost in the rapid downward movement of the soil and rock was converted to seismic waves, which were clearly recorded by seismographs hundreds of kilometers away. Duration of shaking measured on a seismograph 80 kilometers distant was 3 minutes; this is compatible with the speed of the slide, which was about 140 kilometers per hour over the observed slippage of 7 kilometers. (Figure 6.1 shows a seismogram of this event from the Nana seismographic station in Peru.)

A similar collapse earthquake resulted from the greatest landslide to occur in recent history—at Usoy in the Pamir Mountains of Russia in 1911. Prince B. B. Galitzin, a founder of modern seismology, recorded earthquake waves on his seismographs near St. Petersburg that must have radiated from the landslide area. He compared the energy in the earthquake waves with that in the landslide and concluded that an earthquake may have triggered the landslide. More recent work—which makes Galitzin's conclusion doubtful—eventually led to techniques for calculating from seismograms the energy released in an earthquake. It was not until 1915 that an expedition, sent to investigate the Usoy landslide, found that the material involved was 2.5 cubic kilometers!

Humans cause *explosion earthquakes*, or ground shaking produced by the detonation of chemicals or nuclear devices. Underground nuclear explosions fired during the past several decades at a number of test sites around the world have produced substantial earthquakes. When a nuclear device is detonated in a borehole underground, enormous nuclear energy is released. In millionths of a second, the pressure jumps to thousands of times the pressure of the Earth's atmosphere and the temperature locally increases by millions of degrees. The surrounding rock is vaporized, creating a spherical cavity many meters in diameter. The cavity grows outward as boiling rock vaporizes from its surface, and the rock around it is minutely fractured by the shock of the explosion.

The compression of the rock in the fractured region (an area of perhaps hundreds of meters) produces seismic waves that travel outward in all directions. When the first compressive seismic wave reaches the surface, the ground arches upward, and if the wave energy is sufficient, it will blast the soil and rock away, throwing fragments of rock into the air and producing a crater. If the borehole is deeper, the surface may be only cracked and the rock will lift upward for an instant and then "slap down" on the underlying layers.

Some underground nuclear explosions have been large enough to send seismic waves throughout the Earth's interior; waves with amplitudes equivalent to moderate-size earthquakes have been recorded at distant seismographic stations. Some explosions have produced waves that have shaken buildings in distant cities. For example, on April 26, 1968, a nuclear device called Boxcar was tested at the Nevada Nuclear Test Site.* Before the test, Las Vegas citizens — notably the multimillionaire resident and property owner, the late Howard Hughes — expressed misgivings that such an energetic device might cause structural damage and even deaths. Nevertheless the event took place, and people in the surrounding towns felt the shaking, which in Las Vegas, 50 kilometers away, lasted for 10 to 12 seconds. Fortunately, no significant damage resulted.

Of course, people and animals sometimes produce earthquakes (usually small) in other ways. An illustration in a lighter vein is given in the "Earthquake Quiz" at the end of this book.

Geodetic Surveying and the Slow Buildup of Energy

As centuries pass in earthquake country, deep-seated forces beneath our feet deform the rocks steadily and unobtrusively. What are the surface

**This explosion had an energy equivalent to that produced by 1200 thousand tons of TNT!*

manifestations of this crustal warping? What is the evidence for the tectonic forces?

The most obvious manifestations are the great mountain ranges, produced by massive vertical uplift of the Earth's surface above sea level — a process that has taken millions of years (Plate 1). But even crustal movement that has taken place within much shorter periods can be easily detected by careful surveys. In many countries of the world, such geodetic surveys now go back at least to the last century.

In classical surveying there are three major types of geodetic positioning. In two, the degree of horizontal movement is determined. In the first survey type, small telescopes are used to measure angles between markers on the ground surface; this is called *triangulation*. In the second, called *trilateration*, distance between markers on the surface is measured along extensive profiles. Light (sometimes a laser beam) is reflected from a mirror on a distant rise, and the time it takes for the light to travel the two-way path is measured. Because the speed of light varies with atmospheric conditions, in even precise surveys, precision is about 1.0 centimeter over a distance of 20 kilometers.

The third type of classical survey determines the degree of vertical movement by repeated measurements of the difference in *level* across the countryside. Such leveling surveys simply measure differences in the elevation between vertical wooden rods placed at fixed bench marks. Repetition of these surveys reveals any variations in these differences that occur between measurements. Wherever possible, national level lines are extended to the continental edge, so that the mean sea level can be used as the reference point.

Although barely two decades old, there is now a much more precise space-geodetic counterpart of the above ground-based classical ways of obtaining the position of points on the ground surface relative to one another. The one that is being widely adopted to track tectonic deformations and strains along faults is the Global Positioning System, called in brief GPS.

Successor to a U.S. Navy Navigation Satellite System, GPS was developed to provide instantaneous global three-dimensional position and velocity of a ship, with accuracies much higher than previously. It provides the data at any time. Today, the operational GPS system uses at least 18 satellites, with nearly circular 12-hour orbits. Radio signals from each satellite give its own position in real time using a high-accuracy clock. Differences in travel times from a receiver-transmitter at a point on the ground to several satellites enable the latitude and longitude (and, less well, elevation) of any site to be calculated to high precision. Commercially available GPS instruments have become inexpensive (a few hundreds dollars), mainly because they are of enormous value to owners of small boats and airplanes.

At distances of less than 20 to 30 kilometers, along the San Andreas fault for example, short-term accuracies can be obtained with GPS receivers of 1 or 2 parts per 10 million (or better) in the baseline distances. Such capability in geodetic surveying is unprecedented and can be expected to have a broad impact on the measurement of strain along the active fault systems of the world. An added attraction is that GPS geodesy does not require line of sight between measurement points.

All survey methods of observing crustal movements show that in tectonically active areas, such as California and Japan, the Earth's crust moves horizontally and vertically in quite measurable amounts. They also show that in the stable areas of continents, such as the ancient rock masses of the Canadian and Australian shields, change is negligible, at least in the past century.

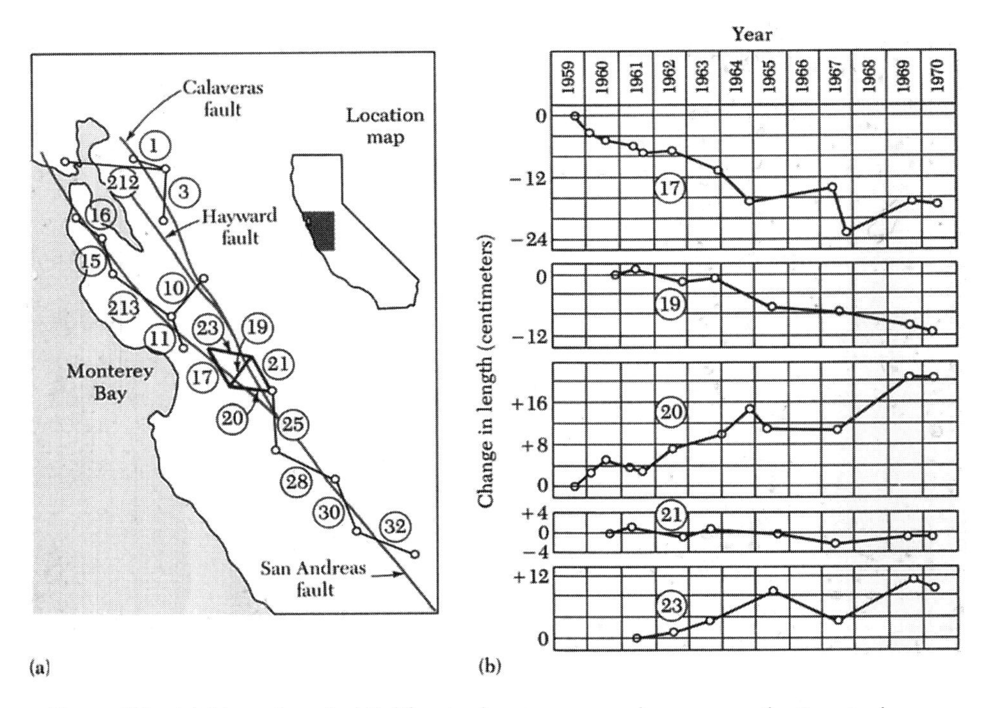

Figure 6.2 (a) Map of central California showing survey lines across the San Andreas fault system as established by the State Department of Water Resources and Division of Mines and Geology. The location of the entire area surveyed is shown in the location map at the right. (b) The temporal changes in lengths of lines 17, 19, 20, 21, and 23 are plotted. [After Bruce A. Bolt and F. Moffitt.]

Geodetic measurements of crustal deformation in the seismically active California region began as early as 1850. In 1959, special observations along the San Andreas fault that could be useful in earthquake prediction were initiated. Some results of these measurements — along five lines in central California from 1959 through 1970 — are sketched in Figure 6.2. Trends in the strain are spectacularly clear. If we examine line 17 in Figure 6.2, which crosses the San Andreas fault near Gilroy, we will see a decrease in length at the rate of about 2 centimeters a year. This is in agreement with right-lateral deformation along the San Andreas fault. Survey lines, such as 21, which do not lie across the major faults show very little change in length.

The displacement occurring along the San Andreas fault enables geologists to make a graphic — if long-range — forecast. The city of Los Angeles, which sits on the Pacific plate side of the San Andreas fault, is grinding northward relative to San Francisco at the rate of about 5 centimeters per year. This means that in 30 million years Los Angeles will have moved an additional 600 kilometers northward to rainy northernmost California!

The most famous geodetic studies related to earthquakes were conducted after the 1906 San Francisco earthquake and led to our basic understanding of earthquake generation. Three sets of triangulation measurements that had been taken across the region traversed by the 1906 break along the San Andreas fault were compared by H. F. Reid (one set for 1851–1865, another for 1874–1892, and the third taken just after the earthquake). These indicated that significant horizontal displacements parallel to the ruptured San Andreas fault had occurred both before and after the earthquake. Reid noticed that distant points of opposite sides of the fault had moved 3.2 meters over the 50-year period prior to 1906, with the western side moving north.* The sudden fault slip that generated the earthquake was measured at a maximum of 20 feet in Marin County (Figure 6.3). These observations led him to his famous elastic rebound theory for the cause of earthquakes (see page 113).

Finally we must consider an exciting new tool for taking the measure of very small crustal movements — one that has already verified the elastic rebound theory in several earthquakes. The method is based on the principle of radar in which reflected radio waves are used to fix the position of airplanes or ships. The new scheme is called SAR (for "synthetic aperture

**From his work, Reid made an approximate prediction of when the next great San Francisco earthquake could be expected. Earthquake prediction is discussed in Chapter 10.*

Figure 6.3 Right-lateral horizontal movement of the San Andreas fault in the 1906 earthquake across the old Sir Francis Drake Highway north of San Francisco, at the southern end of Tomales Bay, Marin County, California. The offset was 6.5 meters (20 feet). [Photo by G. K. Gilbert; courtesy of USGS.]

radar"). The idea is for an airplane (or artificial satellite) to bounce emitted radio waves off the underlying terrain, record the echoes on an inboard radar antenna, and then have a computer program determine the distance range of ground surface reflection, the reflectivity pattern, and a two-dimensional map of the topography.

For very fine resolution of the ground, a physically large antenna (i.e., a large "aperture") is needed. The trick with SAR is to simulate this impractical antenna length from the airborne data recorded sequentially along the airplane flight track. Quite remarkable maps of volcanoes and other key geographic features of the Earth are available as examples on the World Wide Web. In seismology, the first dramatic demonstration of the value of SAR came from published images of the displacement field of the June 28, 1992, Landers earthquake in California (see Chapter 8). Two radar scans were taken under similar conditions just before and after the Landers fault source rupture (see Plate 23). By subtraction, the main topography could be cancelled out, leaving contour fringes of the seismogenic change in range. The fringe widths represent changes of about 1 centimeter in terrain position. Such a geodetic tool is a wonderful technical accomplishment.

Elastic Rebound

The slip that produced the 1906 earthquake is illustrated diagrammatically in Figure 6.4. Imagine this illustration to be a bird's-eye view of straight lines drawn at a certain time at right angles across the San Andreas fault. As the tectonic force slowly works, the line bends, the left side shifting in relation to the right, as indicated by the black arrows. The deformation amounts to a few meters in the course of 50 years or so. This straining cannot continue indefinitely; sooner or later the weakest rocks, or those at the point of greatest strain, break. This fracture is followed by a springing back, or rebounding, on each side of the fracture. Thus, in Figure 6.4 the rocks on both sides of the fault at D rebound to the points D1 and D2 and the elastic forces on the rocks are decreased as they do the work.

This *elastic rebound* was believed by Reid to be the immediate cause of earthquakes, and his explanation has been confirmed over the years. Like a watch spring that is wound tighter and tighter, the more that crustal rocks are elastically strained, the more energy they store. When a fault ruptures, the elastic energy stored in the rocks is released, partly as heat and partly as elastic waves. These waves are the earthquake.

Straining of rocks in the vertical direction is also common. The elastic rebound occurs along dipping fault surfaces, causing vertical disruption in level lines at the surface and fault scarps (see Figure 6.5 and Plate 8). Such scarps are here illustrated for Japan (Figures 5.6 and 6.5), Armenia (Figure 5.9), and the Philippines (Figure 5.12). We will encounter another remarkable example in the Mojave Desert of California of faulting that produced striking offsets in 1992 (see photo on page 80). Vertical ground displacement — produced by earthquakes or other phenomena — can amount to tens of centimeters across wide areas (compare the 1964 Alaska uplift described in Chapter 1).

In two Japanese earthquakes, such vertical movement was quite striking. In the catastrophic Kwanto earthquake of September 1, 1923 — in which over 100,000 persons lost their lives (about 68,000 in fire-ravaged Tokyo) — extraordinary changes in water depths in Sagami Bay, south of Tokyo, occurred. In places the water depth changed by 250 meters, but there is reason to believe that most of this change was due to submarine

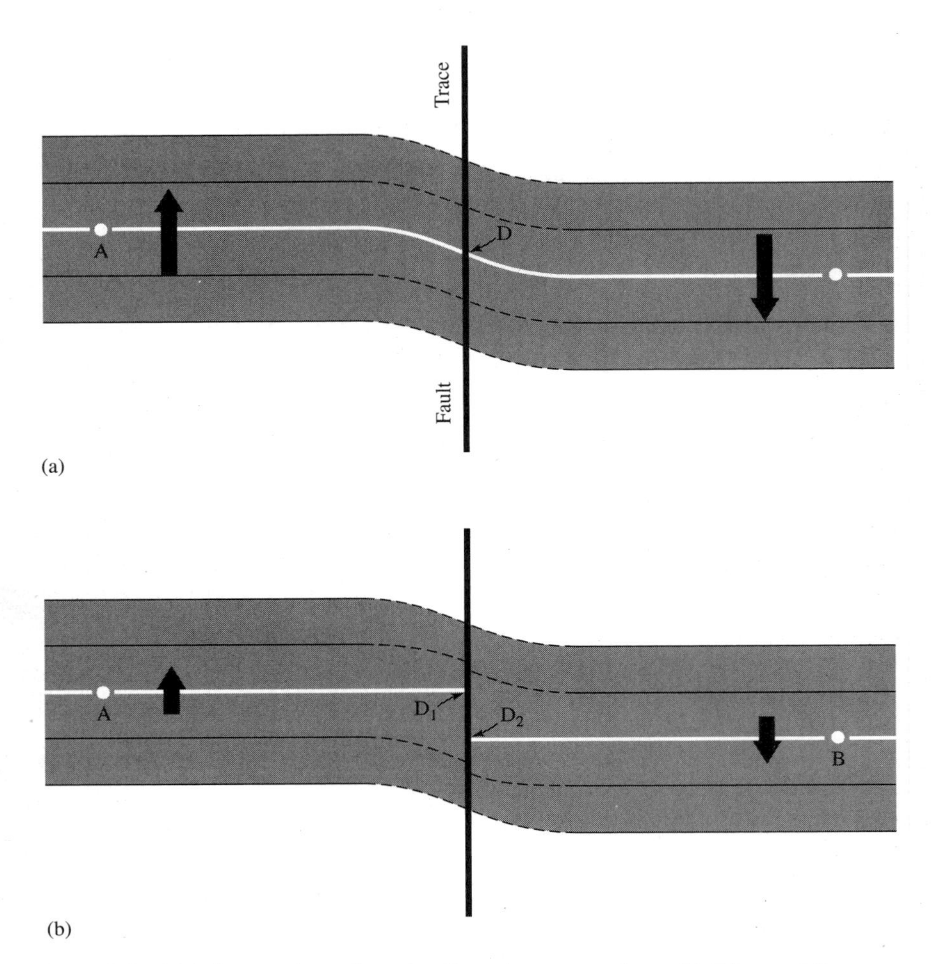

Figure 6.4 A bird's-eye view of marker lines drawn crossing a fault trace at the ground surface. (a) In response to the action of tectonic forces, points A and B move in opposite directions, bending the lines across the fault. (b) Rupture begins at D, and strained rocks on each side of the fault spring back to D1 and D2. [From Bruce A. Bolt, *Nuclear Explosions and Earthquakes: The Parted Veil* (San Francisco: W. H. Freeman and Company. Copyright 1976).]

slides. The Boso peninsula exhibited a number of fault ruptures and an uplift of up to 1.9 meters.

In the Niigata shock of June 16, 1964, vertical ground change in elevation was also noteworthy. Along the west coast of Honshu, bench marks were used to measure the height of the land in relation to the mean sea level for 60 years and just prior to the earthquake in 1964 (see Figure 10.2). From 1898 to 1958, these measurements showed that the coastline of Honshu opposite Awashima Island was rising steadily at a rate of about

PLATE 1
Evidence of earthquakes from prehistoric times. The photos show the shorelines at Turakirae Head, near Auckland, New Zealand, which represent the uplift of the land during at least four earthquakes. The two most recent shorelines were raised by earthquakes in 1460 and 1855. The older terraces have probable earthquake dates of 1100 B.C. and 2900 B.C. [Courtesy of L. Homer, New Zealand Department of Scientific and Industrial Research.]

PLATE 2
Map of the United States showing epicenters of the seismicity for earthquakes of magnitude 5.5 and greater for the period 1700 to 1997. [Courtesy of USGS.]

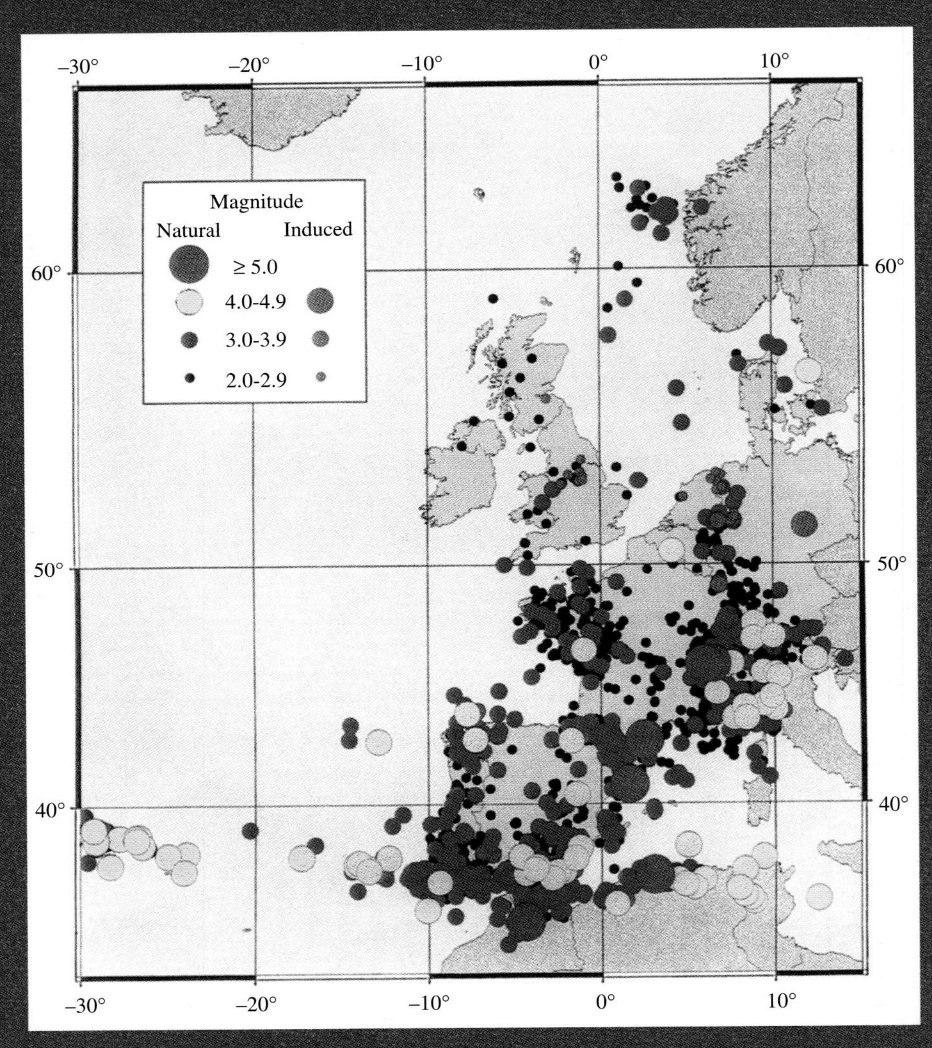

PLATE 3
Circles showing the epicentral locations of natural and induced seismicity recorded in Europe for the period November 1994 to February 1997. [Courtesy of Transfrontier Group.]

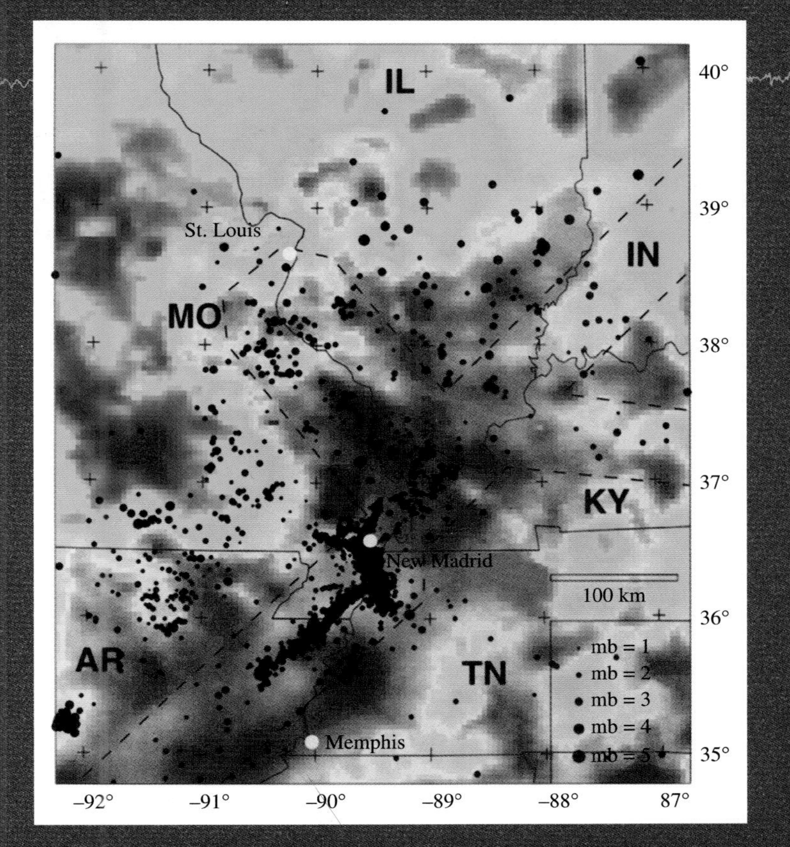

PLATE 4
Epicenters from the New Madrid seismographic network of earthquakes during 1974 through 1994 in the New Madrid seismic zone. Dashed lines show boundaries of the New Madrid rift complex. Colors are measures of the variation in gravitational attraction over the region. mb is the symbol for magnitude. [Courtesy of Braile, Hinze, and Keller, 1997.]

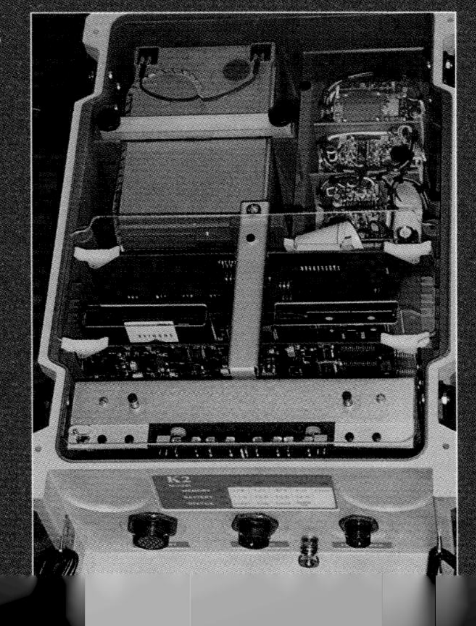

PLATE 5
Photograph of the interior of a modern strong-motion seismograph with electrical circuits with digital seismic ground motion as input. In the upper right are the three motion sensors that record two horizontal and one vertical component of earthquakes without going off-scale in heavy shaking. [Courtesy of Kinemetrics, Inc.]

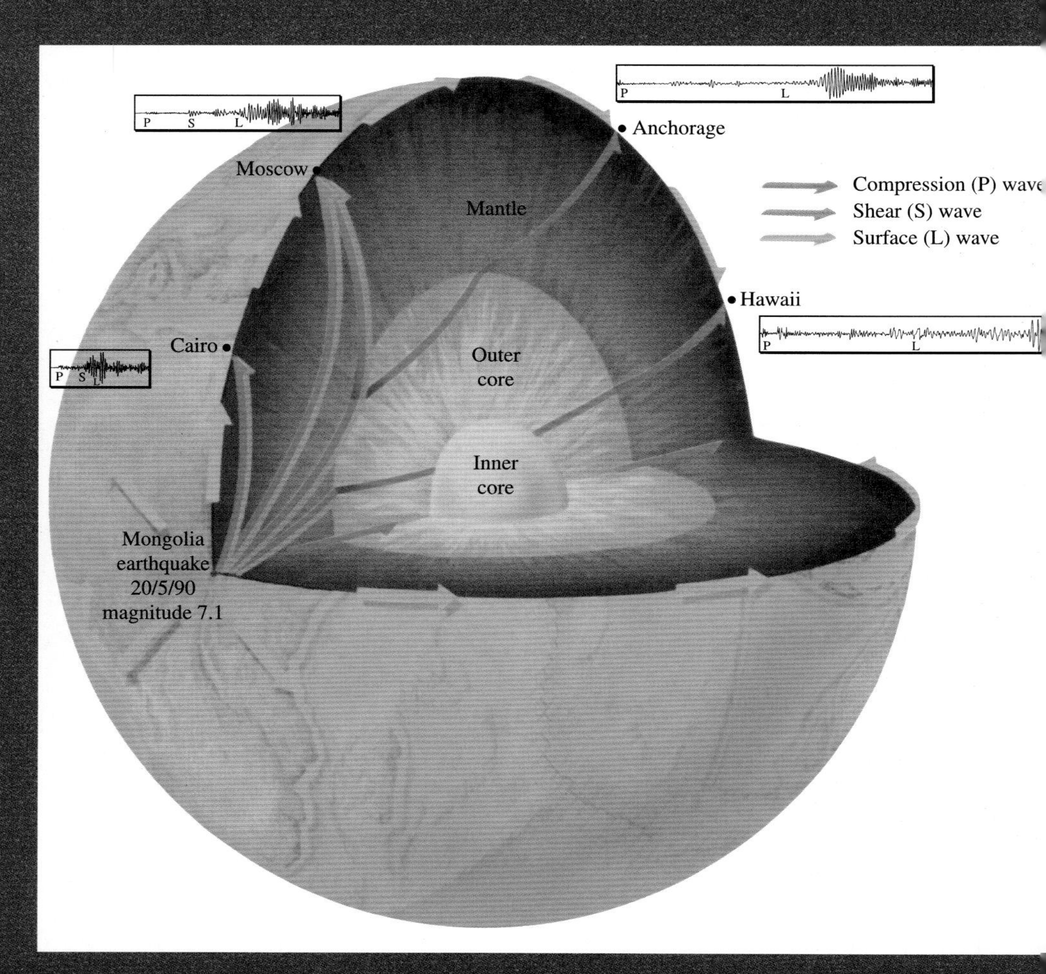

PLATE 6

Diagram showing the main radial structure of the Earth's interior. The figure also shows the actual seismograms recorded at seismographic stations at Cairo, Moscow, Anchorage, and Hawaii from an earthquake in East Africa near Mongalla on May 20, 1990. [Courtesy of British Geological Survey.]

PLATE 7
Destruction of the rail transportation system in the city of Kobe in the 1996 Kobe earthquake. [Courtesy of Michiko Igarashi, Asahi Shinbun.]

PLATE 8
Aerial photograph of the fresh fault scarp formed on the North Island of New Zealand in 1982. Note the lack of damage to the near source buildings. [Courtesy of L. Homer, New Zealand Department of Scientific and Industrial Research.]

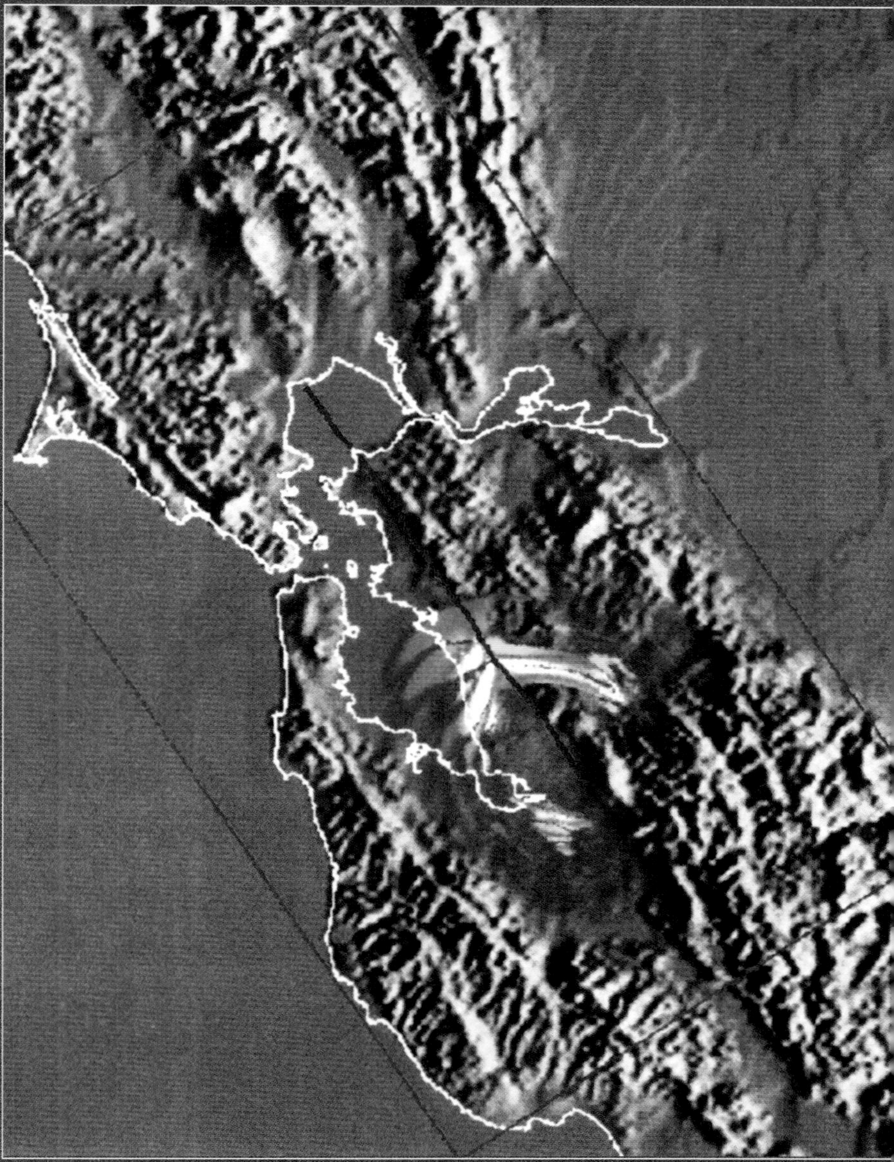

PLATE 9

Plot from a computer simulation of the relative size of seismic waves (white is strongest) after 15 seconds from a rupture of the Hayward fault (red line), California, moving northeast in a theoretical magnitude 7.3 earthquake. The blue patch is the P-wave group speeding ahead of the rupture. Note the geological variation around the fault, producing asymmetry in the strongest wave forms. [Courtesy of S. Larsen and D. Dreger, 1998.]

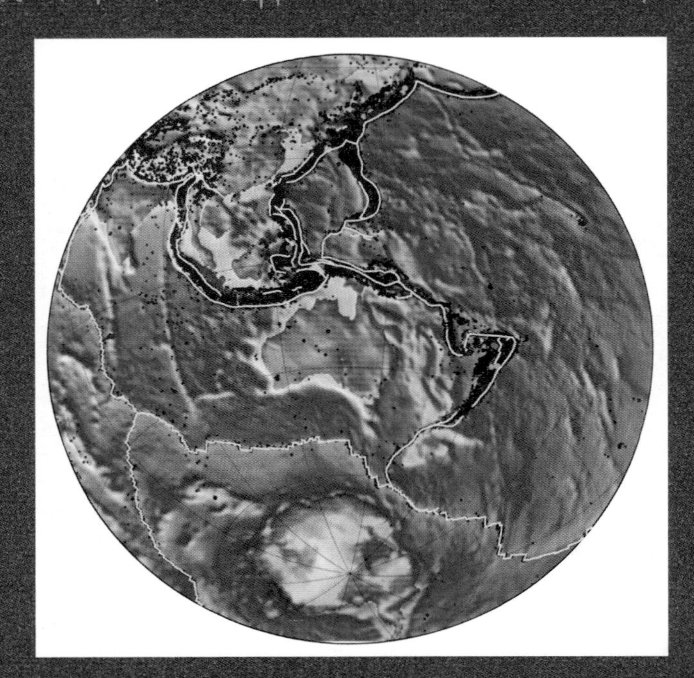

PLATE 10
Seismicity of the Earth, 1964 to 1995 (magnitudes greater than 5.1), with tectonic plate boundaries superimposed. The large circles are major earthquakes with a magnitude over 6.4. This Pacific projection is centered on the Indo-Australia plate. [Courtesy of International Association of Seismology and the Earth's Interior.]

PLATE 11
Projection of the Atlantic hemisphere showing the America and Africa plates, and the mid-Atlantic seismically active ridge. The seismicity and plate boundaries are similar to those in Plate 10. [Courtesy of International Association of Seismology and the Earth's Interior.]

Figure 6.5 Surface horizontal offset of 140 centimeters of the Nojima fault (Japan, 1995) in the right-lateral sense. At this location on Awaji Island the southeastern side (left side of photo) also shows 55 centimeters of ground uplift. [Photo by Aiming Lin.]

2 millimeters a year. The level of the coastline suddenly dropped 15 to 20 centimeters at the time of the earthquake, near Awashima Island.

A controversial case of note, not yet (at the time of this writing) accompanied by a major earthquake, was reported from California in the 1970s. Since the end of the last century, leveling surveys seemed to indicate that significant elevation changes had taken place in the mountainous regions to the north and east of Los Angeles. During the early 1960s, an extraordinary uplift (35 centimeters maximum) of an area of about 12,000 square kilometers was inferred near Palmdale, 70 kilometers north of Los Angeles.* The area of crustal swelling, sometimes called "The Palmdale

**By 1977 further surveys showed that a portion of the uplift had already subsided, and some analyses concluded that a significant part of the "uplift" could have been error in field measurement adjustments.*

Bulge," attracted much public and scientific attention as a possible precursor of an imminent major earthquake.

Changes in the Rocks

Almost everywhere, not many meters below the Earth's surface, the rocks are permeated by groundwater. This water saturates the rocks and fills up the cracks and pores within them. Scientists have now examined what happens to saturated rock samples when, in the laboratory, they are squeezed to high pressures by powerful jacks (similar to those used for lifting automobiles). It has been found that, under some circumstances, wet rocks under shear strain *increase* their volume rather than decrease it. This increase in volume during deformation is called *dilatancy.* There is evidence that the volume increase from pressure arises from opening and extending the many microcracks in the rocks. The groundwater that then moves into the microcracks is much less compressible than air, so the cracks no longer close easily under pressure.

Suppose that we could look into the crust of the Earth with a lens while it was being strained. What sequence of events might we see? First, the slow straining of the crust under the local tectonic forces produces many microcracks throughout the rock. As time passes, water diffuses into the cracks and fills them. During this period, the volume of the region dilates; this process might be detectable at the surface by an upward swelling of the ground, discernible by leveling or tide gauges. Particularly in fault zones, these changes to the rocks first weakens them; then the presence of water in the cracks reduces the restraining forces so that a major crack extends along the fault. In this way the elastic rebound of the strained crustal rocks may begin and spread.

As yet this complete sequence of events has never been directly observed before an earthquake, but there is at least circumstantial evidence that something like it occurs (see Chapter 10). In any event, some speculation on the processes that lead up to earthquakes is helpful when looking for plausible harbingers of earthquakes, such as rapid changes in ground level, ground tilting, and fluctuations of water levels in wells.

The cracks produced by crustal strain provide a reasonable explanation of both foreshocks and aftershocks. A foreshock is caused by an incipient rupture in the strained and cracked material along the fault that did not progress because the physical conditions were not yet ripe. Foreshocks, however, slightly alter the pattern of forces and, perhaps, the movement of ater and the distribution of microcracks. Eventually, longer local fault-rupture commences, producing the principal earthquake. The flinging of the rocks along the major rupture, together with the heavy shaking produced and the local generation of heat, leads to a physical situation very different from that before the mainshock; additional small ruptures occur, producing aftershocks. Gradually, the strain energy in the region decreases, like a clock running down, until, perhaps after many months or even years, stable evolutionary conditions return.

The Effect of Water on Rocks Beneath the Surface

If there were no water in the rocks, there would be no tectonic earthquakes. The reasons are many. First, suppose we calculate the pressure at a depth of 5 kilometers in the Earth's crust due to the pull of gravity on the overlying rocks. We would find that it is equal to the strength of granite or similar rock (that is, the pressure it can sustain without breaking) at the pressure (1300 bars*) and temperature (200°C) appropriate for that depth. At greater depths, because the hydrostatic pressure is already greater than the strength of the rocks, we might expect that they would flow and deform plastically under differential pressures, rather than break through brittle fracture (thus producing an earthquake). Indeed, if a sample of hard granitic rock is squeezed in the laboratory under the appropriate temperature and pressure conditions, generally it will flow and not break. But earthquakes happen, and so we have a paradox.

Experiments on the effect of pressure on minerals containing water of crystallization and on water-saturated rocks do, however, suggest why fracture might occur at large depths in the Earth: water acts in a way that allows a sudden slip to take place, perhaps by providing a kind of lubrication along slide planes, but more effectively by increasing the local pore pressure and, hence, weakening the rock. In these experiments, slips in rock specimens are accompanied by jerks of the pneumatic press that squeezes the rock. These jerks correspond to sudden reductions in the confining pressure. In other words, each jerk signifies an almost instantaneous drop in stress on the slip (fault) surface within the specimen. Out of the laboratory, seismological work on recorded earthquake waves indicates that, in shallow-focus earthquakes, the shearing pressure or stress along faults suddenly drops by amounts ranging from several to a few hundred bars. These rather low stress-drops are much smaller than the strengths of hard rocks,

**1 bar equals 10,000 pascals, which is roughly equivalent to 1 atmosphere pressure. Ten bars is about the tensile strength of concrete.*

which range up to 1 kilobar. It seems possible, therefore, that the water present in rocks along the fault zone weakens them so that only a small amount of shearing stress is removed during the earthquake rupture.

There is other evidence for the effect of water on earthquake mechanisms. Recall from Chapter 5 that the typical geological section in fault zones shows a succession of gouge, crushed and sheared rock, and clays; hydrological conditions in the fault zone commonly produce hydrous, or water-containing, rocks such as serpentinite. At least near the surface, the gouge and clays often show direct evidence of shear slip associated with wet conditions, with successive smooth striated layers called *slickensides*. In fact, the presence of water in springs and deep wells is often quite notable along major faults, and ample groundwater appears to be available at depth in most seismically active zones.

The importance of water in earthquake generation came to the attention of seismologists in 1962 when a series of earthquakes began near Denver, Colorado. Although throughout the years there had been some earthquakes in the area — for example, one of Modified Mercalli intensity VII (see Appendix C) had occurred in 1882, and a few other local shocks had taken place since — the natural seismicity had always been low. Suddenly, there was a change beginning in April 1962, when a succession of earthquakes was felt: between that month and September 1963 local seismographic stations located more than 700 epicenters in the vicinity! The sequence continued for years, the largest earthquakes occurring in 1967. The magnitudes ranged between 0.7 and 4.3 on the Richter scale (see Chapter 8).

Most of these earthquakes were within a radius of 8 kilometers of the Rocky Mountain arsenal northeast of Denver, where weapons were being manufactured by the army. One of the by-products of this manufacture was contaminated water, which was at first allowed to evaporate from surface storage. But in 1961 the army switched to what seemed a more environmentally acceptable method of disposal — to pump the waste liquid down a deep well, bored to a depth of 3670 meters. These wastes were injected under pressure into the borehole from March 8, 1962, to September 30, 1963. Injection ceased for a year and then resumed in September 1964 through September 1965. Subsequently, earthquakes were felt in Denver. Inhabitants complained about the possible relation between the pumping and the outbreak of the earthquakes until they eventually succeeded in halting this method of waste disposal.

The correlation between the amount of water injected and the number of earthquakes was indeed quite strong: a high incidence of local earthquakes occurred in early 1963, followed by a sharp decline in 1964, and then another series of earthquakes occurred in large numbers in 1965 when the amount of water pumped, owing to increased injection pressure, was again maximum. A plausible mechanical explanation is that the increased water pressure at the well produced a flow of groundwater into crevices and cracks in preexisting faults underground. This increase in pore pressure led in turn to a reduction of the shear strength of the rock and gouge material. The condition was then ripe for the tectonic strain in the crust, built up over many years, to be released in a series of slips, producing earthquakes.

The Denver information was uncovered by chance, but it was followed up by a planned field experiment under similar conditions. The necessary

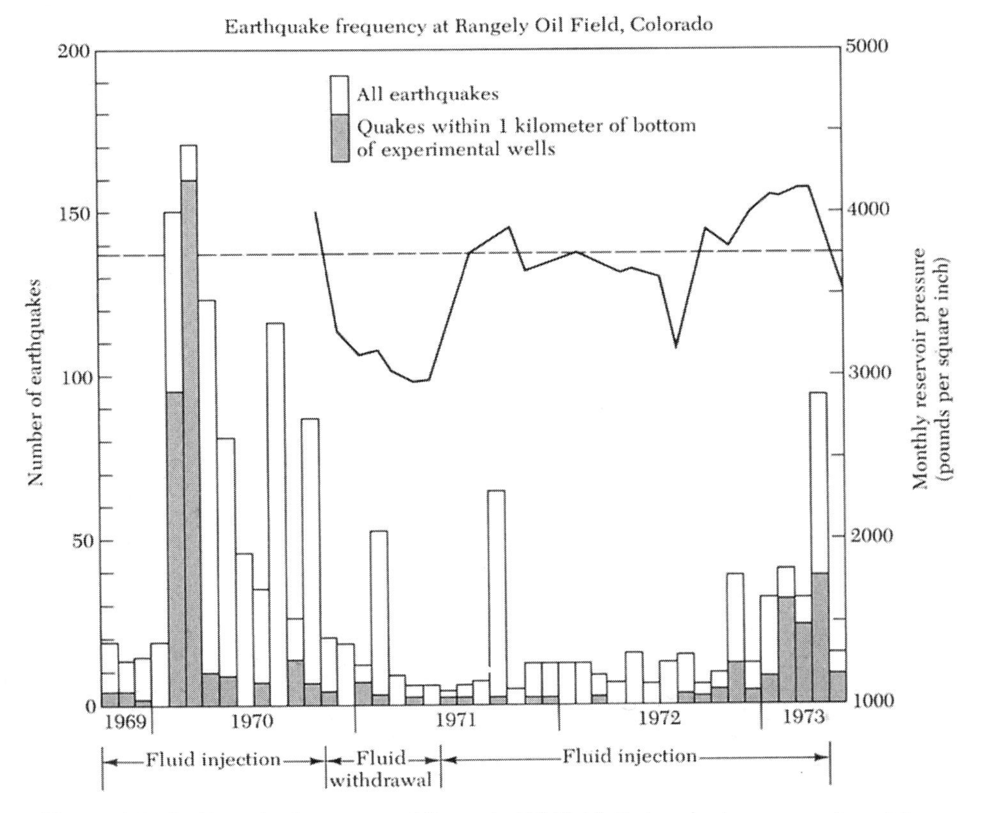

Figure 6.6 Earthquake frequency at Rangely Oil Field, Colorado, in an experiment in which water was alternately pumped in and out of wells. The reservoir pressure is plotted as a solid line. The pressure needed to initiate earthquakes is 3700 pounds per square inch (broken line). Those occurring below that threshold were due to natural causes. Each vertical bar represents the number of earthquakes in 1 calendar month. [Courtesy of USGS.]

work was initiated in 1969 by the U.S. Geological Survey at the Rangely Oil Field in western Colorado. Oil wells were already available at the site, and water, therefore, could be regularly injected into the wells or pumped out of them and the pore pressure in the crustal rock measured. At the same time, a specially sited array of seismographs was put into place to monitor fluctuations in local seismic activity. The results showed an excellent correlation between the quantity of fluid injected and the local earthquake activity, as illustrated in Figure 6.6. When the fluid pore pressure reached a threshold level (1.1×10^4 pascals in this case), the earthquake activity increased. When the pressure dropped as a result of water withdrawal, the seismic activity decreased. Again, it must be emphasized that the wells at Rangely penetrated preexisting faults and that the crust in the region was already under some tectonic strain, as indicated by the occurrence of small local earthquakes over the previous years.

The Denver and Rangely investigations thus demonstrated the crucial importance of water in triggering sudden ruptures deep in the crust. They also led to the idea of earthquake control. One proposal was to pump water through deep boreholes into faults in a region where natural earthquakes might be particularly hazardous. An outbreak of small earthquakes might be thus induced, thereby reducing the amount of strain energy stored in the crust in the vicinity and reducing the probability of a large earthquake. Tampering with the forces of nature in this way is of course in itself hazardous: if control were attempted along a major active fault, the consequences could be especially damaging. But "destraining" of the crustal rocks by water injection at the site of a future critical structure, such as a massive dam, might be worthwhile. Imaginative schemes of this type may be implemented as practical measures at some future time.

Reservoir-Triggered Earthquakes and Dam Safety

In 1971, a great tragedy was averted in the damaging San Fernando earthquake just north of Los Angeles in southern California. The lower Van Norman Dam, less than 10 kilometers from the fault rupture that caused the shaking had been built 30 years before by carrying soil for fill into position by water sluicers. Subsequently, additional hydraulic fill had been placed on the dam. During the 1971 earthquake, a major earth slide took place in the interior portion of the dam, leaving only a meter or so of soil on the downstream side to stop the water from engulfing a densely populated suburban area. Fortunately, the water in the reservoir was not at the allowable maximum height at the time of the earthquake, and the slim earth lip of the dam did not erode but held the water until it could be drawn down. Meanwhile, 80,000 persons were evacuated from the downstream area.

The incident exemplifies the importance of evaluating prospective dam sites for seismic risk. Not only is an earth or concrete dam an expensive structure, but it directly affects the economy of the region through power generation, flood control, and irrigation. As the population grows, structural failure of a large dam poses increasingly greater danger for residents exposed to the sudden inundation of the flood plains. Indeed, in many countries major dams are located in areas that have suffered large earthquakes. The likelihood of damaging earthquakes must be considered — during planning and after construction — to ensure continual safety of downstream habitation. Certainly, geological conditions near the site, including landslides and faulting, must be most carefully studied.

The hazards from normal regional earthquakes aside, however, we must also take account of a curious connection between large reservoirs and earthquakes. There have been at least 13 incidents in different countries in which swarms of earthquakes have occurred very near a large reservoir soon after it has been filled. The circumstantial evidence is that somehow in some conditions reservoirs cause earthquakes. A case history for Oroville reservoir in California and an earthquake nearby is discussed in Chapter 2.

The idea that earthquakes might be triggered by impounding surface water is not new. In the 1870s, the U.S. Army Corps of Engineers rejected proposals for major water storage in the Salton Sea in southern California on the grounds that such action might cause earthquakes. The first detailed evidence of such an effect came with the filling of Lake Mead behind Hoover Dam (height 221 meters) on the Nevada–Arizona border beginning in 1935. Although there may have been some local seismicity before, earthquakes were much more common after 1936. Nearby seismographs in operation since 1940 showed that following the largest earthquake (magnitude about 5) in 1940, the seismicity declined. The foci of hundreds of detected earthquakes clustered on steeply dipping faults on the east side of the lake and had focal depths of less than 8 kilometers.*

In the ensuing years, similar case histories have been accumulated for other large dams, but only a few are well documented. Most of these dams are more than 100 meters high and, although the geology at the sites varies, the most convincing examples of reservoir-induced earthquakes occur in

*A *major study in 1994 in which I participated showed that Hoover Dam is capable of resisting the largest feasible nearby earthquake.*

tectonic regions with at least some history of earthquakes. A few general properties have been suggested: earthquakes at reservoirs are more likely when lake-level changes are comparable with the low water depth; and seismicity is more widespread and deeper for a larger than for a smaller reservoir. It must be stressed, however, that most of the thousands of large dams around the world give no sign of any connection between reservoir filling and earthquakes; of 500 large dams scrutinized in the United States, a poll in 1976 showed that for only 4 percent of them was an earthquake reported with magnitude greater than 3.0 within 16 kilometers of the dam.

Of particular interest are the following four well-studied examples of earthquakes induced by human-made reservoirs. First, Lake Kariba in Zambia began filling in 1958 behind a 128-meter-high dam. Although there is some evidence for minor earthquakes in the vicinity before the construction, from 1958 until 1963 when the reservoir was full, over 2000 local shocks under the reservoir were located with the use of nearby seismographs. The largest shock in September 1963 had a magnitude of 5.8; since then the activity has decreased.

At Koyna, India, an earthquake (magnitude 6.5) centered close to a dam (height 103 meters) caused significant damage on December 10, 1967. After impounding began in 1962, reports of local shaking became prevalent in this previously low-seismicity area. Seismographs showed that foci were concentrated at shallow depths under Shivajisagar Lake. In 1967 a number of sizable earthquakes occurred, leading up to the principal earthquake on December 10. This temblor caused significant damage to buildings nearby, killed over 200 persons, and injured more than 1500. The series of earthquakes recorded at Koyna has a pattern that seems to follow the rhythm of the rainfall (see Figure 6.7). At least a comparison of the frequency of earthquakes and water level *suggests* that seismicity increases a few months after each rainy season when the reservoir level is highest. Such correlations are not so clear in other carefully studied cases.

Another series of earthquakes — which were conclusively induced by a reservoir — occurred in China north of Canton (what is now Guangzhou). The normal regional seismicity is quite low (see Figure 2.2). The Hsingfengkiang Dam (height 105 meters) was completed in 1959. Thereafter, increasing numbers of local earthquakes were recorded, the grand total in 1972 amounting to more than 250,000. Most of the earthquakes were very small, but on March 19, 1962, a strong shock of magnitude 6.1 occurred. The energy released was enough to damage the concrete dam, which required strengthening. Most earthquake foci were at depths of less than 10 kilometers near where the reservoir was deepest, and

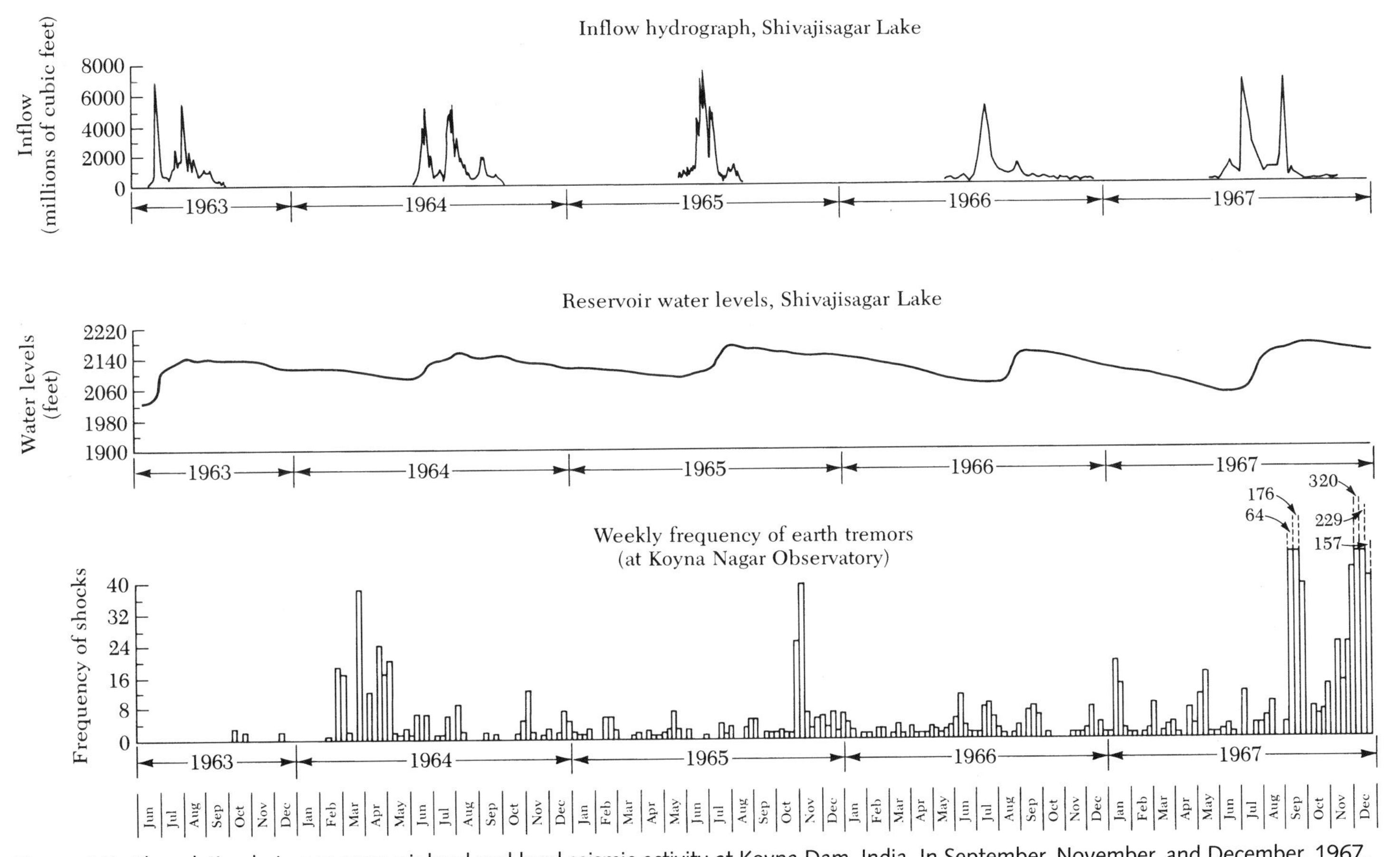

Figure 6.7 The relation between reservoir level and local seismic activity at Koyna Dam, India. In September, November, and December, 1967, earthquake frequency exceeded 40 and, at times, toward the end of the year, amounted to 300 per week. [From data of H. K. Gupta and B. K. Rastogi.]

some of the foci coincided with intersections of the main nearby geological faults.

The final example is the massive Nurek Dam (height 317 meters) in Tadzhikistan, the highest earthfill dam in the world. Even in 1972, before its completion but after water impounding began, increased local seismicity was reported. The full load of stored water was applied onto the crust in 1978; in the years following, no large nearby earthquake has shaken the facility, but many small earthquakes continue to occur.

How does water in a large reservoir stimulate earthquakes? It is hard to believe that it is entirely the effect of the added weight on the rocks because the actual additional pressure a few kilometers below the reservoir is a small fraction of the natural tectonic stresses already present. (Calculations indicate that a few kilometers down the added differential stress available to shear the rock is only about 1 bar (14.5 pounds per square inch); for some geometries of the local geological faults (see Figure 5.5), this added stress may even act to inhibit rather than induce sudden fracture.) A more plausible trigger mechanism is that extra water pressure produced by the reservoir loading spreads out as a pressure wave or pulse through the pores of the crustal rocks. Because of its slow rate of spreading, it may take months or years to travel a distance of 5 kilometers, depending on the permeability and amount of fracturing of the rock. But when the pressure pulse finally reaches a zone of microcracks, it forces water into them and so decreases the forces that are preventing the already present tectonic strain from initiating sudden sliding.

In an area where there is a likelihood of seismic activity, certain preliminary steps must be taken before construction of a dam. First, whether cause for concern is a natural or an induced earthquake, it is essential at the design stages to estimate the intensity of ground shaking that the structure will sustain during its lifetime. Also, preconstruction geodetic surveys of the region are useful for purposes of later detecting any changes in crustal deformations associated with reservoir loading.

Furthermore, in order that earthquake effects can be studied, seismographs and other instrumentation should be installed at an early time. Hydrographs for measuring large water waves (seiches) in the reservoir are also important. In the absence of suitable recording instruments to measure the severity of earthquake motions and of the dam response, the advent of a strong earthquake nearby will pose questions that cannot be answered. If, for example, structural damage has occurred, and no such measurements have been taken, it is impossible to compare vibrational behavior with design earthquake conditions and thus to estimate dam performance for other and perhaps larger shocks, or to make design decisions for repair.

What Produces Seismic Waves?

Rupture has commenced; the ground begins to shake. The rupture begins at the earthquake focus within the crustal rocks and then spreads outward in all directions in the fault plane (see Figures 5.5 and 6.8). The edge of the rupture does not spread out uniformly. Its progress is jerky and irregular because crustal rocks vary in their physical properties from place to place, and the overburden pressure at a particular point in the crust decreases toward the surface. On the fault surface there are rough patches (often called

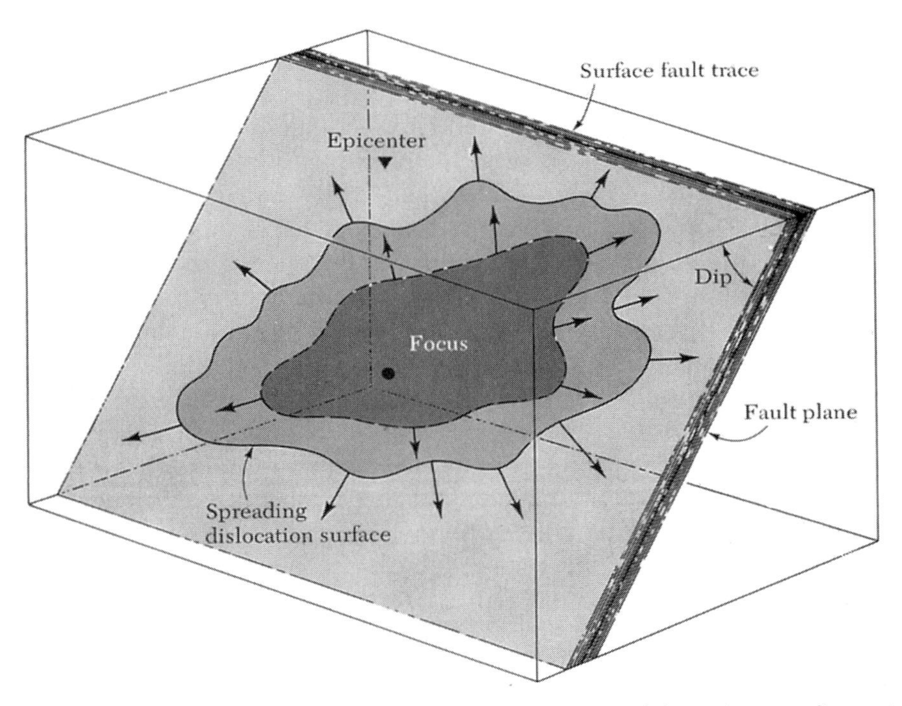

Figure 6.8 Side view into the Earth's crust showing rupture of the rocks spreading out from the focus of the earthquake along the dipping fault plane. Two stages of the rupture are shown. The arrows indicate the direction of the spreading rupture. (The epicenter is the point on the Earth's surface directly above the focus.) [From Bruce A. Bolt, *Nuclear Explosions and Earthquakes: The Parted Veil* (San Francisco: W. H. Freeman and Company. Copyright 1976).]

asperities) and changes in fault direction and structural complexities that act as *barriers* to the fault slip. Thus the rupture front may come almost to a stop; then, because of the rearrangement of elastic forces, it may suddenly break free and swiftly move out to catch up with the rupture on either side of it. If this rupture reaches the surface (as happens in only a minority of shallow earthquakes), it produces a visible fault trace.

The extent of the fault rupture depends on the variation in strain of the rock throughout the region. The rupture continues until it reaches the places at which the rock is not sufficiently strained to permit it to extend further. Then the rupture episode stops.*

After rupturing stops, the adjacent sides of the fault spring back to a less-strained position (see Figure 6.4). During the rupture, the rough sides of the fault rub against one another so that some energy is used up by frictional forces and in the crushing of the rock. The surfaces are locally heated. Earthquake waves are generated at the same time by the rebounding of adjacent sides of the fault at the rupture surface as well as by the rubbing and crushing.

This whole process can be demonstrated in a school laboratory or in the kitchen. To do so, make a model of the elastic crust out of stiff jelly in a shallow mold. Pull the sides of the jelly in opposite directions and then make a small slit in the surface of the jelly. The rupture will spread throughout a plane in the jelly, and the two sides of the jelly will spring back until rupture ceases. While the rupture is tearing through the jelly, the jelly will quiver as elastic waves spread throughout it. Similarly, earthquake waves radiate out from numerous places on the fault plane.

Can this explanation of earthquake genesis be supported by the properties of seismic waves recorded on seismographs around the world? The strength and direction of the seismic waves radiating out from the moving fault slip are much affected by the local geology. This variability is evident from the observed intensity changes and also the theory of seismic waves in diverse elastic rocks (Plate 9). Furthermore, the recorded waves give information on fault motions that may be remote and at inaccessable depths.†

**A scientific "first" was witnessed in conjunction with the Borah Peak, Idaho, earthquake of October 28, 1983. Two elk hunters saw the creation of a prominent fault scarp nearly 2 meters high from about 20 meters in front of their road vehicle. They described, first, dizziness (P waves from the distant focus?) and then, in 2 to 3 seconds, the more-or-less simultaneous perception of the new scarp and violent rocking.*

†Another example of a challenging inverse science problem mentioned in Chapter 4, first solved in this context by Professor Perry Byerly.

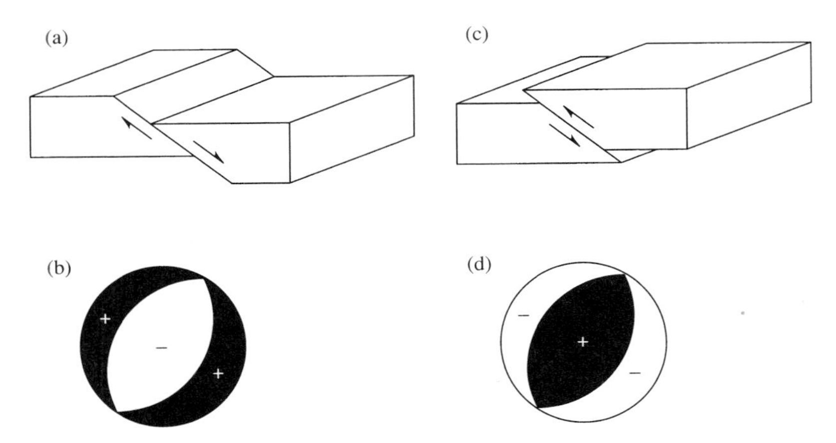

Figure 6.9 Two types of seismological "beach ball." (a) Normal faulting. (b) Fault-plane diagram for the normal faulting in (a). White region represents stations at which the first motion is dilatational (negative); black region represents stations at which the first motion is compressional (positive). (c) Thrust faulting. (d) Fault-plane diagram for the thrust faulting earthquake in (c).

For simplicity, let us consider only the first seismic waves to arrive—that is, the P waves, which are simply pushes and pulls of the rocks in the Earth (see Chapter 1). It thus follows that a P wave will be detected at the surface as either a push or a pull.

Suppose, first, that the source of the recorded P waves is a small explosion at a point in the Earth some distance from the seismograph. Then the first P wave to be generated would, like the air blown into a balloon, push outward on a spherical surface. Seismographs would detect this P wave as a push upward from the ground. This upward movement is referred to as a *compression.*

If the P wave arises from a rupture on a fault (such as in Figure 6.9), a quite different pattern of first P-wave motions must be expected. Now the first P motions, if plotted on a sphere, are not all compressions; nor will they occur in a jumbled way. Rather, the P waves directions will be recorded in a simple pattern on the Earth's surface, depending on the direction in which they first left the fault. Seismographs located at points from which the fault at the focus is moving away will record pulls or *dilatations* for the first P motion. Only those seismographs located at points toward which the fault is moving will record pushes or compressions. The resulting pattern will be alternating compressions and dilatations (see Figure 6.9 and Appendix F) that, when colored black and white, produce the commonly published seismological "beach balls."

Box 6-1

METHOD OF OBTAINING DIRECTION OF FAULTING FROM P-WAVE DIRECTIONS ON SEISMOGRAMS (SEE ALSO APPENDIX F)

In the upper diagram on page 129, F is the focus and E is the epicenter of an earthquake generated at some depth under the Earth's surface by the sudden (right-lateral) slip along a vertical fault plane (bFb′Z). Displacements due to the slip are shown as arrows. The large circle on the surface (A′B′AB) is a projection of the equator (a′b′ab) of a small sphere drawn around the focus. The projection is formed by drawing a cone from below Z through points on the equator.

Consider four paths of travel of P waves from the focus F to four stations on the ground surface S_1, S_2, S_3, and S_4. The arrows on these paths FS_1, FS_2, FS_3, and FS_4 show how the compressions (pushes) and dilatations (pulls) along the fault plane (bFb′Z) are transferred from depth to the surface by the first motion of the P wave.

The lower diagram is a plan view looking down on the circle A′B′AB. For this earthquake, a quadrantal pattern of compressions and dilatations would be observed at the surface. One of the lines AEA′ or BEB′ that separate the different motions will indicate the strike and dip of the fault plane. (In this case, BEB′ corresponds to the direction of the fault.)

The reader can appreciate the elation of seismologists in the 1930s when fault rupture directions, predicted in this way at distant observatories (see Box 6.1), were found to check with field reports from the epicentral region. An early form of remote sensing!

One of the greatest successes of this method of remotely classifying fault movements in earthquakes was its use to map the sense of slip along the oceanic trenches and midoceanic ridges. The fault planes and displacements were found to follow a definite pattern (see Figure 7.5) indicating the global consistency of thrusting, horizontal slipping, and normal faulting.

The discussion thus far has emphasized that regular tectonic earthquakes are caused by the rapid rupturing or slip of the rocks along faults. We must make clear, however, that the speed of rupture varies from place to place along the fault. For some earthquake sources, the speed is close to that of the shear (i.e., S) waves. In others, it slows to a rate comparable with a person's normal walking speed. An illustration of the latter, often quoted, is the rupture in the giant Chilean earthquake of May 22, 1960, which seems to have proceeded along a series of segments, taking about an hour to be completed.

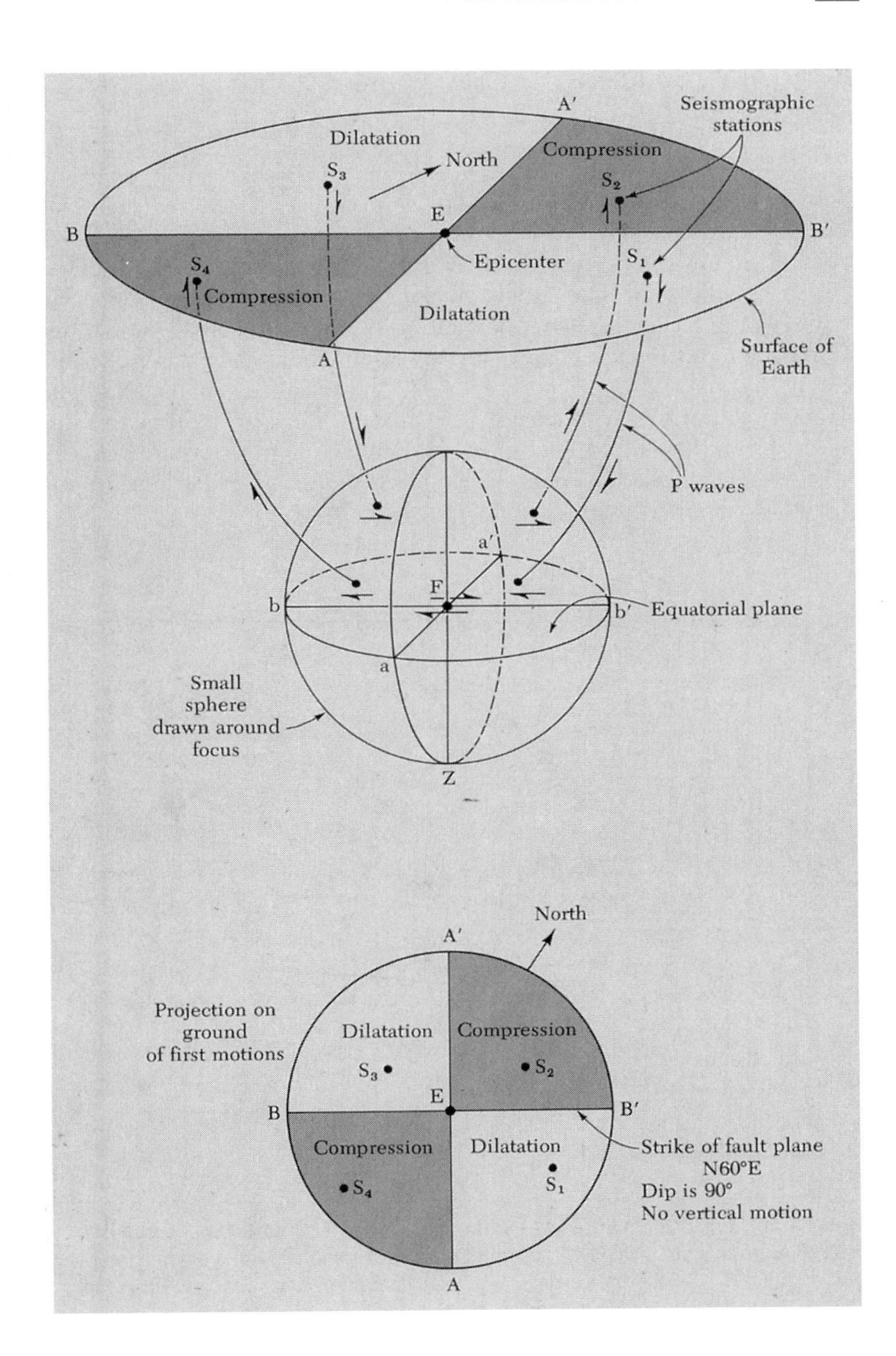

A′
Seismographic stations
Dilatation
North
Compression
S_3
S_2
E
B
B′
S_4
Epicenter
S_1
Compression
Dilatation
A
Surface of Earth
P waves
a′
F
b
b′
Equatorial plane
a
Small sphere drawn around focus
Z
North
A′
Projection on ground of first motions
Dilatation
Compression
S_3
S_2
E
B
B′
Compression
Dilatation
Strike of fault plane N60°E
S_4
S_1
Dip is 90°
No vertical motion
A

Twin Thrust Sources: 1971 San Fernando and 1994 Northridge

Studies of two recent neighboring California earthquakes allow us to understand more clearly how earthquake waves are generated and shake the country rocks and soils. Their sources both occurred under the broad valleys and mountains north of the city of Los Angeles (Figure 6.10). They had almost the same magnitude. In contrast to the strike-slip 1906 San Andreas earthquake, the sources of both were slip on steeply dipping thrust faults over approximately equal surface areas. Each has been carefully studied and has elicited major responses from the scientific, engineering, and planning authorities. Neither was predicted in the short term as to time, place, and size. Both provided relevations concerning the seismic behavior of key structures, such as large dams, electric distribution systems, hospitals, and particularly freeways and high-rise buildings.

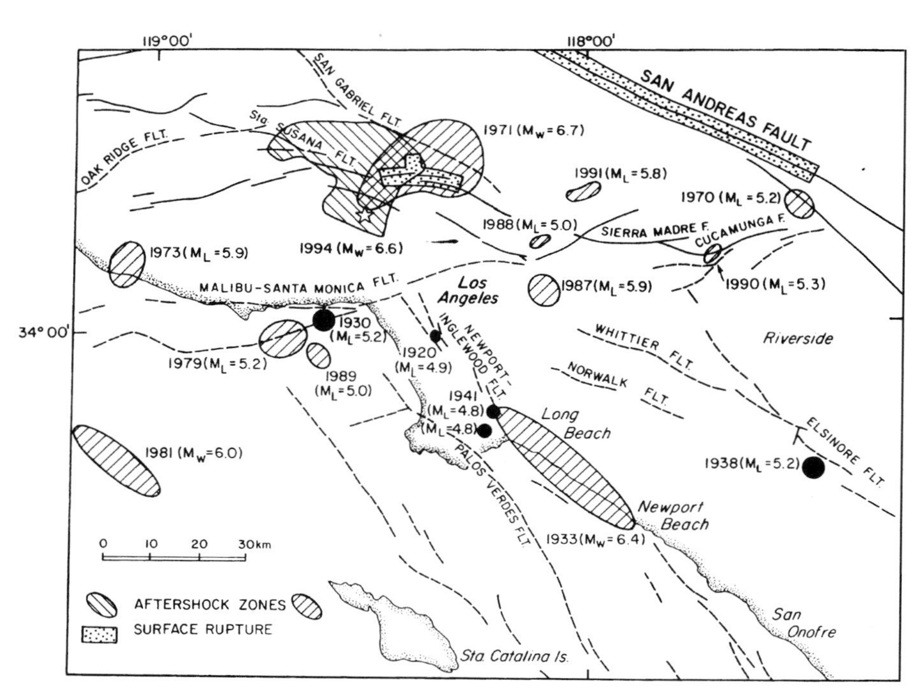

Figure 6.10 Significant earthquakes of magnitude greater than 4.8 that occurred in the Los Angeles area since 1920. Aftershock zones are shaded with cross hatching and include the 1994 Northridge earthquake. Dotted areas indicate surface rupture and include the 1857 rupture along the San Andreas fault (Ft. Tejon earthquake).

Because it occurred in a modern urban/industrial area, the 1971 San Fernando mainshock was one of the most important earthquakes in the evolution of modern seismic design and retrofitting practices — for bridges as well as for residential and commercial buildings. This moderate-sized earthquake caused 53 deaths and damage of just under $1 billion. By comparison, the Northridge earthquake of similar size on January 17, 1994, which mainly affected a neighboring, even more-populated area, caused 56 deaths and an estimated damage of over $15 billion. Each of the earthquakes caused major economic loss through the collapse of buildings, bridges, and other lifelines (see Figure 12.16).

The first of these twin events had a focal depth of about 13 kilometers. The generating fault dipped northeast under the southeastern foothills of the San Gabriel Mountains. The observed surface faulting had both thrust and left-lateral motions, particularly through the city of Sylmar. The total length of the surface trace was about 15 kilometers with some vertical scarps. In contrast, the January 17, 1994, Northridge earthquake had a hypocentral depth of 18 kilometers on a fault plane that dipped this time to the south under the San Fernando Valley, rather than north under the mountains like that of its 1971 seismogenic neighbor. The inferred area of rupture on the fault was approximately 16 kilometers along the strike and 15 kilometers in the direction of slip (see Figure 12.14).

No ground rupture of the thrust fault was observed at the surface in 1994 and the supposition is that the rupture plane lost its structural coherence at depths of about 5 kilometers on the north portion; fault displacements at more than 5 kilometers in depth were perhaps transferred to numerous surficial faults in the Santa Susannah Mountains. Because the faulting did not appear at the surface, the seismic source is termed a *blind thrust*. A network of blind-thrust faults throughout the crust under Los Angeles has long been established by deep drilling and by geological reconstructions of the regional tectonic deformation. Such procedures, however, do not usually lead to unique locations and dimensions of the buried faults. But there can be no doubt that their presence is an ever-present threat to the metropolis. More generally, the danger from blind-thrust faults (often not defined precisely on geological structural maps) is high in many seismic regions of the world. Their detection is often best accomplished by long-term mapping of microearthquake hypocenters in each such region.

7 Earthquakes and Plate Tectonics

The San Andreas fault (plate transform boundary) appears as a gash across the terrain of the Carrizo Plain in central California (see Figure 1.2). Note the offset streams deflected at the fault.

The result, therefore, of our present enquiry is, that we find no vestige of a beginning—no prospect of an end.

James Hutton, 1788

Before proceeding with further descriptions, we should pause to consider the current underlying theory of earthquakes available only since the 1960s. How well does it explain the earthquake properties that have been described in previous chapters?

Driving Forces in the Earth

It is convenient to begin toward the end of the eighteenth century when the Scottish naturalist James Hutton published his *Theory of the Earth* (1795). At that time, the belief was that the Earth was relatively young (measured in thousands of years) and its surface geology formed by regional convulsions of relatively short durations. While Hutton himself never experienced earthquakes, he was greatly concerned with the forces that lift continents and mountains. He became the first influential proponent of the view that geological processes must be embedded in an enormous length of past time, not just thousands of years, as was the theological view at that time.* In a striking phrase describing the grand-scale geological features of the Earth, he concluded, "We find no vestige of a beginning — no prospect of an end." For Hutton, mountains and plateaus could be explained not by a series of deluges like Noah's flood or other catastrophes but by continued application of subterranean heat that caused uplift of the land surfaces above the sea surface. He concluded that the land on which we dwell has been elevated by "extreme heat and expanded with amazing force." He considered volcanic eruptions to be safety valves "in

**The oldest rocks known at present are about 4 billion years old.*

order to prevent the unnecessary elevation of land and the fatal effect of earthquakes."

In the nineteenth century, the views of Hutton were developed and spread by one of the most influential founders of modern geology, the Englishman, Sir Charles Lyell. In the many editions of his *Principles of Geology* (1845–1875), he strengthened Hutton's arguments for geological time spans long enough to allow the present geological strata to form by erosion, volcanic activity, and heat. But he went even further in his radical view by asserting that "the present is the key to the past." In other words, there has been no cranking up of the forces that go to form the striking topographic contrasts of the Earth; rather the causes of mountains, volcanoes, and earthquakes have been uniform and steady over past "deep" time. Not only are the tectonic causes we see now also the past causes, but they have, with small variations, operated at the same rate. "No causes whatever have from the earliest time acted that are not now acting." This principle, called *uniformitarianism,* became the accepted model of the Earth's geological engine. Lyell believed also that the Earth's internal heat, which was the tectonic driving force, was uniform and remained so. We now know that he was largely correct and that the reason for this constant heat is the radioactive decay of mineral elements throughout the Earth's interior. But this heat source was not known to Lyell and his contemporaries.

In his *Principles of Geology,* Lyell included a great deal of descriptive material on earthquakes (see Figure 7.1) and linked them with uplift and other deformations of the Earth's surface. Even now this attractively written textbook can be recommended to the reader for descriptions of many major historical seismic and volcanic events. He describes, for example, large earthquakes along the Chilean coast, which were contemporary with his studies, and he was delighted when Darwin described to him the earthquake he felt near Valparaiso in 1835 mentioned in Chapter 1. He felt that Darwin's observation of the rise of the Chile seacoast associated with the earthquake was confirmation for his view that earthquakes were constructive processes that overcame the destruction of continents by erosion of the land surfaces. In March 1846 Lyell had an opportunity to visit the United States and the region affected by the New Madrid earthquakes of 1811 and 1812 (see Figure 8.1), where he talked with eyewitnesses. His main geological conclusion was that such geological events were examples of "construction," not "destruction," and they contravened the widespread belief that significant changes of relative levels of land and sea had ceased. In sarcastic mode, he writes: "In the face of so many striking facts, it is vain to hope that this favorite dogma will be shaken."

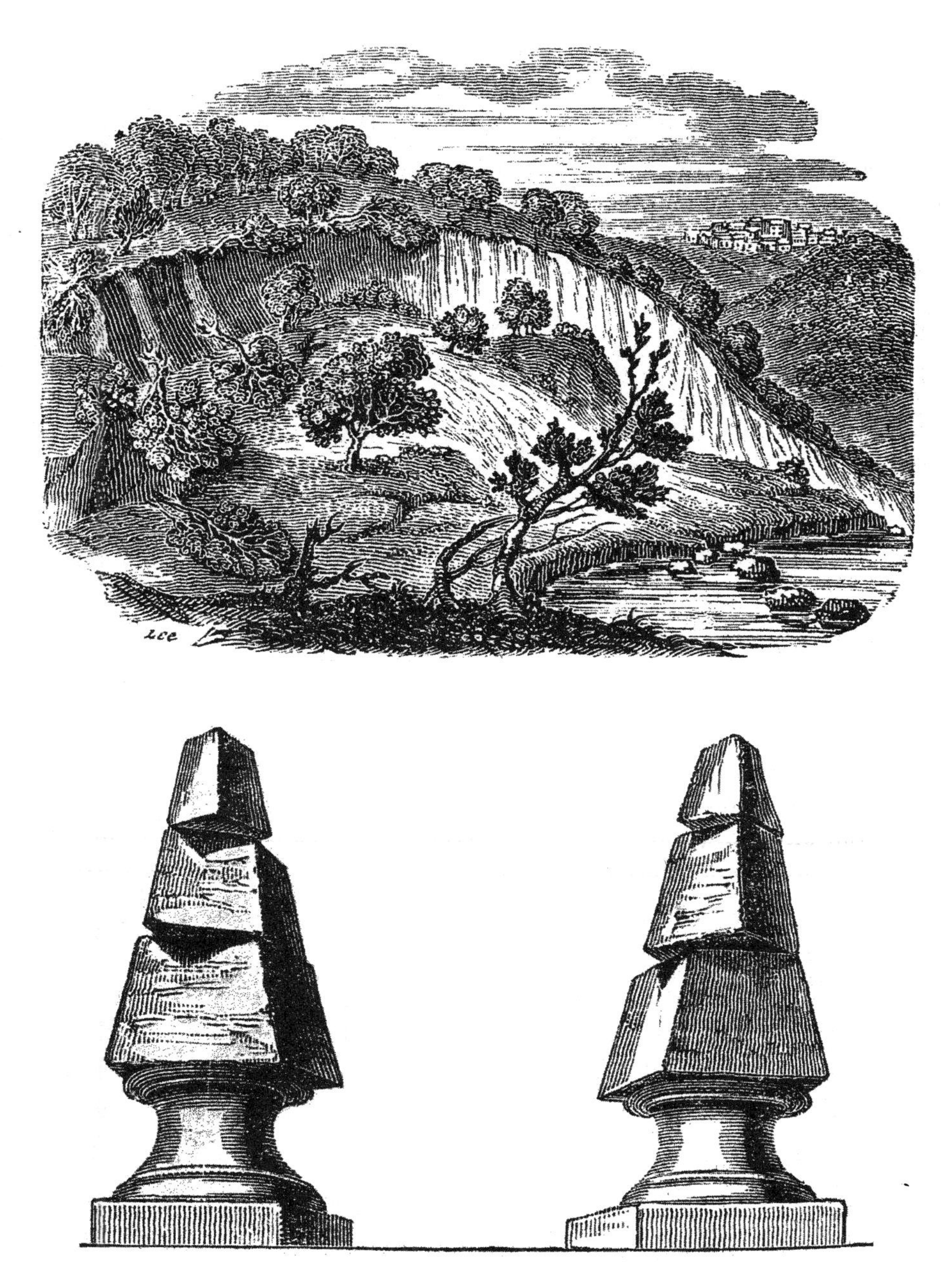

Figure 7.1 Sir Charles Lyell's examples of earthquake effects (1830). Top: The surface at Fra Ramondo in Calabria. Note the breakage of olive trees between up- and down-slid parts of the hill. Bottom: Two obelisks on the front of the convent of San Bruno in Stefano del Bosco, Italy. Each is built of three sections, and repeated earthquakes have loosened the blocks and rotated them to different positions. [From first edition of Lyell's *Principles*.]

Yet for all his insights, Lyell had no theory for the immediate cause of substantial earthquakes. While he described geological faulting accompanying some earthquakes, such as in New Zealand, he did not infer that sudden fault displacement was the source of the seismic waves. Accordingly, he had no way to explain the pattern of earthquake occurrence around the world, or why some earthquakes are contemporaneous with volcanic activity, yet many, such as the New Madrid earthquakes, are not.

It was not until the end of the nineteenth century that evidence began to be published that answered these questions. It became clear that geological faulting seen after earthquakes was not damage to the Earth caused by the earthquake, but was the cause of it. This radical idea was suggested, for example, by Professor Koto after the Japanese earthquake of 1891 in which a high fault scarp (see Figure 5.6) appeared across the rice farms of central Honshu. As has been explained in Chapter 1, the observed rupture of the San Andreas fault in the 1906 San Francisco earthquake provided the clinching evidence. The report of this earthquake convinced most scientists that sudden fault slip, following a slow buildup of elastic strain energy in the rocks of the Earth's crust, was the immediate cause of tectonic earthquakes. The subsequent extension of the 1906 mechanism to every tectonic earthquake was, of course, consistent with Lyell's principle of uniformitarianism.

Gradually, in the first half of the twentieth century, as earthquake recording at seismological observatories allowed remote earthquakes to be detected and mapped uniformly, the long-term pattern of global seismicity was resolved. A remarkable picture appeared — one that cried out for a global explanation (see Chapter 2 and Plates 10 and 11). In addition, both the geological field work and evidence from the recorded seismic-wave patterns (see Chapter 6) indicated that the types of faulting seen at the surface, such as strike-slip or dip-slip displacement, occurred in definite regional patterns. Thus the worldwide distribution of both earthquakes and their fault mechanisms was not random, but systematic. The time was ripe for the emergence of a fresh overall synthesis of this global information.

Until the 1960s, there were only two serious global hypotheses for the formation of mountain ranges and ocean deeps. Either the Earth was cooling and, hence, contracting, or there were large-scale convective movements of the rocks of the Earth's mantle. Neither was specific on the geographic patterns and causes of seismicity and volcanoes.

The Plate Conveyor-Belt Solution

The required comprehensive theory of large-scale tectonic processes, called *plate tectonics*, literally burst on the geophysical scene in the mid-1960s. It was based on the older internal convection engine hypothesis, but it incorporated many of the newly discovered properties of earthquakes and volcanos. It also explained clear-cut patterns of changes in magnetic polarity (i.e., north- or south-pole-seeking directions) of the rocks on each side of the midocean ridges. The polarity in the seafloor rocks had reversed in a regular time sequence, causing a pattern of magnetic "stripes," explainable if the rocks had flowed from a ridge as magma, then solidified to form the oceanic crust which moved away from the ridge.

When we look back on those heady days, the crucial role of seismological research cannot be overemphasized. Certain key scientific abilities had coincided in a decisive way: more precise locations of hypocenters of remote earthquakes by computer computations; the highly reliable estimates of earthquake-source mechanism (normal, thrust, and strike-slip faulting); and consistent focal-depth mapping of dipping Benioff zones. As the mechanical geological model of drifting plates was constructed and refined, certain attributes could be tested against the reality of these seismological results. Thus emerged a new understanding of the Earth's geology.

The basic idea is that the Earth's outermost part (corresponding to the *lithosphere* described in Chapter 4) consists at the present geological epoch of several large and fairly stable slabs of solid and relatively rigid rock called *plates.** These plates evolved to their present pattern during the 200 million years that have elapsed since the breakup of the ancient supercontinent called Pangaea. The energy for the driving mechanisms of these drifting plates comes from heat incessantly produced from the decay of radioactive elements in the rocks throughout the interior of the Earth.

The oceanic and continental crusts we know today make up the top part of the plates. The largest are mapped in Figure 7.2. Each plate extends to a depth of 80 to 200 kilometers; a plate moves horizontally, relative to neighboring plates, on softer rock immediately below. At the edge of a plate, where there is contact with adjoining plates, large deforming (or *tectonic*) forces operate on the rocks, causing physical and even chemical changes in them. At these plate edges, the Earth's geological structure is most affected by the forces of reaction between the plates, and this is where the massive and radical geological changes occur.

The intimate relation between the plate boundary forces and earthquake occurrence is readily apparent in Figure 2.6 along the California margin of the Pacific and North America plates. When the epicenters of small earthquakes are plotted, the regional mosaic is striking. The linear

**The noun "plate" appropriated in this geological sense appears to have become widely accepted about 1970.*

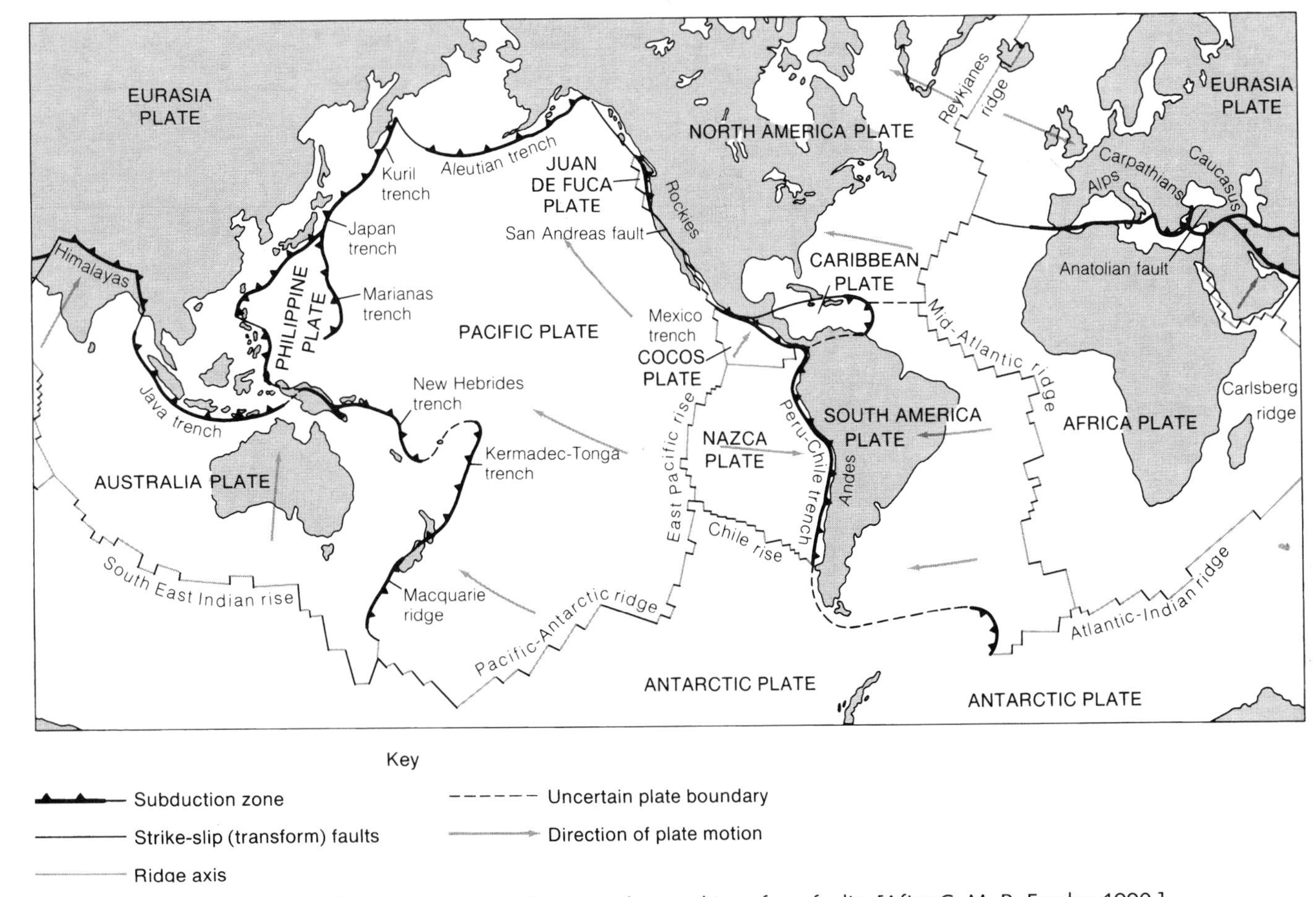

Figure 7.2 The major tectonic plates, midoceanic ridges, trenches, and transform faults. [After C. M. R. Fowler, 1990.]

concentrations of epicenters are seen to coincide with long-active faults such as the San Andreas. Other epicenters of shallow-focus earthquakes scatter throughout the region, indicating minor fracturing of the crust along the plate margin.

Geological evidence shows that plate geometry is not permanent but undergoing constant, gradual change. Magma is continually upwelling at the midoceanic ridges and rises as the seafloor spreads apart. This newly emplaced rock then moves slowly across the Earth's surface as new seafloor on either side of the ridge. In this way, plates extend and move at a uniform speed of several centimeters per year across the planet's surface, like great conveyor belts, cooling and aging as they get farther away from the ridges. For this reason, midoceanic ridges and rises are called *spreading zones*.

These spreading zones have been plotted in Figure 7.2. Notice that none of the lines of ridges appear as unbroken linear trends; they are disrupted by intermittent offsets. These jags coincide with a special kind of horizontal slip between two crustal blocks. At either end, the slip is changed or "transformed" by the emergence of new oceanic floor along the ridge. Such offsets are called *transform faults*, and many earthquakes, all with slip mechanisms that conform to this sense of movement, occur along them (see Figure 7.3). Indeed, in tectonic descriptions involving earthquake generation, the plate transform faults play a unique role.

If new plates, constituting the lithosphere, are constantly being created, what happens to old plates? Because the Earth remains the same size over long periods of geological time, large sections of the moving plates must

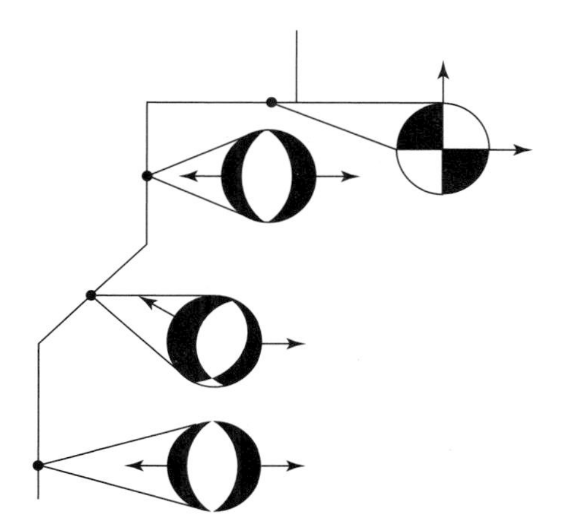

Figure 7.3 Expected fault-plane beach ball patterns for earthquakes (•) along a plate boundary (solid line). Arrows indicate the azimuth of the horizontal slip vectors. The region is in extension with strike-slip and normal faulting. [After C. M. R. Fowler, 1990.]

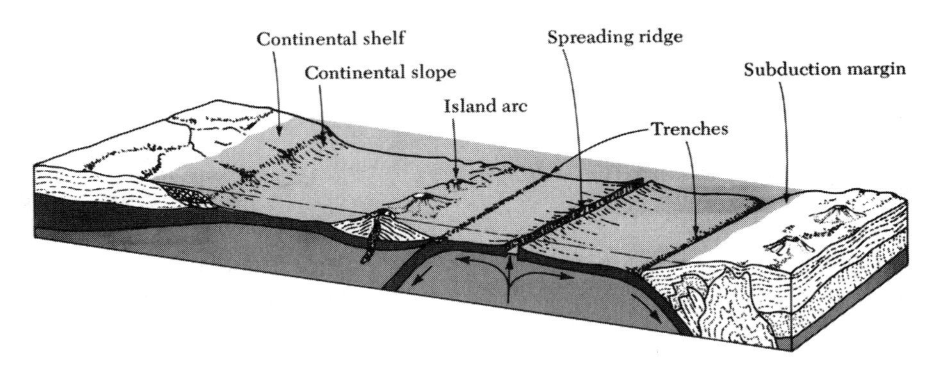

Figure 7.4 Generalized cross section through the ocean crust showing the uppermost layers of rocks under the ocean spreading out from the midoceanic ridge and sliding down under the volcanic island arc and active continental margin along the deep trench. Earthquake foci are concentrated at the tops of the descending layers and along the ridge. Magma rises upward above the subduction zones, and the erupting lava builds spectacular volcanic cones.

also be absorbed at some place. The burial ground of plates is believed to be the ocean trenches associated with advancing continents or island arcs (see Chapter 2). At these places, called *subduction zones*, the surface layers of rock plunge into the Earth's interior (see Figure 7.4). At greater depths, temperature and pressure increase, and the sinking lithosphere is gradually reworked until it becomes mixed and absorbed into the rocks of the deeper interior. At present, the plates containing Africa, Antarctica, North America, and South America are growing, while the Pacific plate is getting smaller.

From this perspective of an evolving, mobile Earth, earthquakes occurring along the midoceanic ridges are produced by processes arising from the growth of the plates. Along these submarine mountain chains are many surface ruptures and downdropped blocks. When these breaks occur in the rocks, energy for local earthquakes is released. In contrast, massive mountain chains (such as the Himalayas and the Alpine belt of the Mediterranean) arise when plates collide "head-on," producing almost continuous earthquake activity.

As a plate bends downward at the ocean troughs, such as the Japan trench, it fractures, generating shallow earthquakes. In the process of its downward movement, additional force is generated, causing further deformation, fracturing, and mineral dehydration, thus giving rise to deep-focus earthquakes (Plate 13). The deep earthquakes occurring along and within this descending plate (called the *downgoing slab*) define the remarkably regular dipping seismic zone — the Wadati-Benioff zone — that we discussed earlier (see Figure 2.4). Finally, at depths of 650 to 680 kilometers, either the plate is altogether absorbed into the rocks of the mantle or its brittle properties are altered enough so that it can no longer release energy as earthquakes.

Explanation of Inter- and Intraplate Earthquakes

Plate tectonics provides the general geological theory needed to understand global and regional earthquake occurrence. Many of the tantalizing observations described in previous chapters now become explicable. First, many more earthquakes will occur along the edges of the interacting plates *(interplate earthquakes)* than within the plate boundaries *(intraplate earthquakes)*. Earthquakes occurring where plate boundaries converge, such as at trenches (see Figure 4.5), contribute more than 90 percent of the world's release of seismic energy for shallow earthquakes, as well as most of the energy for intermediate- and deep-focus earthquakes. Most of the largest earthquakes, such as the 1960 and 1985 Chile earthquakes, the 1964 Alaska earthquake, and the 1985 Mexico earthquake, have originated in the subduction regions as a result of the thrusting of one plate under another. In contrast, about 10 percent of the world's earthquakes occur along the divergent ocean-ridge systems; these contribute only about 5 percent of the total seismic energy of earthquakes around the world.

We have noted that within the various seismic zones there are impressive similarities of fault types. In particular, there are consistent networks of active dip-slip, strike-slip, and oblique faults, already discussed in terms of their geometry and forms of slip (see Figure 5.4). The spreading ridges and subduction zones are predominantly the sites of normal and thrust earthquake-source mechanisms, respectively. But at many places along the margins of the patchwork of plates, there are offset ridges and offset subduction slabs. These offsets are the transform faults mentioned earlier. These faults connect the otherwise dead-ends of midoceanic ridges and subduction zones into a continuous network of earthquake-rich belts around the globe. A set of possible transform types is drawn in Figure 7.5. The reader might like to characterize the tectonic type of the San Andreas fault using the Pacific plate boundary geometry in Figure 7.2.

However, as the epicenters on the map in Figure 2.2 show, shallow-focus earthquakes also take place within plates, and the theory does not explain these in an obvious way. Such *intraplate earthquakes* must arise from more localized systems of forces in the crust, perhaps associated

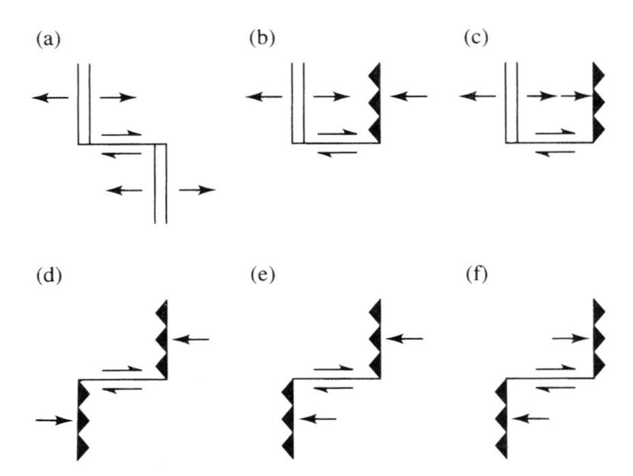

Figure 7.5 The six types of right-lateral transform faults (arrows give direction of relative slip). (a) Ridge-ridge fault; (b) and (c): ridge–subduction zone faults; (d), (e), and (f): subduction zone–subduction zone faults. Double lines indicate ridges. [After C. M. R. Fowler, 1990.]

with ancient geological structural complexities or with anomalies in temperature and strength of the lithosphere. Because the directions of forces on plates (see Figure 7.2) vary across them, the mechanism of the sources of earthquakes and their distribution differ in different parts of a plate (see Chapter 6).

In the United States, a number of large intraplate earthquakes has occurred since the European settlement. The most important were the great earthquakes that struck the New Madrid area of Missouri in 1811 and 1812 (see Figures 8.1 and 12.1). They caused heavy damage in the area and were felt as far away as Washington, D.C., New England, and Montreal, Canada. These earthquakes are produced by faults in the upper part of the North America plate now buried beneath alluvial deposits extending along the Mississippi River system from the Gulf of Mexico to the vicinity of New Madrid. It is likely that we are seeing a shearing reactivation of an ancient attempt to rift apart the North America plate, perhaps with intrusions of dense rock thrust upward in this region.

Other seemingly anomalous concentrations of earthquakes apparent in world seismicity maps need special explanations. For example, Figure 2.2 shows a cluster at the Hawaiian Islands remote in an aseismic seafloor. This and similar earthquake sources are associated with long-lived vigorous volcanic activity remote from subduction zones. We will take up this especially interesting case in Chapter 9.

The reader can now use plate properties to explain the mechanics of many famous historical earthquakes that were previously mysterious. Thus we can speculate that the 1755 Lisbon earthquake (see Chapter 1) was caused by sudden slip of the transform fault running from the mid-Atlantic ridge near the Azores through the Straights of Gilbraltar. Darwin's 1885 coastal elevation in Concepción occurred when the Nazca plate abruptly subducted under the Chile Andes, and Lyell's New Zealand earthquakes were the result of trench-trench collision of the Pacific and Indo-Australia plates. The right-lateral San Andreas fault slip in 1906 was clearly a consequence of the northeast movement of the Pacific plate.

Let us turn to a more recent case: the Philippine 1990 earthquake discussed in Chapter 5. This fault rebound resulted from the forces between two major tectonic plates (see Figure 5.11): the Pacific plate pushes the Philippine Sea plate beneath the eastern side of the Philippine archipelago at a rate of about 7 centimeters per year. The oceanic portion of the slower-moving Eurasia plate is being subducted along the western side of the islands of Luzon and Mindanao at a rate of about 3 centimeters per year. The role of the Philippine fault zone can then be seen as decoupling the northwestward motion of the Philippine Sea plate from the mismatched southeast motion of the Eurasian plate. The result is superposition of tectonic forces which gives rise to the frequent release of strain energy on many major faults throughout the region.

Predictions Using the Plate Tectonic Theory

What about making forecasts from geological plate theory? No scientific theory is worth its salt if it cannot make predictions beyond the evidence that was accumulated to form the theory. Thus, the theory of Isaac Newton describing gravitational forces, modified later by Einstein's extensions for very large masses and light-warp speeds, is regarded as top rank; it enables the theoretical equations to be used to predict the location of a planet in its orbit or the point on the moon where a spacecraft will land with great precision and confidence.

Unlike Sir Charles Lyell and all geologists up till about 1970, we now have, in tectonic plates and the mantle convection currents assumed to drive them, a general theory of the Earth's dynamical engine. But if this theory can be even comparable to the power of Newton's gravitational theory, various elements of it must be found useful in forecasting future earthquakes, tsunamis, volcanic eruptions, and seafloor mountain uplifts. An intriguing test comes from studying the quiet segments in major interplate earthquake zones.

The idea is as follows: the grand scale of the plate pattern and the steady rate of plate spreading imply that along a plate edge the slip should,

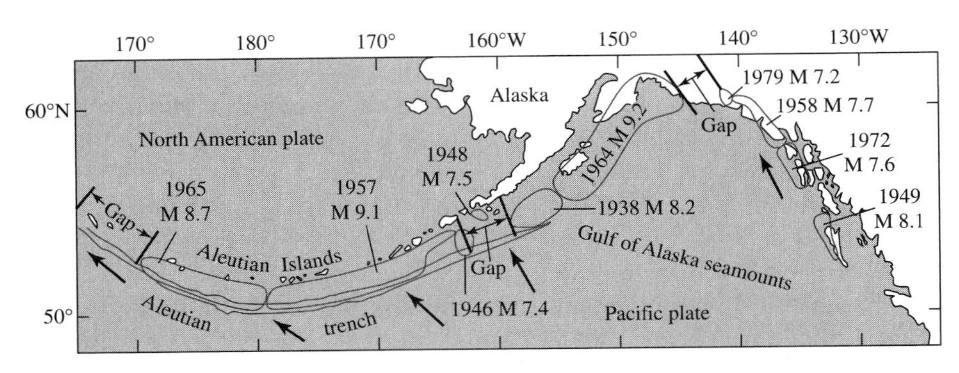

Figure 7.6 Rupture areas of large, shallow earthquakes (with approximate magnitudes) from 1930 to present and seismic gaps along the Alaska-Aleutian arc. The Yakataga gap is at longitude 143°W, and the Shumagin gap is at 160°W. The dark arrows show the direction of the motion of the Pacific plate relative to the American plate. [After J. C. Savage, M. Lisowski, and W. H. Prescott, *Science,* 231, 585, 1986.]

on average, be a constant value over many years a new form of uniformitarianism. Thus, if two fault slips that are some distance apart along a trench produce earthquakes, we might expect that a similar-sized slip will occur between them in due course. This idea suggests that the historical patterns of distance and time intervals between major earthquakes along major plate boundaries provide an indication of places at which large earthquakes might next occur. With this in mind, a *seismic gap* has been defined as an area in an earthquake-prone region where there has been a below-average level of seismic energy release, thought to be temporary.

Consider, for example, the plate boundary of the Alaska-Aleutian arc, as illustrated in Figure 7.6. The sites of the inferred seismic-energy release in some recent large earthquakes are indicated in the figure by the contours. If all such earthquake locations for the last 50 years are plotted, many sections of the arc are covered. There remain, however, some gaps (indicated by the heavy lines), which could be likely areas for plate slip and thus for major earthquakes in the future.

At the center of Figure 7.6 is the Shumagin gap, for which there is evidence of rupture in 1788, 1847, and perhaps 1903. The Yakataga section, at the north of the arc, was the source of an earthquake in 1899. Surveys indicate that the Pacific plate converges into the subduction zone at a rate of about 16 millimeters per year in a N15°W direction, roughly perpendicular to the Alaska arc. Also, strain accumulation has been measured from distances between ground markers in both these seismic regions since about 1980. With this evidence, the two regions may be the most likely

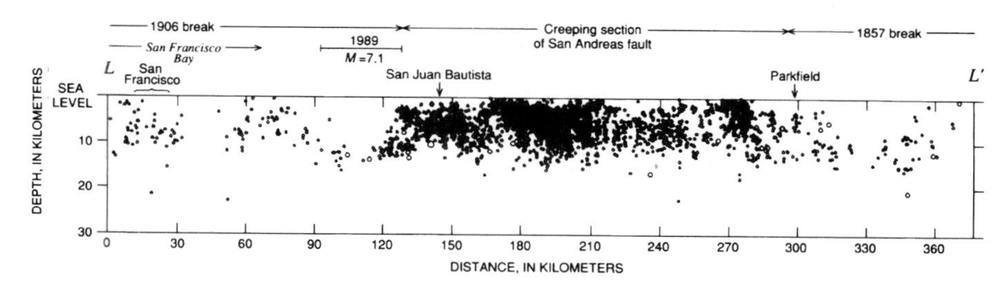

Figure 7.7 Cross section along the San Andreas fault zone from north of San Francisco to south of Parkfield. Background seismicity for 20 years prior to 1989. North of San Juan Bautista the fault had been virtually aseismic since 1906. On the Loma Prieta segment, the seismicity outlined a U-shaped area (Loma Prieta gap). Aftershocks of the 1989 main shock filled the former quiet zone of the Loma Prieta gap. [From Plafker and Galloway, 1989.]

sites for the next great thrust earthquakes along the Alaska-Aleutian arc. Yet the surveys have not detected any significant crustal deformation in the Shumagin gap, raising the speculation that subduction is sometimes episodic, with long intervals of strain accumulation occasionally interspersed by intervals of rapid accumulation. In the Yakataga gap, the surveys indicate that the rocks are being strained at an appropriate rate for the eventual occurrence of another great earthquake.

In California, there is a seismic gap along the San Andreas fault south of the fault rupture that caused the 1857 Fort Tejon earthquake (see Chapter 10). Another example of a seismic gap is given in Chapter 12 in the discussion of the tragic 1985 Mexico earthquake, which occurred when part of the subduction zone under the Pacific margin of Mexico slipped. The Loma Prieta earthquake discussed in Chapter 1 also fits the "deficiency of seismicity" theory. Pre-1989 mapping of foci of small earthquakes along the San Andreas fault south of San Francisco (Figure 7.7) highlighted a sparse region about 80 kilometers long centered on Loma Prieta. The mainshock and myriad of aftershocks in 1989 had foci that neatly coincided with this seismicity gap.

We must be cautious, however, about simple applications of a seismic-gap theory to short-term earthquake-hazard estimation because there are known exceptions. For example, in 1979, a moderate earthquake in the Imperial Valley of California was produced by rapid energy release along the Imperial fault. But the same section had slipped to generate an earthquake of similar size in 1940 (see Chapter 3). Thus, relatively frequent repetition of earthquakes from the same fault section cannot be ruled out.

8 The Size of an Earthquake

Sand boils from liquefaction effects on Marina Green, San Francisco, after the 1989 Loma Prieta earthquake. [Courtesy of Timothy Barker.]

The earthquake began everywhere with tremors . . . ; then the great shove of the destructive shock arrived, in some places rather before, in some a little after, the moment of loudest sound, and it died away suddenly into tremors again.

—*Robert Mallet, on The Neapolitan (Italy) earthquake, December 16, 1857*

So destructive was the earthquake in southern Italy in December 1857 that local communications were disrupted and almost a week went by before news of its extent reached foreign parts. Immediately, Robert Mallet, an engineer, applied to the Royal Society in London for a travel grant and proceeded to the Kingdom of Naples, where he spent 2 months making the first scientific, perceptive field studies of the effects of a great earthquake. Mallet's methods included detailed mapping and tabulation of felt reports and damage to buildings and geological movements. In this way, he sought to measure the strength and distribution of the ground motion.

Intensity of Shaking

By drawing lines on a map between places of equal damage or of equal *intensity* (he called these *isoseismal lines*), Mallet determined the center of the earthquake shaking and hence identified the source of the seismic waves. Also, the patterns of isoseismal lines indicated to Mallet the rate at which the shaking effects diminished with distance and provided him with an estimate of the relative size of the earthquake.

In the decades that have followed Mallet's work, seismologists have used earthquake intensity as their most widely applicable yardstick of the size of an earthquake. Intensity is measured by means of the degree of damage to structures of human origin, the amount of disturbance to the surface of the ground (see Plate 14), and the extent of human and animal reaction to the shaking. The first intensity scale of modern times was developed by M. S. de Rossi of Italy and Francois Forel of Switzerland in the 1880s. This scale, which is still sometimes used in describing an earthquake, has values ranging from I to X. (The published intensities of the 1906 San Francisco earthquake are based upon it.) A more refined scale, with 12 values, was constructed in 1902 by the Italian seismologist and volcanologist G. Mercalli. A version of it, called the abridged Modified Mercalli Intensity Scale, is given in Appendix C. It was developed by H. O. Wood and Frank Neumann to fit construction conditions in California (and most of the United States). Figure 8.1 illustrates how the scale was used to evaluate the shaking in the New Madrid, Missouri, earthquake of December 16, 1811.

Anyone who lives in regions where building and social conditions are similar to those of California can estimate the strength of a local earthquake on the Modified Mercalli scale (refer to Appendix C).* For example, suppose the earthquake is felt by all, most people are frightened, and many run outdoors. Suppose too that it shifts heavy furniture (such as refrigerators, large television sets, or sofas), causes plaster to fall, and damages some chimneys. Then it rates VI on the Modified Mercalli scale. Alternative intensity scales have been developed and are widely used in other countries, notably in Japan and in the countries of the former Soviet Union, where building conditions differ from those in California.

The assessment of earthquake intensity on a descriptive scale depends on actual observations of effects in the meizoseismal zone, not on measuring the ground motion with instruments. The descriptive scale continues to be important, first because in many seismic regions there are no seismographs to measure strong ground motion, and second, because the long historical record from seismically active countries is founded on such descriptions. However, the method has one problem that can affect the accuracy of the intensity rating: at a particular town or village the effect reflecting the *greatest* intensity is often chosen, thus increasing the local rating of the earthquake. A particular difficulty is the use of landslides caused by earthquakes. The Modified Mercalli scale gives landslides a rating of intensity X, but the fact is that landslides are common in many regions — even nonseismic areas — and quite small seismic shaking is known to be an effective landslide trigger.

When a study of the intensity of an earthquake is made nowadays, questionnaires (related to the description in Appendix C) are often

**After an earthquake in the mountains of northeast California, a householder said that effects in the earthquake "felt like a bear was on the roof." What would be the Modified Mercalli intensity value?*

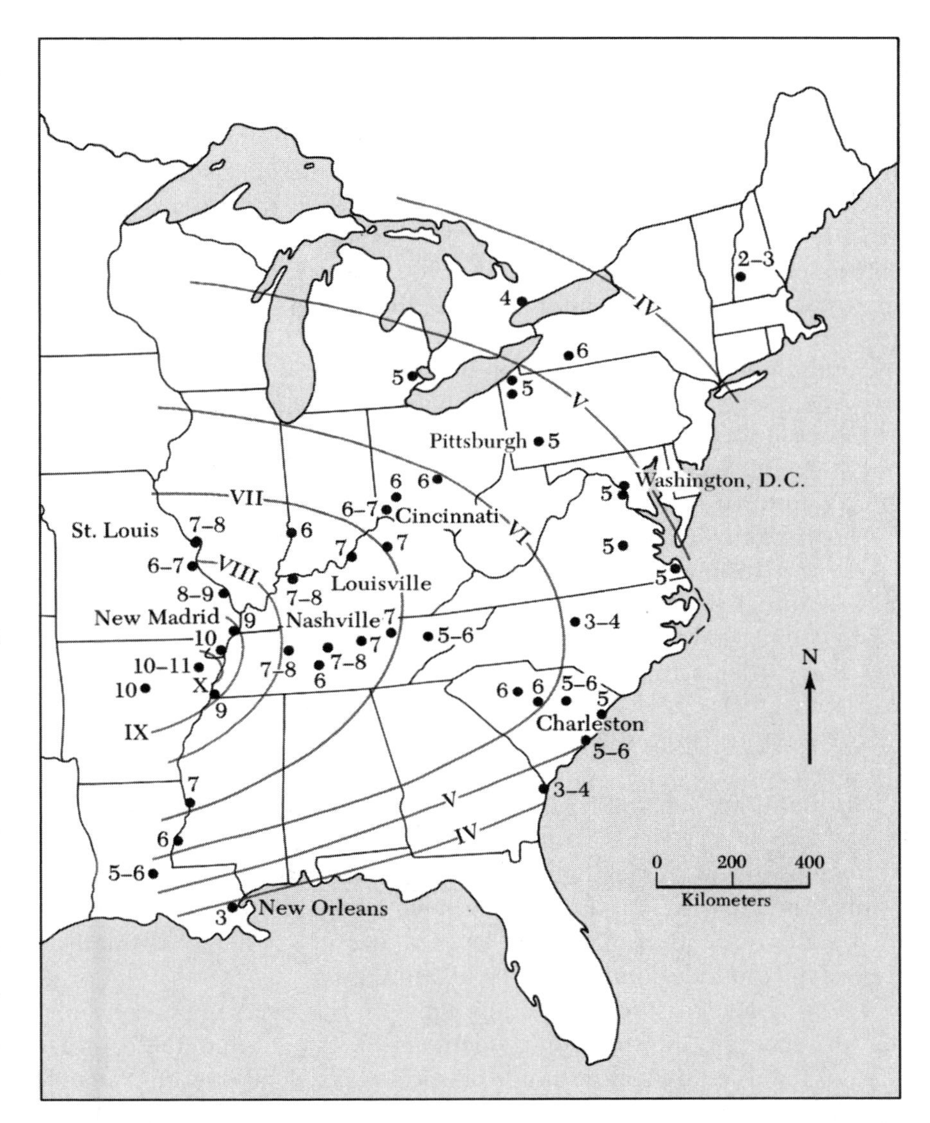

Figure 8.1 Isoseismal lines of intensity (Modified Mercalli scale) in the New Madrid, Missouri, earthquake on December 16, 1811. The felt radius of the earthquake extended to the East and Gulf coasts. Intensity in the then-sparsely populated area west of the epicenter is unknown. Intensity values at specified points are given in Arabic numerals, and the isoseismals are labeled by Roman numerals. [Courtesy of O. Nuttli and *Bull. Seism. Soc. Am.*]

circulated to inhabitants of the affected region.* Based on the responses to these questionnaires, a map such as that shown in Figure 8.1 can be drawn. Then areas of equal intensity are separated by isoseismal contours. The resulting isoseismal maps provide crude but valuable information on the distribution of the ground shaking away from the earthquake source, and also they may indicate the effect of the underlying irregular rock layers and surficial soil on the intensity of shaking. The relation between the rock type of San Francisco and the intensity of the 1906 earthquake is commonly cited. Figure 8.2 shows the correlation between the strength of shaking and damage (part *a*) and the rock and soil conditions (part *b*). Clearly, the harder rock in the hills (Kjf) coincides with an area of rather low damage to structures (many chimneys did not fall), whereas high intensities occurred on the filled lands (Qal) around the Bay shore. It is of interest to compare the intensities and isoseismals for the 1989 Loma Prieta earthquake (Figure 12.12) with those in 1906.

A commonly asked question is, Why does the intensity of earthquake shaking vary so much from place to place even in a local area? Of course, changes in rock type is one reason (see Plate 9), but we saw already (Figure 8.2) that the Modified Mercalli intensity in San Francisco on April 18, 1906, was particularly severe in the low-lying areas of soft soil and Bay sand. Yet, on the rocky hills relatively moderate shaking occurred. There was much the same contrast observed in the 1989 Loma Prieta, California, earthquake.

Furthermore, soft soils, particularly when water saturated, often slide, slip, or lose their cohesive strength when shaken, resulting in building foundations moving, cracking, or subsiding. But, in addition, the incoming seismic waves that are shaking the underlying rock are modified by the different elastic properties of any soil layers above it. The effect is usually to increase the ground motions that shake buildings.

The physical reason for this amplification has been understood for a long time. Simply, the soils are less dense and less rigid than the basement rocks, so that the physical restraints on ground accelerations and velocities are less. An analogy is jelly on a dessert plate that develops large undulations as the rigid plate is moved to and fro by even a small amplitude. The layering of the soil further boosts the amplifying effect as seismic shear (S) waves are bounced up and down between the layer boundaries. This

**Future scale revisions should perhaps include the ubiquitous hamburger and gasoline stations as special items! Estimated intensity maps of earthquakes in some localities can now be seen, soon after the shaking, at certain World Wide Web sites. See references at end of book.*

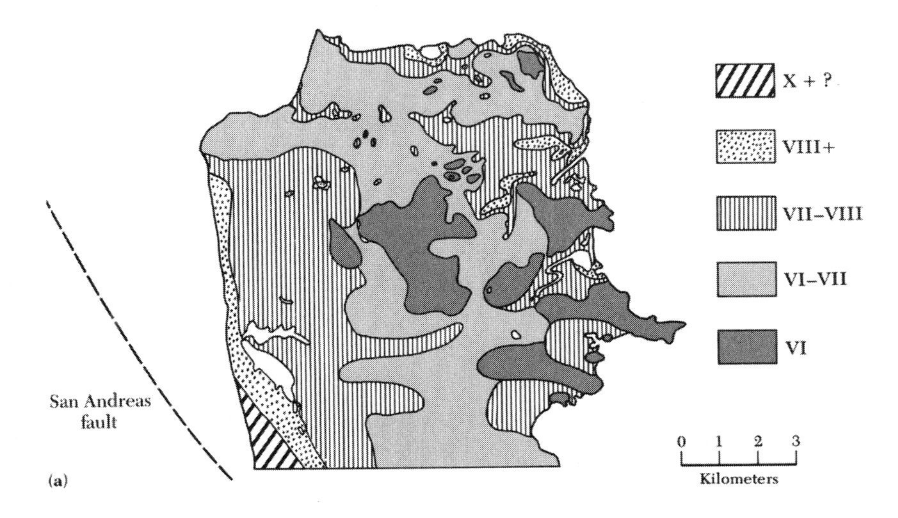

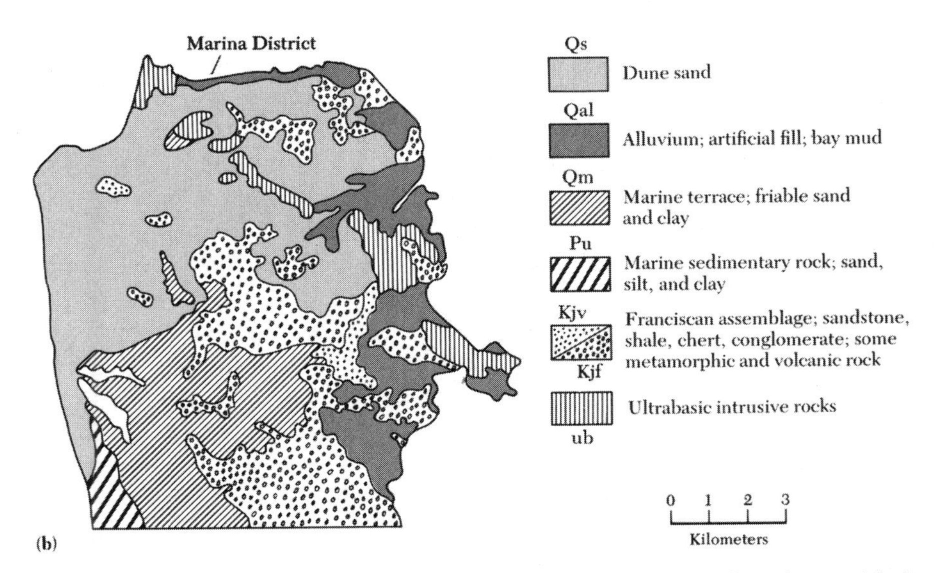

Figure 8.2 (a) Isoseismal lines on the San Francisco peninsula (based on the Modified Mercalli scale) drawn by H. O. Wood after the 1906 San Francisco earthquake. (b) A generalized geological map of San Francisco peninsula. Note the correlation between the geology and the intensity.

theoretically predicted soil amplification has been confirmed by recordings of down-hole seismographs as well as by the innumerable cases of enhanced intensity of shaking at soil sites in past earthquakes. The effect partly explains the many historical accounts of severe earthquake damage to buildings in towns situated in alluvial valleys.

How to Calculate Earthquake Magnitude

If the sizes of earthquakes are to be compared worldwide, a measure is needed that does not depend (as does intensity) on the density of population and type of construction. A strictly quantitative scale of size that can be applied to earthquakes in both inhabited and uninhabited regions was originated in 1931 by K. Wadati in Japan and developed by the late Professor Charles Richter in 1935 in California. The scheme is to use the wave amplitudes measured by a *seismograph*. This idea is similar to that of astronomers who grade the size of stars using a stellar magnitude scale based on the relative brightness seen through a telescope.

Because the size of earthquakes varies enormously, the amplitudes of their ground motions differ by factors of thousands from earthquake to earthquake. It is therefore most convenient to compress the range of wave amplitudes measured on seismograms by using some mathematical device. The most common is to replace numbers by their logarithms. In other words, instead of using amplitudes the powers of these numbers are used. Consider first a nearby earthquake source. Richter defined the magnitude of a local earthquake as *the logarithm to base 10 of the maximum seismic-wave amplitude (in thousandths of a millimeter) recorded on a standard seismograph at a distance of 100 kilometers from the earthquake epicenter.* When the logarithmic scale is used, every time the magnitude goes up by 1 unit, the amplitude of the earthquake waves increases 10 times (i.e., for magnitude 4 and 5, $10^5/10^4 = 10$).

The seismograph used as standard for magnitude determinations of local shocks is a simple type with torsion suspension of the mass developed by H. O. Wood and J. Anderson (or its equivalent). The farther the earthquake source is from the seismograph, the smaller the amplitude of the seismic wave, just as a light appears dimmer as the observing distance from the source increases. Because earthquake sources are located at all distances from seismographic stations, Richter further developed a method of making allowance for this attenuation with epicentral distance when calculating the Richter magnitude of an earthquake. The procedure, presented graphically in Box 8.1, shows how with this scale anyone can easily calculate the magnitude of an earthquake.

Box 8-1

EXAMPLE OF THE CALCULATION OF THE RICHTER MAGNITUDE (M_L) OF A LOCAL EARTHQUAKE

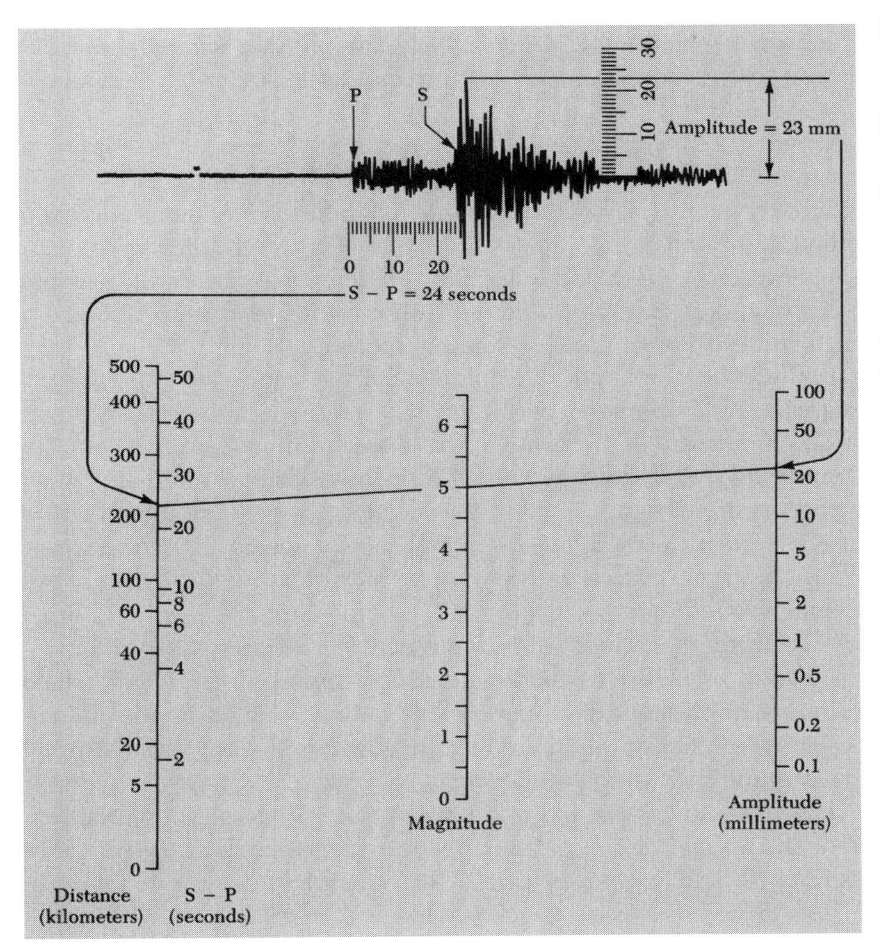

Procedure for calculating the local magnitude, M_L

1. Measure the distance to the focus using the time interval between the S and the P waves (S − P = 24 seconds).
2. Measure the height of the maximum wave motion on the seismogram (23 millimeters).
3. Place a straight edge between appropriate points on the distance (left) and amplitude (right) scales to obtain magnitude $M_L = 5.0$.

The idea behind the Richter magnitude scale was a modest one at its inception. The definition of local magnitude (M_L) was for southern California earthquakes only. The type of seismic wave to be used was not specified; the only condition was that the wave chosen — whether P, S, or surface wave — be the one with the largest amplitude. Richter wrote, "I did the work to provide a purely instrumental scale for rough separation of large, medium and small shocks."

Today, the use of magnitude has expanded beyond recognition from these modest beginnings. The convenience of describing the size of an earthquake by just one number, the magnitude, has required that the method be extended to apply to a number of types of seismographs throughout the world. Consequently, there are a variety of magnitude scales, which are based on different formulas for epicentral distance and ways of choosing an appropriate wave amplitude.

Earthquake magnitudes are used in three main ways. First, they are recognized by the general public, as by scientists, engineers, and technicians, as a measure of the relative size of an earthquake; people correlate a magnitude, at least roughly, with the severity of an earthquake. Second, magnitudes are of significance in the ongoing efforts to monitor a comprehensive nuclear test ban treaty; research has indicated that comparison of different kinds of magnitude is one of the best ways to distinguish between an underground nuclear explosion and an earthquake due to natural causes.* Third, magnitudes of previous earthquakes are used in an approximate way to predict what the greatest acceleration of the ground shaking may be in an earthquake at a site of an important structure (see Chapter 12). The information is then used by the engineer to design a structure that will withstand such strong motion.†

The magnitude scale has no upper or lower limit, although the largest size of an earthquake is certainly limited by the strength of the rocks of the Earth's crust. In this century, two or three earthquakes recorded on Wood-

**Details of this seismological detective work are given in Bruce A. Bolt,* Nuclear Explosions and Earthquakes: The Parted Veil *(San Francisco: W. H. Freeman and Company, 1976). A comprehensive test ban treaty banning tests of nuclear weapons, both above and below ground, was at last agreed on internationally in 1996. Seismological tools play a large part in monitoring this long-sought agreement.*

†This practice is somewhat unfortunate, because near an earthquake source there is no strong correlation between earthquake magnitude and maximum peaks of acceleration.

Anderson seismographs have had Richter magnitudes of 8.9. The great 1960 Chile earthquake of May 22 had a Richter magnitude of 8.25, and the enormous 1964 Alaska megathrust registered a Richter magnitude of 8.6 on the Berkeley standard seismograph. Their actual relative size was much greater. It turns out that such Richter values for the relative size of large near and distant earthquakes can be quite misleading.

At the other extreme, highly sensitive seismographs can record earthquakes with a magnitude of less than minus 2. The energy release in such events is about equivalent to that produced when a brick is dropped from a table onto the ground. Generally speaking, shallow earthquakes have to attain Richter magnitudes of more than 5.5 before significant damage occurs near the source of the waves. Smaller earthquakes, however, can cause damage to weak structures.

The current practice at earthquake observatories is to use two or more magnitude scales, all different from the original Richter scale. One modification is used because earthquakes that have deep foci give very different seismograph recordings from those having shallow foci, even though the total amount of energy released in each event might be the same. In particular, deep-focus earthquakes (see Chapter 2) have only small or insignificant trains of surface waves. It is therefore desirable, when dealing with all global earthquakes, to be able to calculate a uniform magnitude that does not depend on the presence or absence of surface waves.*

Richter's original definition is simply changed. Let us examine the seismogram in Appendix G. An uncomplicated earthquake record clearly shows a P wave, an S wave, and a train of Rayleigh waves. (The seismogram shows only the vertical component of the ground motion.) Now, if Richter's procedure for determining local magnitude were followed, we would measure the amplitude of the largest of the three waves and then make some adjustment for epicentral distance and the magnification of the seismograph. But it is just as easy to measure the maximum amplitude of any one of the three waves and hence to find three magnitudes — one for each type of wave.

It has become routine in seismology to measure the amplitude of the P wave, which is not affected by the focal depth of the source, and thereby determine a P-wave magnitude (called m_b). For shallow earthquakes (such as that recorded in Appendix G) a surface wave train is also present. It is

**This problem might be avoided by using a new measure of earthquake strength called* seismic moment *(see page 158). Research on earthquake size using this measure has suggested revisions in previous estimates of the relative magnitudes of great earthquakes.*

common practice to measure the amplitude of the largest swing in this surface wave train that has a period near 20 seconds. This value yields the surface-wave magnitude (M_S). Neither of the magnitudes (m_b or M_S) is the Richter magnitude, but each has an important part in describing the size of an earthquake.

For the shallow-focus earthquake shown in Appendix G, the measurements yield a body-wave magnitude of 5.3 and a surface-wave magnitude of 5.0. Many measurements of this kind for shallow earthquakes have suggested an approximate relation between m_b and M_S. This empirical relation allows the conversion of one type of magnitude into another, at least for moderate-size earthquakes. It turns out, however, that M_S correlates much more closely with our general ideas of the size of an earthquake than does m_b. For example, the 1964 earthquake in Alaska, which was a very strong shallow earthquake, had a surface-wave magnitude M_S of 8.6, whereas the body-wave magnitude m_b turned out to be only 6.5. Thus for this particular earthquake, the magnitude of the P wave (due to its short period) was not a good description of the Alaskan earthquake as a whole, but the M_S value was a better (but still not completely satisfactory) measure of overall size.*

Additional magnitude scales, such as the *moment magnitude* (M_W) have been introduced to improve further the uniform coverage of earthquake size (see the section on seismic moment on page 157).

Energy in Earthquakes

Today everyone is giving some thought to the energy resources available to humankind. Oil, coal, wind, and the sun, as well as nuclear energy, are all used as sources of energy. In order to discuss energy quantitatively, we must recognize that energy is the measure of the work that can be done by some machine; in the metric system, common units of energy are ergs. The present total consumption of energy in the United States per annum is about 10^{26} ergs.

From a global perspective, such an amount of energy is really quite small. The amount of heat that flows out of the Earth as a whole, to be lost through the atmosphere into space each year, is about 10^{28} ergs. Earthquakes, too, emit a great deal of energy. As we discussed in Chapter 6, they are the result of the sudden release of strain energy stored previously

**The interested reader might try to explain why this is so. The answer is related to the often repeated observation: the longer the surface fault rupture, the greater the surface-wave magnitude (see Appendix G).*

in the rocks in the Earth. From measurements of the seismic-wave energy produced by the sudden fracture, it is estimated that each year the total energy released by earthquakes throughout the world is between 10^{25} and 10^{26} ergs. When seismograms from various stations around the world are used to calculate the energy in the reported waves, an earthquake of Richter magnitude 5.5 turns out to have an energy of about 10^{20} ergs. By way of comparison, the energy that nuclear physicists calculate was released in the atomic bomb blast of Bikini in 1946 was about 10^{19} ergs.

It is tempting to correlate the energy release of an earthquake with its size, as measured by the earthquake magnitude scale (see Appendix I). Although the correspondence is not a precise one, it is nevertheless useful for estimating the amount of energy actually released in earthquakes. The relationship that seismologists suggest prevails between magnitude and energy release is given in Appendix G. This logarithmic relation indicates that an increase of 1 unit in magnitude M_S increases the amount of *seismic energy* E released by a factor of about 30 (not a factor of 10, as is sometimes quoted).*

Earthquake waves, of course, carry energy, and when they encounter buildings some of it is transferred to vibrate the structure. Nowadays, the amount of seismic energy per unit area per second transmitted to the shaking building can be measured by seismographs.

What Is the Seismic Moment?

As the science of seismology develops, more and more precise terms are needed to describe fully the size of an earthquake. As was mentioned at the beginning of this chapter, the first such term was the *seismic intensity* introduced as early as 1857 by Mallet in his study of the destructive earthquake near Naples, Italy. Although in modern forms the intensity scale is still of use, it is not a true mechanical measure of source size, like force or energy. Rather it indicates the relative strength of the shaking locally.

The first index of the size of an earthquake based on the measured wave motion was *earthquake magnitude*, which has been described above. This method refers the maximum single ground-wave amplitude projected back to the earthquake focus by allowing for attenuation (see Box 8.1). Such peak values, however, do not directly measure the overall mechanical

**If the amount of energy in a magnitude 2 earthquake were represented by the volume of a golf ball, the amount of energy released in the 1906 San Francisco earthquake would be represented by a sphere of radius 33 meters.*

power of the source — just as the strongest wind gust is not a reliable measure of the overall force of a windstorm.

In seeking a physically meaningful measure of the size of the earthquake source itself, seismologists turn to the classical theory of motion of mechanical systems in which movement results from the application of forces. Such forces do work by exerting pressures on the machine, which begins to move with a certain energy.

As outlined in the last section, one way of describing the overall size of the source is by the *seismic energy* radiated as waves through the rocks. There is a problem with this measure of size, however. The energy of the shaking is absorbed by fracture and friction in the rocks so that the recorded motion is always less than if the earthquake "machine" were a perfect one. Correction for such damping of the seismic motions has to be made in order to estimate the true total energy released. Numerous attempts using many earthquakes to develop consistent formulas for estimating energy from the wave motion measured by instruments have not been completely successful so far.

Nowadays there is a more robust procedure available. Seismologists now favor a measure called *seismic moment* in estimating the size of seismic sources. Like energy, the concept has been adopted from the theory of mechanics and has been found to yield a consistent scale of earthquake size. It has become more and more widely used by seismologists and earthquake engineers because of the greater reliability of absolute size thus provided in seismicity catalogs.

The underlying mechanical concept of moment can be described in terms of a simple personal experiment. Place both hands on the edge of a heavy table and push on one while pulling on the other in a horizontal direction (see Figure 8.3). The more widely separated the hands, the easier it

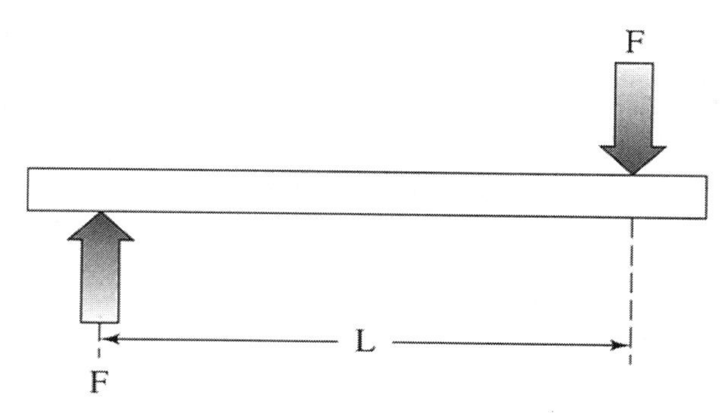

Figure 8.3 The two equal but opposite forces F, distance apart L, rotate the bar under a moment FL.

is to rotate the table. In other words, the effort required to produce the rotation is reduced by increasing the leverage of the forces exerted by the two arms even if the force of the hands remains the same. These two equal and opposite forces are named the force *couple*. The size of this couple is called its *moment*, and its numerical value is simply the product of the value of one of the two forces and the distance between them. The basic idea can be extended to any system of forces that produces sudden slip on a fault. As is seen in Figure 6.4, it is easy to apply the moment description to the size of an earthquake source, because the elastic rebound along a rupturing fault can be thought of as being produced by force couples along and across it.

There is little doubt that it will become more and more commonplace to see every earthquake identified by a numerical value of its moment. A bonus is that this measure is quite independent of any frictional dissipation of energy along the fault surface or as the waves propagate away from it. Further, the estimation of moment can be made either by using field measurements of the dimensions of the fault source slip (see Glossary) or from the characteristics of the seismic waves recorded at the ground surface.

Seismic moments range over many orders of value from the smallest to the largest earthquakes. For example, between magnitude size 2 and magnitude 8 earthquakes, the seismic moment would typically range over 9 orders of scale (powers of ten). The moment of the 1906 San Francisco earthquake is estimated to have been over 10 times that of the 1989 Loma Prieta earthquake.

It is unfortunate that the unfamiliar physical unit in which seismic moment is measured (see Appendix G) limits the widespread adoption for public use of seismic moment to replace earthquake magnitude. However, moment values can be correlated with the magnitude to produce yet another variety of magnitude which differs from those mentioned in the previous sections. This magnitude is written M_W and called the *moment magnitude* (see Appendix G). For comparison, the 1989 Loma Prieta earthquake had estimated magnitudes of 7.1 M_S and 6.9 M_W. For scientific purposes, M_W is now tending to supersede Richter magnitude and the surface-wave magnitude M_S.

The reason is simple. As we have noted, the seismic moment upon which M_W is based is a measure of the whole dimension of the slipped fault which for great earthquakes may have lengths of hundreds of kilometers. In contrast, the seismic waves used to estimate M_S have wavelengths of about 100 kilometers for all earthquakes of moderate to major size. Because such a wavelength samples only a fraction of the slipped fault area, the M_S value calculated is saturated at some upper threshold.

Acceleration of Ground Shaking

Ground acceleration is another measure of size. We have all felt the considerable forces that arise in a rapidly accelerating or braking car, in a jet aircraft as it takes off, or in the roller coaster at an amusement park. Some have felt similar forces during the hectic moments of an earthquake. The notion of acceleration is of key importance when trying to measure any type of varying motion such as strong ground shaking. Indeed, as mentioned in Chapter 3, we can think of each portion of the earthquake waves as being associated with a certain acceleration of the ground.

For many years, a widely asked related question has been, "How fast and by what amount does the ground move during an earthquake?" Definite answers were hard to find until the development of the modern strong-motion seismograph, designed to operate near to the source of an earthquake in such a way that it would not go off-scale during the strongest shaking (see Chapter 3). Such recorders measure seismic ground acceleration directly.

It is useful to scale acceleration against a value with which everyone is familiar, because experience does not commonly give us a feel for the magnitude of accelerations when stated in physical terms (often expressed as a centimeter per second per second, or cm/sec^2, for example). This base reference is the *acceleration due to gravity;* that is, the acceleration with which a ball falls if released at rest in a vacuum (to eliminate wind resistance). We will call this acceleration 1.0g.* It is quite a sizable rate of increase of speed. In terms of automobile accelerations, for example, it is equivalent to a car traveling 100 meters from rest in just $4\frac{1}{2}$ seconds.

Although acceleration of seismic motion is important, a full understanding of vibratory effects also requires an understanding of the velocity and displacement of the ground and of such wave properties as frequency. (The relations between acceleration, velocity, and displacement, as well as the concepts of wave period, wave frequency, and wavelength, are explained briefly in Appendix H).

The farther the waves travel, the more the high-frequency waves are attenuated in comparison with the long-period ones. For example, in 1964, long surface waves from the Good Friday earthquake in Alaska were recorded at the seismographic station in Berkeley, California, with periods of 17 seconds and maximum ground displacement of 1 centimeter. Yet the length between the wave crests was so long (approximately 50 kilometers)

*$g = 980$ *cm/sec*^{2} $= 980$ *gal, approximately, the acceleration due to gravity.*

that nobody in Berkeley was aware of moving up and down during the passage of the waves.

Strong-motion seismographs (Plate 5), called accelerographs, have now provided hundreds of records of seismic shaking, both away from and within buildings, in many countries of the world. Measurements from accelerograms indicate that the highest acceleration in the shaking of firm ground in most moderate earthquakes, at places a few tens of kilometers from the seismic source, lies in the range of 0.05g to 0.35g (compare Figure 12.2).

Some peaks of high-frequency waves reach accelerations equal to gravity. This usually occurs when the ground motion is measured on firm ground or rock very near the source of the waves.* As discussed in Chapter 1, both vertical and horizontal ground motions are measured in earthquakes. Generally, the vertical acceleration is less than the peak horizontal acceleration. The reason is that most vertical motion is due to P waves and the horizontal motion is due to S waves produced by the energetic fault rebound. An average value from many of the available California accelerograms suggests an average ratio of about 50 percent for the vertical compared to the horizontal motions. One of the largest horizontal accelerations recorded thus far was on the abutment of Pacoima Dam in the damaging 1971 San Fernando earthquake centered north of Los Angeles — it reached 1.15g. The vertical acceleration was recorded with a peak of about 0.70g.

Studies indicate that damage is often much more attributable to the speed of the back-and-forth motion of the foundation than to its peak acceleration. In general, the higher the seismic intensity, the higher the average velocity of the shaking. Nevertheless, the mean accelerations have much bearing on the forces affecting a structure (see Appendix C). Consequently, in designing to avert earthquake damage, engineers have come to rely heavily on the estimates of ground acceleration that a structure might be expected to experience in its lifetime.

Because buildings are built to withstand the pull of gravity, even when no special earthquake code is followed, they will usually withstand substantial accelerations in a vertical direction during earthquake shaking. In contrast, experience has shown that it is the horizontal motions of the ground

**Such as in the Bear Valley, California, earthquake on September 4, 1972 (M_L = 4.7), the Ancona earthquake in Italy on June 21, 1972 (M_L = 4.5), and the Northridge, California, earthquake on January 17, 1994 (M_L = 6.7). In such recordings, horizontal accelerations of 0.6g to over 1.0g have been recorded independent of the magnitude of the earthquake.*

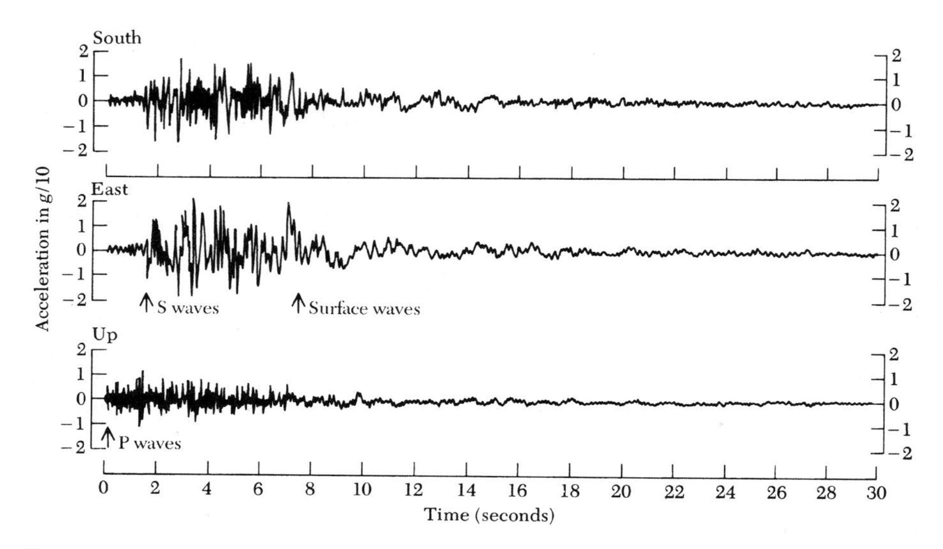

Figure 8.4 Three components of strong ground acceleration recorded in the parking lot outside the Hollywood storage building about 20 kilometers from the fault rupture in the 1971 San Fernando earthquake, California.

that topple structures and even throw people to the ground. Many types of buildings, such as adobe buildings in South America and the Middle East, are not able to withstand even 10 percent of the acceleration of gravity in a horizontal direction.

Typical accelerations of the ground recorded during the 1971 San Fernando earthquake are reproduced in Figure 8.4. Three components of the ground acceleration were recorded: the bottom trace, from left to right, indicates the vertical shaking; the top two traces show the two horizontal components — north-south and east-west ground shaking. At the bottom, a time scale is given in seconds. Acceleration of the shaking has been recorded as a fraction of gravity.

Consider in Figure 8.4 the structure of the wave form, running from left to right. First of all, the instrument triggered with the arrival of a P wave and, during the first second and a half, a rather high-frequency but small-amplitude P wave shook the ground. The bottom trace shows that the vertical component shaking did not increase much after this time, reaching a maximum of a little over 0.1*g*. In sharp contrast, however, after about 2 seconds the horizontal components of the ground acceleration became markedly larger with the arrival of S waves and surface waves. On the east-west component, the arrows show the onset of the S and surface waves; the horizontal ground shaking reached a maximum of a little more than 0.2*g* after 3 seconds elapsed.

Although "peak," or maximum, acceleration values are important, another point to recognize is that the damage to structures may be occurring throughout the entire period of strong ground shaking. Indeed, the overall damage may be more closely correlated to the total duration of the strong motion than to any particular peak on the record. For this reason, the second parameter of importance in acceleration records is the duration of strong shaking. A useful measure of duration is called the *bracketed duration*. This is the duration of shaking above a certain threshold acceleration value, commonly taken to be 0.05*g*, and is defined as the time between the first and last peaks of motion that exceed this threshold value. In Figure 8.4, the bracketed duration above 0.05*g* acceleration is only about 6 seconds.

Landers, 1992: The Largest U.S. Earthquake in 28 Years

Many factors are really needed to define earthquake size adequately. In the last decade of the twentieth century, two powerful earthquakes with unexpected characteristics occurred in California. First, in 1992, a long series of faults, mapped in the remote Mojave Desert of California on the state geological map of major faults, suddenly slipped, causing strong shaking over much of southern California and as far away as Denver, Colorado. The surface-wave magnitude of the principal tremor was measured to be 7.5 and the moment magnitude 7.3.

The greater event struck on Sunday, June 28, at 4:58 a.m.; 3 hours later, the second earthquake (M_S = 6.5) was generated by a separate fault about 45 kilometers from the first, near the town of Big Bear. Right-lateral surface ruptures occurred on an *en echelon* series of strike-slip faults with step-over of the slip at the end of segments. The fault rebound progressed in this segmented way from the hypocenter to the northeast in a unilateral direction. These faults were known to have slipped in Quaternary time and some even in this century. What was surprising was that the 1992 rupture continued through three mapped fault segments that were not believed to be parts of a single throughgoing fault. In this light, the large size of this seismic source was not predicted.

Surface slip ranged from an average of about 2 meters near Landers (see Figure 1.2) to as much as 5.5 meters along the northwestern part of the rupture (see the figure facing the start of Chapter 5). A surprise was the substantial vertical ground offset along limited parts of the main fault rupture, with scarps amounting to 1 meter at fault bends.

The epicenter was located between the towns of Landers and Yucca Valley, approximately 30 kilometers northeast of the San Andreas fault zone. The area is lightly populated, but high felt intensity was reported. Mr. Jerry Gobrogge lost the side wall of his bowling alley in Yucca Valley. Soon after he described the motion: "It was terrible. It was just terrible. It never quit, it just kept shaking and it hasn't stopped." This earthquake, officially named the Landers earthquake, was in fact the largest earthquake in California since the often-cited seismic event in Kern County in 1952. It resulted in 1 death and 25 serious injuries. More than 77 homes were destroyed and 4300 damaged, with losses estimated at about $50 million.

There were no significant surprises in the response of the relatively few buildings in this desert area to such a large earthquake — an indication of improvements in building codes and practices. Some structures astride or adjacent to the extensive fault-related ground disruption were, as might be expected, structurally impaired. Generally, however, the ground shaking produced widespread but mainly nonstructural damage and damage to building contents.

In agreement with Mr. Gobrogge's impression, the Landers earthquake was notable for its long duration, consistent with its high magnitude as discussed on page 159. The strong shaking lasted on some instrument records for 30 seconds or more at epicentral distances of 20 to 30 kilometers. The amplitudes of acceleration of the ground were generally as expected from past earthquakes, with the highest recorded horizontal acceleration reaching 0.86g at a site near the fault rupture.

Upthrow

In earthquake engineering, recent studies suggest that certain structures respond significantly to vertical shaking of the ground, and that design methods that take into account only the horizontal motion are deficient in important ways. The vertical component appears to be particularly significant in calculating seismic designs for massive dams and foundations of such surface structures as pipelines. In a number of earthquakes, strong-motion accelerograms have indicated that ground accelerations (both vertical and horizontal) at the basement level of buildings are magnified by considerable factors in the upper floors of the buildings. For example, in the March 22, 1957, San Francisco earthquake (Richter magnitude 5.3), accelerograms in a building in San Francisco showed an amplification in the maximum vertical acceleration of a factor of 3 between the basement and the fourteenth floor.

One measure of extreme vertical shaking is the "upthrow" of objects. Reports of such upthrow in large earthquakes, from around the world, have depended upon field observations after the shaking. If the reports are true, do they indicate that earthquake accelerations in the vertical direction exceed the acceleration of gravity? In fact, instrument measurements of vertical ground accelerations greater than gravity have been obtained. For example, near the source of the Gazli earthquake ($M_S = 7.0$) in the Uzbek Republic of the former Soviet Union, on May 17, 1976, a vertical acceleration of 1.3g was recorded.

One of the earliest dramatic accounts of upthrow is by R. D. Oldham* on the great Assam, India, earthquake of 1897. He reports that loose stones were tossed in the air at Shillong and elsewhere, "like peas on a drum." All the available reports indicate that particularly violent shaking occurred. People were thrown to the ground and injured by the shock. Boulders were displaced, leaving cavities in the earth where they had lain. The ejection was so abrupt that the sides of the cavities were almost unbroken. There were also stories of posts coming out of their holes without cutting the edges of the surrounding soil.

Oldham also writes about the disturbance of the surface soil:

> In the western portion of the southern spur, and all around the civil surgeons' quarters to the distance of a mile down the Nankachar Road where the soil is sandy and the surface fairly level, the ground looks as if a steam plow had passed over it, tearing up the turf and throwing the clods in every direction, some uphill and some down, and in many cases, turning the sods completely over so that only the roots of grass are visible.

Some Japanese earthquakes have also produced strong upward motion. In the two Imaichi earthquakes (Richter magnitude about 6) of December 26, 1949, diverse pieces of evidence are mentioned. Near Ochiai village, close to the source, a stone implement called an *ishiusu*, about 50 centimeters in diameter, was said to have been tossed upward about 20 centimeters a few times, like a rebounding rubber ball. In nearby Imaichi, published studies report that although objects upon the shelves did not fall, iron kettles were thrown off their hooks.

A particularly interesting observation on dynamical systems comes from reports of the Kwanto earthquake of September 1, 1923 (compare Chapter 6). The disturbances in alluvial material were particularly marked; in fact, near Manazuru Point, the soft ground was so shaken that potatoes were extruded onto the ground. Nearby, large trees sank in the soft soil

**In 1906, Oldham discovered the Earth's massive core using earthquake waves (see Chapter 4).*

until only their tops were visible. No doubt liquefaction of the sandy soil was a dominant factor (see Chapter 11).

In California earthquakes, too, vertical separation of objects has occurred. The rupturing in the 1971 San Fernando earthquake was thrust faulting, and it produced surface fault offsets as great as 3 meters. This faulting was an efficient generator of seismic waves, because large ground accelerations were recorded across a fairly wide area. At the Los Angeles County Fire Station 74 in North Dexter Park, San Fernando, a fireman on duty was tossed out of bed onto the floor and the bed fell on top of him. The receiver of a standard wall phone came off its hook and every object in the building was upset. The building was shifted off its foundations and outside "rocks were thrown off the ground and large cracks appeared in both soil and rock."

At the same station, the movements of a fire truck — which must have been due at least in part to the elastic springing system of the vehicle — were also recorded by B. J. Morrill:

> A 20-ton fire truck enclosed in the garage moved 6 to 8 feet fore and aft, 2 to 3 feet sideways without leaving visible skid marks on the garage floor. The truck was in gear and the brakes were set. Marks which appear to have been made by the right rear tire were found on the door frame, three feet above the floor, while the metal fender was not damaged. The fender extends several inches out beyond the upper portion of the tire. Four feet above the floor, the hose rack was broken by the rear step of the truck. The step was bent up while the hose rack was broken downwards.

Even more striking evidence of vertical separation was the displacement of the building in which the firefighters were housed. As Figure 8.5 shows, the building was displaced from the foundations in such a way that the bottom row of shingles — which, prior to the earthquake, overlapped the foundations by 10 centimeters — was undisturbed. This evidence is hard to explain as sliding of the building, but rocking, perhaps in a large ground displacement pulse, remains an alternative cause.

Disturbed surface soil was also observed after the 1971 earthquake in several places. For example, along flat ground at the top of one ridge, the surficial soil was considerably shattered, giving the appearance of plowed land. One explanation is that the soil was overturned during vertical accelerations that exceeded 1.0g during several cycles of the seismic waves.

Similar reports had been published already from field studies of numerous other earthquakes around the world. Although some individual reports can be discounted, apparently certain types of objects do separate vertically

Figure 8.5 Northwest corner of the quarters building of the Los Angeles County Fire Station in Kagel Canyon, San Fernando, after the 1971 earthquake. Note that a line of shingles that had overlapped the foundation is undamaged. [Photo by B. J. Morrill, 1972; courtesy of USGS.]

and rotate (see Figure 7.1) in earthquakes, indicating that the acceleration of at least the localized motions exceeds that of gravity. It is possible that one explanation is not sufficient for all phenomena reported; certainly, a significant proportion of apparent vertical motions and offsets (for example, of gravestones) can be explained as the consequences of rocking and rotation set up by the seismic waves.

9 Volcanoes, Tsunamis, and Earthquakes

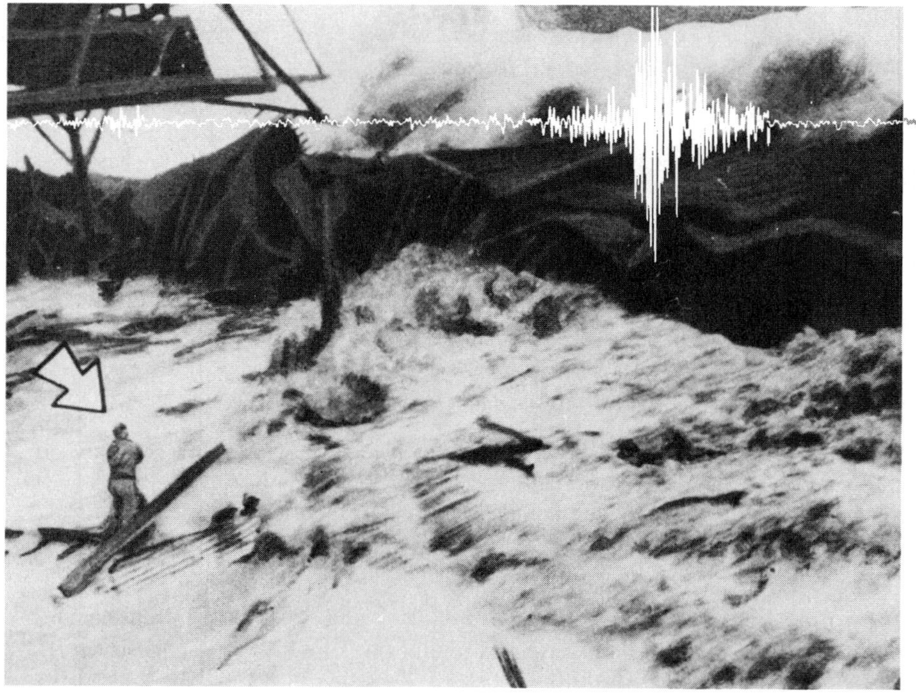

Seismic wave crashing over the pier at Hilo, Hawaii, in the great 1946 tsunami generated by an earthquake in the Aleutian Islands. Note the man in the path of the waves. He was never seen again. Photograph by an unknown seaman on the S.S. *Brigham Victory*. [Courtesy of R. L. Wiegel.]

For nearly forty million years the first island struggled in the bosom of the sea, endeavoring to be born as observable land. For nearly forty million submerged years its subterranean volcano hissed and . . . spewed forth rock, but it remained nevertheless hidden beneath the dark waters of the restless sea . . . a small climbing pretentious thing of no consequence.

— *James A. Michener, Hawaii*

Earthquakes and volcanoes often, but by no means always, accompany each other in certain tectonic regions, particularly midoceanic ridges and near deep ocean trenches (see Figure 9.1) as explained in Chapter 7. At the *subduction* zones, that is, along the trenches, the mechanical link is the downward movement of the lithospheric plates into the Earth. This movement is illustrated in Plate 13. As the surface rocks bend and thrust downward at a deep trench, they become strained and finally fracture, thus producing earthquakes.

At the same time, the temperature rises and local melting of the dipping plate occurs; the chemical composition of the rocks changes, and a molten fraction of the rocks makes its way to the surface, where it may be stored for a time in *magma chambers* — huge reservoirs underneath volcanic vents. From these pressure chambers, magma moves upward from time to time to issue forth as lava. In contrast, at places where two tectonic plates are moving together along a transform fault (such as the San Andreas fault) or collision zone (such as at the Himalayas), volcanoes are usually absent (see Figure 2.6).

Earthquakes and volcanoes are equally hazardous. For example, some of the extreme volcanic hazards of the world are posed by the stratovolcanoes of the Antilles Islands arc (see Figure 9.1), where subduction separates the North America and Caribbean tectonic plates. In 1902, an infamous explosive eruption of Mont Pelée struck the island of Martinique, producing a glowing avalanche of hot gases, steam, and rock debris (called a *pyroclastic* flow), and causing over 28,000 deaths. There were no large earthquakes before the event to warn of the likely coming of the violent blasts that plunged the city of Saint Pierre into darkness and rained scalding mud and hot gas on the houses. Many explosive episodes followed for over a year, producing widely felt shaking.

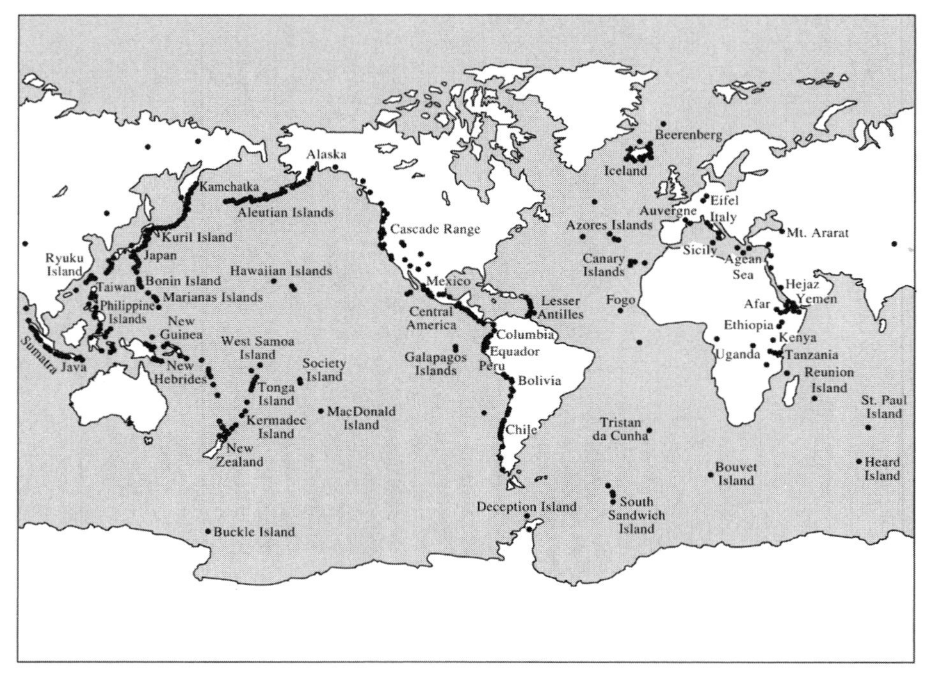

Figure 9.1 Volcanic activity of the Earth. Dots show the locations of volcanoes known or believed to have erupted within the past 12,000 years, as determined by radiocarbon dating, geological evidence, and other techniques. Each dot represents a single volcano or a cluster of volcanoes. [Data originally compiled by the Smithsonian Institution, Washington, D.C. Based on a map published in 1979 by World Data Center A for Solid Earth Geophysics at the Geophysical and Solar-Terrestrial Data Center, National Oceanic and Atmospheric Administration, Boulder, Colorado.]

In the hope of providing early warning of these catastrophic eruptions, a seismographic network has been operated in the volcanic islands for decades. The detection in 1979 of earthquake swarms under La Soufriére volcano on the island of Saint Vincent was a crucial ingredient in the decision to evacuate the threatened population. (*Soufriére* is a name given to several volcanoes that give off sulfurous gases.) As a result, when the 1979 eruption began it caused few casualties. More recently, the Soufriére Hills vents of Montserrat began explosive eruptions in July 1995. Like many, but not all, such hazards this volcanic crisis began with precursor seismic activity and this increased in intensity from 1992 to late 1994. In late 1996, about one-third of the volcanic dome collapsed, followed by huge flows of lava and an ash column that rose to 14,000 meters. The pyroclastic and lava flows in 1996 and 1997 devastated a heavily settled part of this popular resort island. Fortunately, the population had been evacuated well before on the basis of continuous monitoring at the Montserrat Volcano Observatory of the seismicity, geochemical changes, and deformation.

Volcanism also occurs within the tectonic plates far from plate boundaries, such as at Yellowstone National Park in Wyoming and, somewhat differently, in Iceland. The upwelling of magma at such places appears to be the result of fissuring or fracturing of the solid lithosphere as it progresses slowly across the mantle. Very much like the vertical plume of hot gas, steam, and ash from a factory smokestack on a still day or the eruptive plume of a volcano, so buoyant molten rocks move upward into the rifts in the plates, forming *hot spots* with fixed roots in the mantle below the lithosphere. The Hawaiian Islands chain is a prime example of this intraplate volcanism. The track of the islands extends progressively in age from the very active young volcanoes on Hawaii, in the north, westward to the long-extinct volcanoes of Midway. The particular location of intraplate volcanoes generally limits the size of earthquakes associated with this type of tectonic activity. Nevertheless, many small earthquakes, and sometimes quite substantial ones, result from the eruptive forces. The hot spot engine explains some of the isolated clusters of earthquakes discussed on page 31.

Eruptions in Hawaii

Let us look more closely at earthquakes and volcanoes of the Hawaiian archipelago type, where there is no present evidence for a downgoing tectonic slab (see Figure 7.2). The most severe earthquake to strike the Hawaiian Islands since 1868 occurred in the early morning (4:48 a.m., Hawaiian Standard Time) of November 29, 1975. The focus was about 5 kilometers below the surface, and the epicenter was approximately 45 kilometers south of Hilo, on the big island of Hawaii's southeastern coast (see Figure 9.2). About 1 hour earlier, a smaller, precursor earthquake, with nearly the same focus, had shaken the vicinity.

Two companion events occurred. The fault movement under the ocean that had caused the earthquake also generated damaging ocean waves, or tsunamis (see the section on page 180). These sea waves ran up to heights of 6 or more meters at isolated localities on Hawaii and inundated camp sites at Halape Beach where 36 people were spending the Thanksgiving holiday weekend. At least one man was killed and others

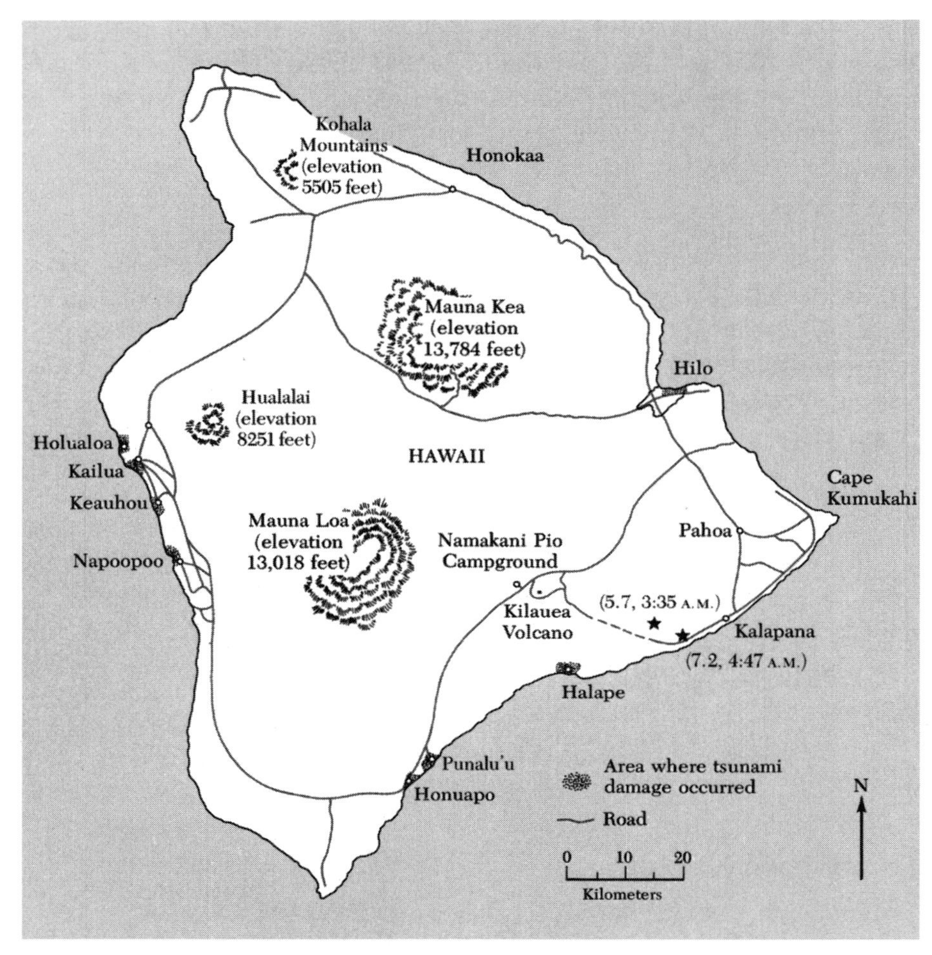

Figure 9.2 Map of the island of Hawaii showing Kilauea and Mauna Loa volcanoes. The stars denote the epicenters of the earthquakes of November 29, 1975. The magnitude and time of occurrence for each are given in parentheses. [Courtesy of USGS.]

were hospitalized with minor injuries. (The tsunami should not have come as a surprise. An earthquake in the same region on April 2, 1868, was followed by a tsunami that ran up almost 3 meters high at Hilo.)

The second striking event, the same day, was an eruption of Kilauea volcano. Lava erupted on the floor of Kilauea's caldera less than an hour after the main earthquake. Lava issued from fissures, and lava fountains rose as high as 50 meters into the air. New fissures also opened on the east wall of Halemaumau crater within the caldera. By the following morning,

November 30, the volcanic activity had ceased, less than 18 hours after it began.

The earthquake itself produced significant damage on Hawaii amounting to more than $4 million. Roads were cracked, and some were rendered impassable as a result of rock slides and slumping. Electrical power was cut off in some areas. In Hilo, the urban center closest to the earthquake source, light-to-moderate damage from shaking was prevalent. Plate-glass windows were broken in some old wood-frame shops, furniture was shifted or overturned, and a few older homes partially collapsed. Fortunately, strong-motion seismographs were available to record for posterity the actual ground shaking during the earthquake; these showed that strong seismic waves lasted about 14 seconds.

This conjunction of three types of energy release — earthquake, tsunami, volcanic eruption — on Hawaii in 1975 provides us with a context for exploring the links between them.

Volcanic activity and local earthquakes occur together in two ways. First, often before, during, and just after an eruption, minor seismic events — called *volcanic earthquakes* — increase in the vicinity of the volcano. Some kilometers below the volcanic vent, very hot viscous magma moves sluggishly under great steam pressure through a network of tubes and pipes from one storage chamber to another. As this motion takes place, various parts of the surrounding rock become hotter and more strained as the magma pushes through them. These forces fracture the neighboring rocks, and strain is relieved by the elastic rebound mechanism discussed in Chapter 6.

Sometimes, fault rupture precedes the motion of magma and eruption of lava. As in the episode of November 1975 in Hawaii, earthquake waves from a nearby rupturing fault may shake up the molten material in the storage reservoirs beneath the volcano. In a way similar to the violent shaking of a bottle of soda pop, steam and gas — which previously dissolved in the magma — may then begin to boil off, forming bubbles of superheated steam that accelerate the escape of the lava from surface tubes and then escape as gaseous material. In turn, this release of superheated steam and gas disturbs the unstable equilibrium of the magma below the vent, thereby producing further flow in the subterranean tubes and the stimulation of local volcanic earthquakes.

As can be imagined, this mixture of properties — local strain in the rocks, mobile magma, and dissolved gases — provides many mechanisms for diverse seismic-energy release. Sometimes the earthquakes recorded in a volcanic environment look ordinary with separate P- and S-wave onsets (see Chapter 1). Sometimes there is no distinct S-wave part, as would be the case when the seismic waves encountered liquid zones. A very distinct seismic result of volcanic activity is a more-or-less continuous shaking of the ground that persists for hours or days. This *volcanic tremor* can usually only be detected very near the vents but can sometimes be felt nearby as high-frequency harmonic trembling of the ground, as I remember well from a field study of the 1965 eruption of Mount Taal in the Philippines. These volcanic harmonic tremors are useful in forecasting probable imminent eruptions.

Volcanic Hazards

Near active volcanoes, the population is subject to the definite danger of lava flow onto fields and property or, in some areas, of cataclysmic ejection of gas and superheated water, mud, and other materials. As with any geological hazard, each individual must decide whether the risks posed by a nearby volcano are reasonable ones. On Hawaii, most people believe that the risks are not excessive and are no worse than the natural hazards of other regions of the world — such as tornadoes. Although volcano insurance is available in some parts of the world, it is usually expensive.

Volcanic hazards have two mitigating features. First, even though lava may cover the land surface, the land can be used again, at least for certain types of agriculture, after a few decades. Also, when a volcano begins to erupt, warning signs often occur soon enough to allow evacuation from the threatened area. To be sure, specific prediction of a damaging eruption is not yet often possible and may never be common in a practical sense. Nevertheless, there are often clues of an impending eruption: rising water temperature in fumaroles and the changing composition of erupting gases; also, deformation of the ground surface around volcanoes sometimes precedes eruptions. One possible explanation is that the swelling occurs when the magma reservoirs beneath the vents are filling, indicating that a lava flow is likely to occur. Seismographs have also been tried as a predictive tool, because networks of these instruments around active volcanoes sometimes detect significant changes in background seismicity. For example, sometimes the foci of earthquake swarms migrate to shallower depth, perhaps indicating the movement of magma upward.

This predictive idea is not without verification: one example is a 1974 eruption of Mauna Loa on Hawaii. Workers at the Hawaiian Volcano Observatory had detected signs of increased restlessness beginning in April, when there was a marked increase in the number of small earthquakes near the summit of the big volcano. Subsequently, little change in the earthquake frequency occurred until July 1975. Then, the recorded seismicity noticeably increased and became shallower. A glow was reported from the erupting lava before midnight on July 5, and by the early morning, observers in aircraft saw a line of red-hot fountains extending along the summit caldera. Mauna Loa had ended a 25-year period of dormancy, since its last great eruption of June 1950. The eruption lasted 2 days and covered approximately 13.5 square kilometers of land with new lava. Although a warning was issued to residents that might be threatened, the lava flows did not reach developed areas.

Another extraordinary example of the efficacy of this application of geodesy was associated with the devastating eruption of Mount Pinatubo in 1991 in the Philippines. Explosive eruptions of this dangerous, but long dormant, subduction-zone volcano had occurred about 500, 3000, and 5500 years ago. Seismic swarms, felt locally in March, led to seismographic and geodetic monitoring. By May and June, larger and shallower earthquakes were occurring with large emissions of gas and ash. There was thus growing evidence for a dramatic finale. By June 15, the majority of people had been convinced enough to evacuate, avoiding the climactic eruption of an ash cloud 400 kilometers wide and 34 kilometers high, with pyroclastic flows spreading over 15 kilometers from the summit. There is little doubt that the 600 people killed would have been magnified manyfold but for the seismographic evidence.

Explosion of Mount Saint Helens, 1980

A violent volcanic eruption occurred in the United States at Mount Saint Helens, Washington, on May 18, 1980 (see Figure 9.3). It was one of the largest natural releases of energy since the catastrophic volcanic eruption at Krakatoa in 1883.

Mount Saint Helens belongs to the row of towering volcanic peaks that extends along the Cascade range from northern California to southern Canada. These volcanic cones consist of alternating layers of tephra (ejected fragments of older rock) and lava flows, producing what are called *composite volcanoes.* At their summits is a *caldera* — a crater produced by rock collapse after a great eruption.

In their mature stages, composite volcanoes erupt explosively, throwing out many cubic kilometers of glassy ash and pumice. Such cataclysms are like the one at Vesuvius in 79 A.D., in which the Roman scholar, Pliny the elder, died; thus they are called Plinian eruptions. They are potentially a great hazard.

Figure 9.3 The eruption of Mount Saint Helens (2950 meters high), Washington, USA, May 18, 1980. The eruption cloud rose to 20 kilometers. [Courtesy of USGS.]

For many decades, volcanologists had been aware of the hazards from the Cascade volcanoes; predictions based on geological volcanic deposits along the range indicated that there might be a large eruption before the twenty-first century. In March 1980, an earthquake swarm near Mount

Saint Helens began a sequence that led to the giant eruption 2 months later. By the end of March, seismographs in the vicinity recorded swarms of hundreds of earthquakes each day. By mid-April, a bulge was observed developing on the snow-covered summit, and it continued to enlarge through early May 1980. These events were interpreted as the emplacement of a large volume of magma near the top of the volcano. As a result, the forest and resort regions on the north flank of the volcano were evacuated.

The sequence of events leading to the most violent phase of the eruption started just after 1530 Universal Time on May 18, 1980, with a small earthquake (magnitude 5.5) at a focal depth of about 2 kilometers beneath the mountain. This ground shaking triggered a massive landslide on the north slope of Mount Saint Helens, releasing about half a minute later a powerful blast of superheated steam and rock northward and upward, as evident from Figure 9.3. The cause of the eruption was probably the unloading of the overburdened rock mass by the landslide and the consequent exposure of a magma or hydrothermal reservoir. A few minutes later, a second but vertical eruption occurred in the north part of the caldera before the first one was fully developed. This eruption was accompanied by another earthquake.

The extent of the blast zone in the first and lateral eruption has been measured from the devastated area in which all fir trees were blown over. It covers a nearly semicircular northward section of the mountain, 30 kilometers across from west to east and extending outward 20 kilometers from the peak. The eruption killed 65 people, one of whom was a professional volcanologist who was observing gas flow and rock deformations at his post about 10 kilometers from the summit.

The blasts strongly disturbed the atmosphere: airwaves spread out as very long pressure waves around the Earth. At some weather stations and observatories, the instruments, called barographs, recorded the fluctuations in air pressure in the waves; these measurements have been used to compute the rapidity of the eruption.*

Figure 9.4 shows the recorded airwaves that followed the direct path southeast from Mount Saint Helens to Berkeley, California, followed by the waves traveling the long path northwest completely around the Earth. The onset of the wave around the antipodes (the far side of the Earth) appeared 33 hours later, corresponding to an air speed of 314 meters per second.

**In 1963, I installed at the seismographic station at the University of California, Berkeley, a special barograph to measure the air-pressure fluctuations that might be generated in very large earthquakes. This barograph could also record the atmospheric-pressure variations after volcanic eruptions.*

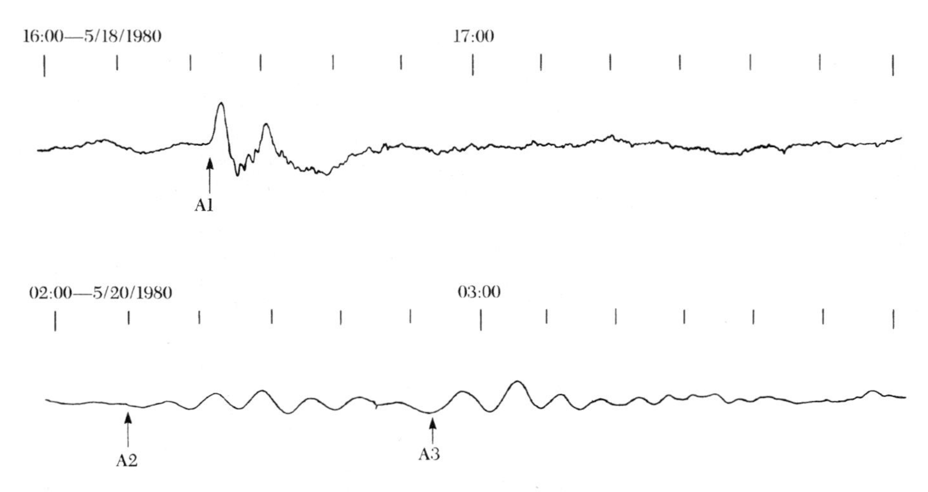

Figure 9.4 Fluctuations in air pressure at Berkeley after the Mount Saint Helens eruption, May 18, 1980. The top record shows the direct (A1) atmospheric waves and the bottom shows the two overlapping wave trains (A2 and A3) through the antipodes.

Mammoth Lakes Earthquake Swarm, California

Another well-studied association between volcanic activity and earthquakes is along the eastern front of the Sierra Nevada, California. Here is one of the most spectacular mountain areas of the world. The geological processes that created this scenic beauty have operated for millions of years and are still ongoing today. This mountain range — a huge mass of the Earth's crust, 650 kilometers long and over 30 kilometers wide — has broken free along the great fault system on its eastern front and has tilted westward. Notable along this fault system is the recent volcanism in the Long Valley region, within which is situated the popular ski and holiday resort of Mammoth Lakes, Mono County (see Figure 1.2).

In May 1980, three earthquakes of magnitude 6 or greater occurred in the vicinity of Mammoth Lakes. By August 1, a sequence of over 600 small earthquakes (see Figure 9.5) had shaken the region. During the previous year, several earthquake swarms had occurred along the southern boundary of the Long Valley caldera, and this had prompted seismologists to install seismographs and other geophysical devices in the area.

Although the larger earthquakes caused a relatively minor amount of damage from shaking in the town, they led to widespread concern that the seismic sequence heralded a flareup of major volcanic eruptions in

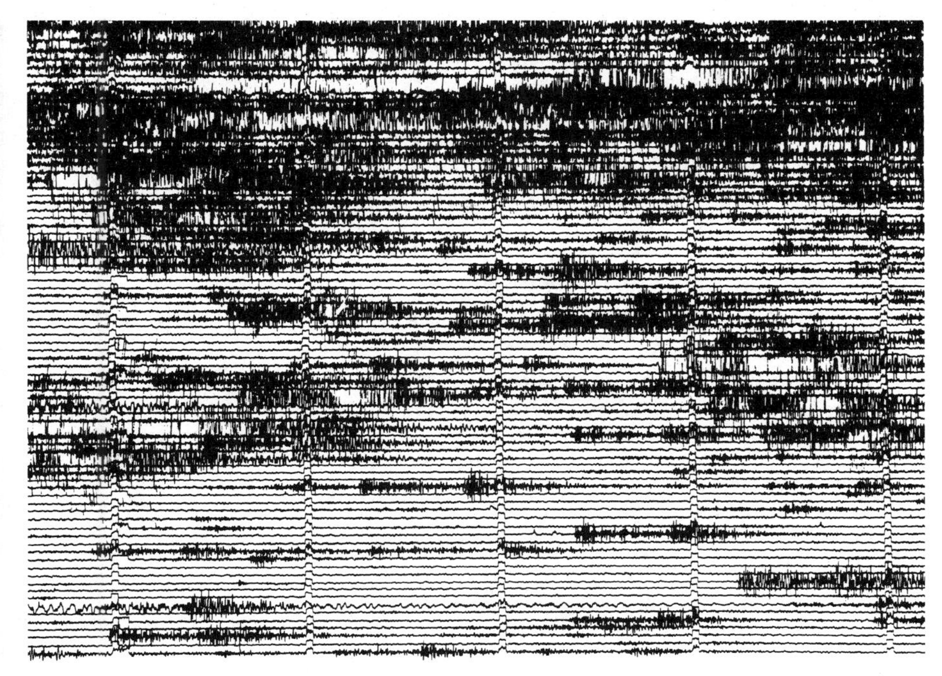

Figure 9.5 Seismogram recorded at the University of California station at Priest, California, showing a swarm of hundreds of microearthquakes from the Mammoth Lakes area on May 26, 1980. The epicentral distances are about 300 kilometers, and the time between trace offsets is 1 minute.

the Long Valley caldera.* Because of limited access roads to the town of Mammoth Lakes, such sudden violent eruptions could constitute a grave risk. Consequently, steps were taken by government groups to monitor closely the geological conditions in the Mammoth Lakes area with the aim of building a geological model that could both explain and predict the tectonic events. People asked: Was the 1980 seismic episode a forerunner of a future eruption?

The region of Long Valley and adjacent Mono Basin is one of four areas in California for which there is well-documented evidence of multiple volcanic eruptions in the past 2 millennia. To begin with, the Long Valley caldera itself was formed by a gigantic eruption about 760,000 years ago, involving ejection of magma with total volume of about 600 cubic kilometers. Such colossal eruptions are infrequent. None have occurred

**This led to a heightened nationwide awareness of volcanic hazards. The cataclysmic eruption at Mount Saint Helens occurred in the same month.*

during the written history of humankind, but study of age-dated volcanic ash in the area indicates that at least 20 regional eruptions have occurred in the last 2000 years.

The most recent volcanic eruption in the region occurred on August 23, 1890, witnessed by only a few people in this remote place. Earthquakes occurred at the same time. A quotation from the local newspaper, *The Homer Mining Index*, gives a colorful perspective:

> Remarkable earthquakes at Mono. The southern end of Mono Lake was considerably agitated last Sunday and dwellers in that shaky locality were much perturbed. Steam was issuing from the Lake as far as could be seen, in sudden pops, and the water was boiling fiercely, while high waves rolled upon the beach and receded, leaving the sand smoking. In a moment, the air was thick with blinding hot sulfurous vapor, and subterranean moans and rumblings made the witness think that the devil was holding highjinks down below. Fences wobbled up and down and sideways.

In 1980, to help with modeling the tectonic process, measurements were made of the tilt and elevation of the ground in the vicinity of Long Valley. The foci of many aftershocks were pinpointed in relation to faults in the area by a local seismograph network. Ground rupture was mapped east of Mammoth Lakes along a known fault and its northwestern branches in a zone 20 kilometers long. Aerial photographs showed significant zones of cracking and ground failure that coincided with other faults. As well, there was landsliding along zones of weakness on the steep slopes and older scarps. For a time there was speculation that an episode of dangerous volcanic activity was likely, causing concern among local inhabitants. But earthquake frequency and steam vent activity began to diminish and, by 1985, the Mammoth Lakes area was again quiet.

Almost a decade went by quietly until in 1989 and again in 1997 the geophysical scenario repeated with many microearthquakes and uplift of the land surface. Portable seismographs have been deployed during these seismic episodes to map the extent of the magma chamber and the deformation processes. One thing is sure, because of the ongoing tectonic deformation, structures should be built in this area to resist substantial earthquakes and with appropriate concern for volcanic hazards.

Tsunamis

Along seacoasts, another disaster may follow large earthquakes. The sudden offset of a major fault under the ocean floor shoves the water as if it were

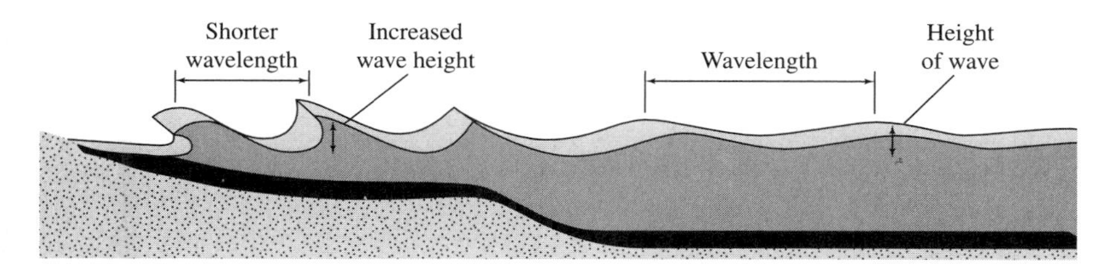

Figure 9.6 Terminology used in describing characteristics of tsunamis (seismic sea waves).

being pushed by a giant paddle, producing powerful water waves at the ocean surface. These water waves spread out from the vicinity of the earthquake source and move across the ocean until they reach a coastline; and sometimes they run up on land and in rivers many hundreds of meters (Plate 15). Just offshore, their height sometimes increases greatly, as shown in Figure 9.6, and they surge and run up on the shore with disastrous effects, as shown in Figure 9.7. These are the "great waves" that Charles Darwin wrote about (see Chapter 1).

Figure 9.7 Aftermath of a 1983 tsunami that killed 104 people near Minehama, Japan. [Courtesy of Japan Meteorological Agency.]

As mentioned previously, a common way to refer to these long water waves in English is "tidal waves." The name is not accurate. They do not arise from the attraction of the moon and the sun on the water of the ocean. In Spanish, the word *maremoto* fits neatly, but, for want of a better word, the names *seismic sea wave* and *tsunami* are used by scientists in English. Use of the Japanese word is particularly appropriate because the islands of Japan have suffered greatly from the destructive effects of tsunamis, some of which have come from as far away as South America.

Box 9-1

NOTABLE TSUNAMIS OF THE WORLD

Date	Source region	Visual run-up height (meters)	Location of report	Comments
1500 B.C.	Santorini eruption		Crete	Devastation of Mediterranean coast
Nov. 1, 1755	East Atlantic	5–10	Lisbon, Portugal	Reported from Europe to West Indies
Dec. 21, 1812	Santa Barbara Channel, Calif.	Several meters	Santa Barbara, Calif.	Early reports probably exaggerated
Nov. 7, 1837	Chile	5	Hilo, Hawaii	
May 17, 1841	Kamchatka	Less than 5	Hilo, Hawaii	
April 2, 1868	Hawaii Island	Less than 3	Hilo, Hawaii	
Aug. 13, 1868	Peru-Chile	More than 10	Arica, Peru	Observed in New Zealand; damage in Hawaii
May 10, 1877	Peru-Chile	2–6	Japan	Destructive in Iquique, Peru
Aug. 27, 1883	Krakatoa eruption		Java	Over 30,000 drowned
June 15, 1896	Honshu	24	Sanriku, Japan	About 26,000 people drowned
Feb. 3, 1923	Kamchatka	About 5	Waiakea, Hawaii	
March 2, 1933	Honshu	More than 20	Sanriku, Japan	3000 deaths from waves
April 1, 1946	Aleutians	10	Wainaku, Hawaii	159 deaths in Hawaii, 5 deaths in Alaska

One of the most massive tsunamis in history hit the eastern, or Sanriku, coast of Honshu following a great earthquake centered out at sea on June 15, 1896. The best guess is that this earthquake was produced by a wide area of ocean floor rebounding up or down along a submarine fault of the Japan trench. The seismic sea wave washed onto nearby land as much as 25 to 35 meters above high-tide level. Entire villages were engulfed. More than 10,000 houses were washed away, and 26,000 people were killed. To the east, the tsunami waves spread across the Pacific Ocean (see Figure 7.2)

NOTABLE TSUNAMIS OF THE WORLD (Continued)

Date	Source region	Visual run-up height (meters)	Location of report	Comments
Nov. 4, 1952	Kamchatka	Less than 5	Hilo, Hawaii	
March 9, 1957	Aleutians	Less than 5	Hilo, Hawaii	Associated earthquake magnitude 8.3
May 23, 1960	Chile	More than 10	Waiakea, Hawaii	
March 28, 1964	Alaska	6	Crescent City, Calif.	119 deaths in Alaska and Calif. and $104 million damage from tsunami
Nov. 29, 1975	Hawaii Island	About 4	Hilo, Hawaii	
May 26, 1983	Honshu, Japan	14	Minehama, Honshu	104 deaths along the western coast of Japan; associated earthquake magnitude 7.7
July 12, 1993	Hokkaido, Japan	20	Okushiri Island	Over 200 deaths; magnitude 7.8
June 3, 1994	Java	11	Malang	222 people killed; magnitude 7.2
Jan. 1, 1996	Indonesia	5	Sulawesi	9 deaths; magnitude 7.7
July 17, 1998	Papua, New Guinea	More than 10	Aitape	About 2500 deaths Local earthquake source magnitude 7.0

and were recorded at Hilo, Hawaii, with an amplitude of 3 meters. The waves then continued to the American coast, where they were reflected back toward New Zealand and Australia. The devastation was repeated on March 2, 1933, when a tsunami, with crests as high as 25 meters, again washed ashore along the Sanriku coast, killing about 3000 people.

References to the devastation of tsunamis can be found throughout recorded history. The earliest description is of a damaging sea wave near the north end of the Aegean Sea in 479 B.C. In the ensuing centuries other tsunamis occurred along the coastal plains and offshore islands of the Mediterranean. Within the last 200 years of the modern era, about 350 tsunamis have produced fatalities. Details on some of the most devastating tsunamis are given in Box 9.1. In 1998, the most recent dreadful tsunami tragedy wiped out whole villages in New Guinea, killing over 2500 people.

A comparison of historical earthquakes and tsunamis shows that a large tsunami inundating a stretch of populated coastline is likely to be much more destructive than the shaking from all but exceptionally large earthquakes, or from moderate earthquakes in densely populated areas with poor construction — such as in the 1988 Armenia earthquake (see the description of the 1964 Alaska earthquake in Chapter 1). Only the really great earthquakes, such as the 1556 and 1976 earthquakes in China and the 1908 disaster in Messina (see Appendix A), cause as many casualties as the largest and most damaging tsunamis. On the average, about one major tsunami occurs each year somewhere in the world; they can occur in most oceans and seas but are particularly frequent in the Pacific, Indian, Mediterranean, Atlantic, and Caribbean. They have also arisen in large inland seas, such as the Caspian and Black seas in Central Asia.

While the evidence is that most great sea waves are caused by fault rupture with vertical displacement along a submerged fault, there are also other causes. One example is a submarine landslide, such as occurred in Sagami Bay in Japan in the devastating earthquake of 1923 (see Chapter 6). These underwater landslides may themselves be triggered by a nearby earthquake, as in the 1923 earthquake. Sometimes a landslide or avalanche of soil and rock from a mountain into a bay, a large lake, or even a reservoir can produce a local water wave that is deadly.

A famous landslide-induced sea wave occurred at Lituya Bay, Alaska, after a local large earthquake on July 9, 1958. Water waves rushed onto the opposite shores of the bay as far as 500 meters, stripping vegetation in their path. More recently, a giant water wave was produced by a landslide (not triggered by an earthquake) into the Vaiont Reservoir in Italy in October 1963. A large volume of water overtopped the Vaiont Dam by 100 meters and swept down the valley of the Piave River, killing almost

3000 people. Such incidents are warnings that towns and marinas located around lakes, bays, and reservoirs that may be affected by earthquakes and landslides should take defensive planning measures.

The only other known terrestrial source of great tsunamis is a major volcanic eruption.* The classical example was the wave following the collapse of the top of Krakatoa volcano in 1883, one of the most violent geological paroxysms in historical times. During the summer, numerous earthquakes and considerable volcanic activity had occurred on Krakatoa Island, in the Sunda Strait between Java and Sumatra in the East Indies, with its peak standing to a height of 2000 meters. At the end of August, a series of violent eruptions took place, with great masses of ejecta streaming out from the volcano vent. (By August 28, the cataclysm was essentially over; a total of about 16 cubic kilometers of ash and pumice had been ejected.) On August 27, the central vents caved in: where the island had stood there was now ocean water 250 meters deep. This sudden collapse produced an enormously energetic tsunami. The wave was not high enough in the deep water to sink ships present in Sunda Strait, but when it reached shallow water along the coast, it washed away 165 villages without a trace and killed more than 36,000 inhabitants. The wave height was said to exceed 35 meters along the shore, and when it reached Port Alfred in South Africa later, it was still in excess of 30 centimeters. Eventually, it made its way around the African continent into the English Channel, where it was observed to have a surge of 5 centimeters.

In the open ocean, the distance between the crests of a tsunami may be greater than 100 kilometers, and the elevation is seldom more than 1 meter in height. Such waves themselves cannot be detected by ships at sea. As the water depth decreases, the speed of the waves slows down (see Box 9.2). When the tsunami approaches the shoreline, sometimes the water level will first fall, denuding beaches and leaving fish stranded. For example, in 1923, during this phase of a tsunami at Hilo, Hawaii, some persons were drowned by the first wave crest when they unwisely rushed onto the exposed flatlands to pick up the fish.

The local height of a tsunami is affected by the topography of the sea bottom and the continental shelf and by the shape of the shoreline. For example, in open and hook bays, the tsunami can cause the water level of

**The oceanic impact of a large comet or asteroid (bolides) would also produce one. As an extreme example, a giant bolide hit 65 million years ago near Yucatan, Mexico, producing a 45-kilometer radius crater, called Chicxulub, leaving widespread sedimentary deposits around the Caribbean. Some geologists hold that this impact terminated the age of the dinosaurs.*

Box 9-2

PROPERTIES OF A TSUNAMI

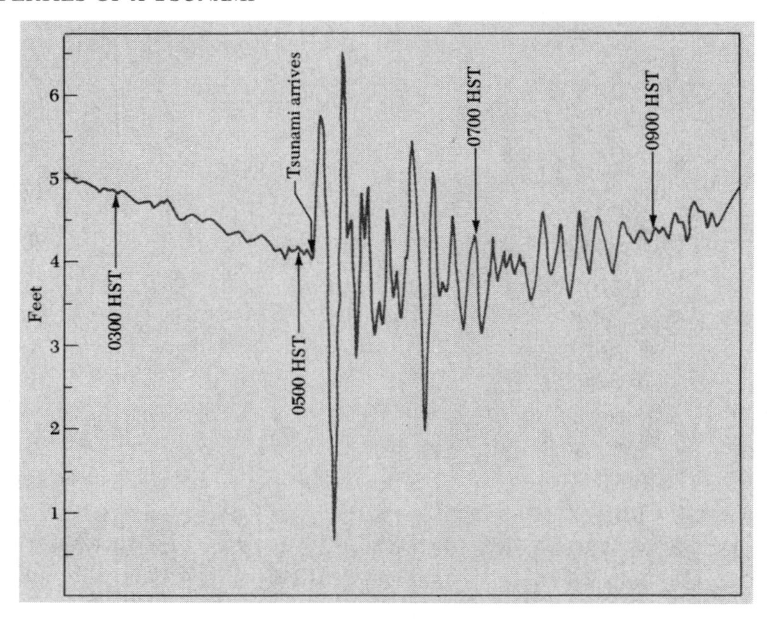

Record of a tsunami in Kuhio Bay, Hilo, Hawaii, on November 29, 1975. (HST = Hawaii Standard Time.)

The record shows the following properties of the water waves.

1. The onset was an upward swell of $\frac{1}{2}$ meter.
2. The water level then fell 1 meter (3.3 feet) below normal.
3. The period between tsunami crests was about 15 minutes.
4. The duration of the tsunami wave action was over 4 hours.

In the deep ocean of depth d, the speed v of a tsunami (long water wave) is equal to $\sqrt{gd}$, where g is the acceleration of gravity (980 centimeters per second squared). For

$$d = 5 \text{ kilometers}$$

$$v = \sqrt{980 \times 5 \times 10^5} \text{ centimeters per second}$$
$$\simeq 800 \text{ kilometers per hour}$$

For a period $T = 15$ minutes from (3) above, the wave length, λ, was:

$$\lambda = v \times T$$

$$= 800 \times \tfrac{1}{4} \text{ kilometers}$$

$$= 200 \text{ kilometers}$$

one side of the bay to rise dramatically, while the other side is sheltered and shows little change. Coastal regions — such as the Pacific side of the Japanese island of Honshu — that face the source of the tsunami usually suffer the highest run-up of the water, but the lee side of promontories and peninsulas can provide shelter. There has been some confusion on the names for tsunami inundation. The elevation above the tide level (at the time of the tsunami) reached by the water is called the *run-up height*. This vertical distance is not the same as the tsunami water-wave height offshore or the horizontal distance of water run-up from the normal water's edge.

The tsunami onrush is sometimes amplified in a bay or river mouth, producing an almost vertical wall of water, called a *water bore*. The photograph in Figure 9.8 shows the bore at Hilo, Hawaii, that formed during the onslaught of the tsunami produced along the deep Aleutian trench in the great 1946 Alaska earthquake. In the river estuary at Hilo a nearly vertical wave front, about 7 meters high, rushed across the estuary and churned across the roadway and bridge, finally reaching a power plant at the south end of the bay, where it short-circuited the electrical system, plunging most of the island of Hawaii into darkness.

Long-period movements of water can also be produced in lakes and reservoirs by large, usually distant, earthquakes, and sometimes by strong winds. As a classical example of the latter kind, it had been known for

Figure 9.8 Bore from a tsunami (of April 1, 1946) racing into the mouth of the Wailuki River, Hilo, Hawaii. Note that part of the bridge has already been destroyed— by an earlier wave of the same tsunami. [Plate 8, from G. A. Macdonald, F. P. Shepard, and D. C. Cox, "The Tsunami of April 1, 1946, in the Hawaiian Islands," *Pacific Science*, vol. 1, no. 1.]

centuries that Swiss lakes were apt to rise and fall rhythmically by a few centimeters. The duration of a complete oscillation of the Lake of Geneva, for example, is 72 minutes. In the late nineteenth century, the Swiss professor F. A. Forel made a systematic study of this type of water wave, which he called a *seiche* (pronounced *sāsh*). The term *seismic seiche* was coined by Anders Kvale in 1955 to describe oscillations of lake levels in Norway and England caused by the earthquake of August 1950 in far off Assam, India. More recently, the 1964 Alaska earthquake generated water oscillations in wells in the United States, 4000 kilometers away along the Gulf of Mexico.

Tsunami Alerts; the 1993 Hokkaido Disaster

The central coast of California and the northern coast of Europe are not as susceptible to tsunami damage as the coasts of Japan, Alaska, and South America. One reason is that no subduction zone is present to cause major thrust faulting (but compare the tsunami hazard in the Cascadia subduction zone of northwest contiguous United States described in Chapter 10). In California the motions along the San Andreas transform fault system are predominantly horizontal, so that where it is under the ocean (as off the Golden Gate Bridge at San Francisco) even *major* horizontal offsets, as in the 1906 earthquake, do not push the seawater enough to generate a seismic sea wave of any consequence. Nevertheless, a tsunami generated by an earthquake source far away can strike almost any coast and can be severe enough to cause heavy damage.

The Seismic Sea Wave Warning System was set up in the Pacific after the devastating Aleutian tsunami of April 1, 1946, in order to reduce the danger from Pacific tsunamis. The system is international: evidence of a large earthquake is immediately cabled, radioed, or phoned in to the Tsunami Warning Center in Honolulu from earthquake observatories in the United States, Japan, Taiwan, the Philippines, Fiji, Chile, Hong Kong, New Zealand, Samoa, and elsewhere. Then information on variation in water level, obtained from tide gauges at various observatories in the region of the earthquake, is provided. On the basis of this information, tsunami alerts may be issued.*

**One inherent danger in a tsunami alert is people's high curiosity. For example, an alert was issued at Berkeley, California, during the 1964 Alaska earthquake. Afterward the police chief complained to me that not only did the tsunami alert cause local people to go to the waterfront to watch the wave come in — but some of his police did the same thing!*

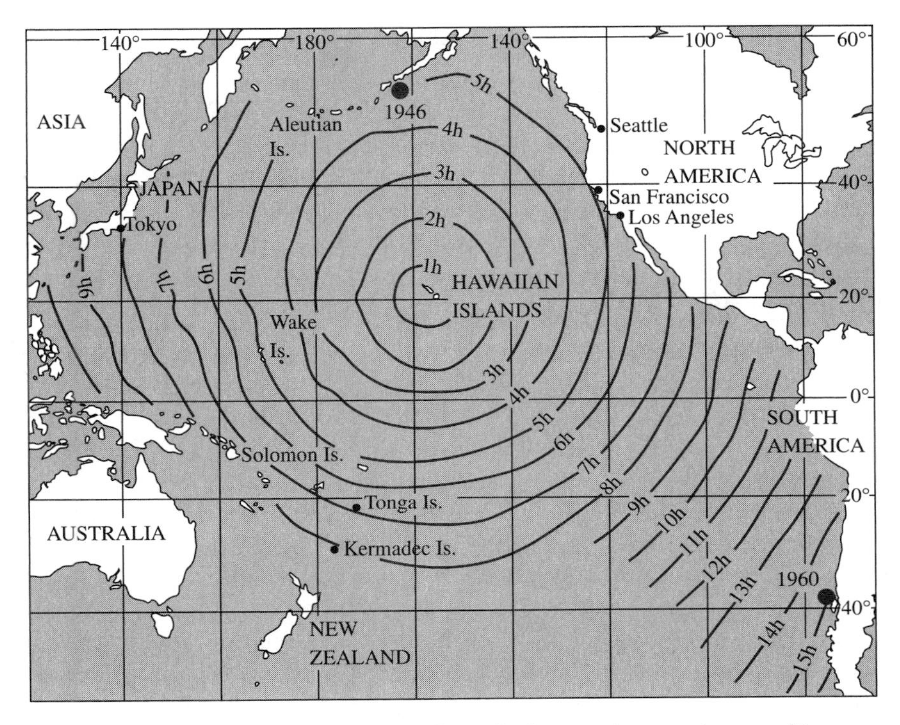

Figure 9.9 Contours showing the time of travel of tsunami waves from positions around the rim of the Pacific Ocean to the Hawaiian Islands. The large dots mark the earthquake source origins of the disastrous tsunamis of 1946 and 1960.

At more distant places, although tsunamis travel swiftly (see Figure 9.9), there is usually ample time to issue warnings. For example, the travel time of a tsunami from the coast of Chile to Hawaii is about 15 hours, and from the Aleutian Islands to northern California, about 4 hours. It is crucial to realize, however, that coastal regions subject to large submarine faulting with vertical displacements may be exposed to large wave run-ups, with arrival times of high surges only minutes after the earthquake. Such is the case in Alaska, Japan, and the Cascadia subduction zone of North America (see Chapter 10). There is only one safety reaction in such areas — get to high ground immediately after the heavy shaking is felt.

Again, the shape of the coastline and the adjacent seafloor topography have a direct bearing on the degree of tsunami hazard. The parts of California that have the highest risk are the northernmost coast, above Cape Mendocino, the area near the entrance to San Francisco Bay, Monterey

Bay, and several points in southern California from San Diego northward to Point Buchon.*

The Japanese islands with their dense coastal populations are particularly prone to devastating tsunamis generated very near to the shore. Consequently, much effort has been made to reduce the hazard. Early-warning networks of seismographs and tide gauges are now linked to radio and television stations. The system was urgently upgraded in 1994 following public criticism of the disaster on the west coast of Okushiri Island and Hokkaido on July 12, 1993 (see Figure 10.1 and Plate 16). Undersea fault displacement north of Okushiri Island produced a magnitude 7.8 earthquake and a tsunami with run-ups reaching over 20 meters. Two hundred and thirty-nine people were killed with great harbor destruction, and over 500 houses were destroyed. Varying degrees of damage also occurred along the coasts of Russia and Korea.

An understanding of the inundation propensity of a particular coastline or harbor is of the greatest help in preparing for and protecting against tsunamis. The ability of modern high-speed computers to simulate theoretically the passage of tsunami waves across the ocean from their source to a coastal site has allowed reliable maps of inundation to be prepared (Plate 16). The interested reader can now actually follow the progression of tsunamis generated by computer simulations at a number of World Wide Web sites [see World Wide Web (Internet) Resources at the end of this book].

Some protection can be achieved in such cases by building breakwaters and barrier walls to block the inundation of the sea, but this defense has been developed to a significant degree only in Japan.

If you live in a coastal area subject to tsunamis, there are a number of steps you can take to minimize the danger from tsunami run-up: if you feel a strong earthquake, move quickly away from tidal lowlands to higher ground; if you hear that an earthquake has occurred under the ocean or a coastal region, be prepared to move to higher ground; tsunamis often signal their arrival by a precursory rise or fall of coastal water, and this warning should be heeded; do not go down to the beach or waterfront to watch for a tsunami; and follow the advice of your local emergency organization.

**Historically, a few small-to-moderate tsunamis have been generated by sudden seismic movements of the California seafloor — as on December 21, 1812, Santa Barbara Channel, and November 4, 1927, Point Arguello.*

Tragedy at Crescent City, California, 1964

The most recent tsunami of importance along the Pacific coast of California occurred after the great Alaska earthquake of March 1964 (see Chapter 1). In San Francisco Bay, a wave height of about 1 meter was recorded near the Golden Gate Bridge, and some boats moored at bay marinas were battered. Total damages along the California coast were about $10 million, the largest amount in a century. Of this sum, almost three-quarters was borne by Crescent City, on the far north coast of California.

During the 1964 Alaska earthquake, which occurred at 3:36 a.m. Universal Time, the first alarm was sounded at the Honolulu Observatory at 3:44 a.m. One hour later the position of the focus of the earthquake and its magnitude were accurately determined, but the main communication channels with Alaska had been severed. In California, an advisory bulletin of a possible "tidal wave" was received from Honolulu by the State Disaster Office at 5:36 a.m. UT. A more definite advisory warning was received at 6:44 a.m. UT and was sent to sheriffs, police chiefs, and civil defense directors of coastal cities and counties. At Crescent City, the county sheriff received the warning at 7:08 a.m. UT, and he notified people in the low-lying areas to begin evacuation. During this time of preparation, long sea waves had been moving across the Pacific Ocean, and they arrived at Crescent City about $4\frac{1}{2}$ hours after they had been generated in Prince William Sound in the Gulf of Alaska. If a large circle is drawn on a terrestrial globe at right angles to the underwater fault from Prince William Sound to Kodiak Island, also in the Gulf of Alaska, it will be found to head toward northern California (see Figure 7.2). The seismic sea waves that arrived at Crescent City were particularly high because the topography of the seafloor along the continental shelf at Crescent City is so shaped that the wave heights were amplified.

Several seismic sea waves inundated Crescent City harbor, and the third and fourth waves damaged the low-lying area around the southward-facing beach. The third wave washed inland more than 500 meters. Thirty city blocks were flooded, damaging or destroying one-story wood-frame buildings. After the first two smaller waves had struck, some people returned to the flooded area to clean up. Seven, including the owner and his wife, returned to a tavern at the shore to remove valuables. Because the sea seemed to have returned to normal, they remained to have a drink and were trapped by the third wave, which drowned five of them. This story illustrates the problem inherent in warning systems: too many alerts cause people to become blasé; too few foster ignorance of safety procedures.

The Crescent City disaster had one positive aspect. The town was rezoned afterward and the waterfront area developed into a public park. Businesses previously in the low-lying area were relocated on higher ground. The city is now a more attractive and safer place to live than before the killer tsunami of Easter 1964.

Atlantis and Santorini

Everyone is familiar with the story of the lost continent of Atlantis, the haunting legend of a great island civilization destroyed by a natural catastrophe. The legend comes to us from the writings of Plato, and many scholars have tried to separate the facts from poetic fiction.

Modern scholarship now leans toward the view that if there ever was a highly developed island culture that sank beneath the sea, it was not in the Atlantic Ocean but much closer to Greece and Egypt. In this century, archaeologists have rediscovered the highly accomplished Minoan civilization centered around the island of Crete and flourishing from before 2000 B.C. There is strong reason to believe that this "first civilization of Europe" suffered a sudden eclipse, and some overwhelming disaster or series of disasters, such as major earthquakes and volcanic eruptions, must be sought for an explanation.

At this stage of the puzzle, geophysical and geological work has brought to light evidence that the simultaneous abandonment and devastation of many Minoan palaces and villages were due to a powerful geological event, not to foreign marauders. The source of this natural violence is thought to be the volcano Santorini on the island of Thera, situated about 120 kilometers north of Knossos on Crete (see Figure 9.10). The geological evidence indicates reasonably gentle activity of the volcano about 1500 B.C., with increased violence proceeding to a cataclysmic stage. Indeed, it is suggested that the final eruptive stages of Santorini were similar to the outburst of Krakatoa, Indonesia, in 1883. Minoan cities on the island of Santorini were buried and their ruins preserved to the present time in the pumice and volcanic ash. Few human remains have been found in the ruins, indicating that the populace was probably warned by the early stages of the eruption — perhaps by accompanying earthquakes.

As at Krakatoa, it is highly likely that tsunamis were generated that were large enough to cause destruction on the nearby shores of Crete and to flood lowlands around the eastern Mediterranean. These speculations are supported by direct evidence of the effect of tsunamis in recent years in the Greek islands. For example, a large 1956 earthquake centered near the southeast coast of Amorgos in the Cyclades islands of Greece, was followed

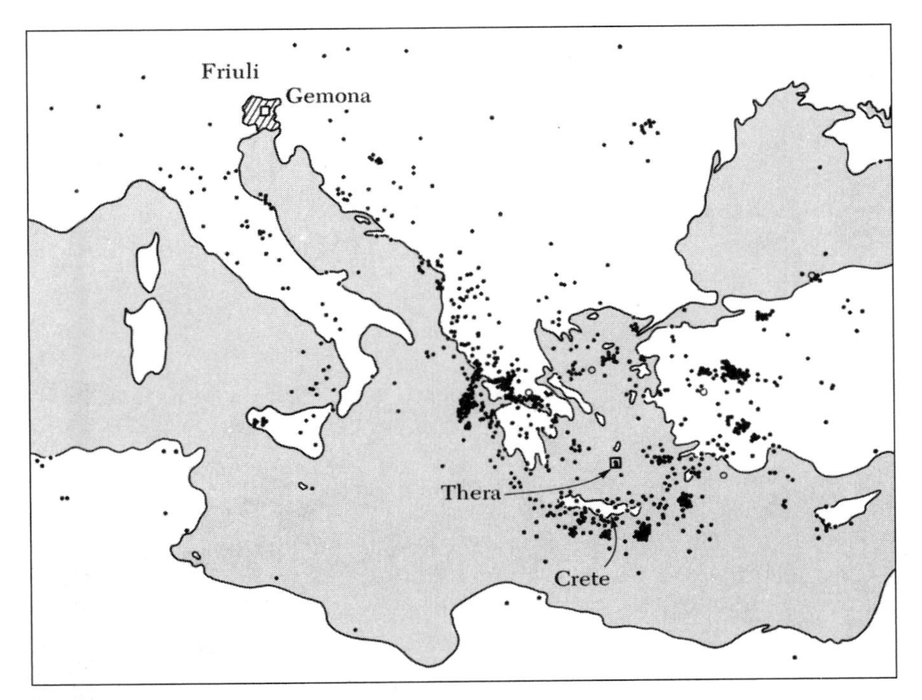

Figure 9.10 Seismicity of the eastern Mediterranean region. The dots represent epicenters for all earthquakes with magnitudes of 4.5 or more that occurred between 1965 and 1975. [From the hypocenter data file, National Geographic Solar Terrestrial Data Center, Boulder, Colorado. Courtesy of W. Rinehart.]

by a tsunami that caused damage on Amorgos and on many surrounding islands, including Patmos, Crete, and Milos. More than 80 small ships and boats were wrecked and one person was drowned. On the coast of Amorgos, wave heights of 25 to 40 meters were reported.

Another famous case at the other end of the Mediterranean region occurred around the Iberian peninsula after the great 1755 Lisbon earthquake with a submarine source in the mid-Atlantic Ocean (see Chapter 1). Then, a series of high ocean waves washed ashore along the west coasts of Spain, Portugal, and Morocco and increased the death toll. The water wave at Lisbon reportedly reached 5 meters above high-tide level. In the Mediterranean Sea it soon died out, but in the North Atlantic it disturbed British, Dutch, and French harbors hours later. Unfortunately, for the mitigation of present-day hazards in smaller oceans and seas like the Mediterranean, the travel time of a tsunami from its source is not long enough for an early-warning system to be as effective as it has proved to be in the Pacific Ocean.

10 Events That Precede an Earthquake

Collapsed portion of the elevated Hanshin expressway in the 1995 Kobe (Hyogoken-Nanbu) earthquake (see Chapter 5). [Photo by F. Seible.]

. . . after the wind an earthquake; but the Lord was not
in the earthquake!
And after the earthquake a fire; but the Lord was
not in the fire; and after the fire a still small voice.
—1 Kings 19:11

Can earthquakes be predicted? A variety of prediction methods have been used for centuries, ranging from accounts of "earthquake weather" to arrangements of the planets and jumpiness of animals. All have been unsuccessful. The growth of a small fault rupture into a large one in a given case seems to depend on a myriad of geological details. Indeed, there are physical arguments that the precise onsets of individual earthquakes are, in general, inherently unpredictable.

Efforts at Earthquake Forecasting

Beginning in the 1960s, a major scientific effort at earthquake prediction grew rapidly in seismic lands, particularly in Japan, the former Soviet Union, the People's Republic of China, and the United States. The aim was to establish at least as much reliability in earthquake forecasting as there is in weather forecasting. Publicity was given to the prediction of the date and place of a damaging earthquake, especially in a very short time interval.

Some seismologists, dubious of the reality of such hopes (of which I was one), stressed another aspect of earthquake forecasting: the prediction of *seismic intensity* at a particular site. This is the factor that is paramount in choosing the sites of important structures such as dams and hospitals and constructing hazard maps (see Plate 17). Moreover, in the long run, it provides the almost complete mitigation of earthquake hazards. In this chapter, we will examine attempts at scientific prediction of the time and place of earthquakes and defer discussion of prediction of strong ground shaking to Chapter 12.

Recall from Chapters 2 and 7 that studies of world seismicity patterns and global tectonics have made it possible to predict the probable place at which a large majority of damaging earthquakes can be expected to occur. However, this record does not enable us to forecast their precise times of occurrence. Even in China, where between 500 and 1000 destructive

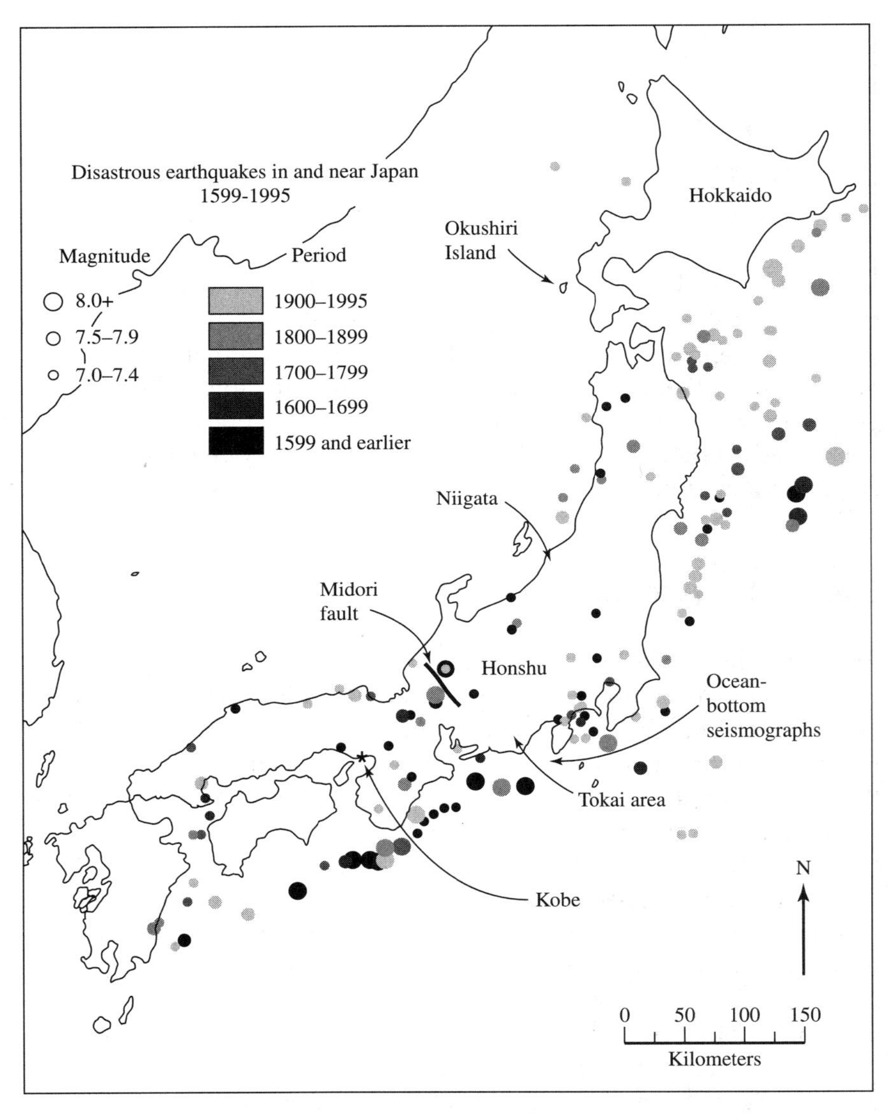

Figure 10.1 Seismicity of Japan. Hokkaido is the northern large island. [Courtesy of Japan Meteorological Agency.]

earthquakes have occurred within the past 2700 years, statistical studies have not convinced skeptics of periodicities between great earthquakes, although they do indicate that long periods of quiescence can elapse between them.

In Japan, where there are also long-term earthquake statistics (see Figure 10.1), hopes for earthquake prediction are deeply rooted, and vigorous earthquake-prediction research has been underway since 1962 but so far with only failures. (The damaging earthquake near Kobe in 1995 and the tsunami-producing earthquake of 1993 were not forecast definitively.) Nevertheless, the Japanese program, drawing on the contributions of hundreds of seismologists, geophysicists, and geodesists, has produced a variety of information and possible clues.

An often-quoted claim came from the west coast of Honshu, Japan. Here, geodetic measurements of ground elevation, plotted in Figure 10.2, suggested that for about 60 years (before 1964) steady uplift and subsidence of the coastline had taken place in the vicinity of Niigata. This rate slowed at the end of the 1950s; then, at the time of the infamous Niigata earthquake of June 16, 1964, mentioned in Chapter 6, a sudden subsidence of more than 20 centimeters was detected to the north adjacent to the epicenter. However, this pattern (see the graphs in Figure 10.2) was discovered only after the earthquake, and recent checks of the measurements throw much doubt on the reality of what was claimed to be an anomalous uplift.

From the 1960s in Japan, special studies of historical earthquake cycles in the vicinity of Tokyo, together with local measurements of present crustal deformation and seismicity, suggested to some Japanese seismologists that a repetition of the great 1923 Kwanto earthquake (see Chapter 6) is not now imminent but that earthquakes in neighboring areas are much more likely. In particular, the Tokai area (see Figure 10.1) is being closely watched. As the new millennium approaches, however, opposition to the short-term prediction program in Japan that has lasted through the last decades of this century has grown, with many Japanese seismologists calling instead for a research focus on the underlying causes of earthquake onset.

Over the years, many types of trigger forces that might initiate fault ruptures have been postulated. Some of the most serious proposals are severe weather conditions, volcanic activity, and the gravitational pull of the moon, sun, and planets.* Numerous catalogs of earthquakes, including

**In 1959, an absurd claim was even published in a scientific journal that the tiny gravitational attraction of the distant planet Uranus induced periodicities in earthquakes.*

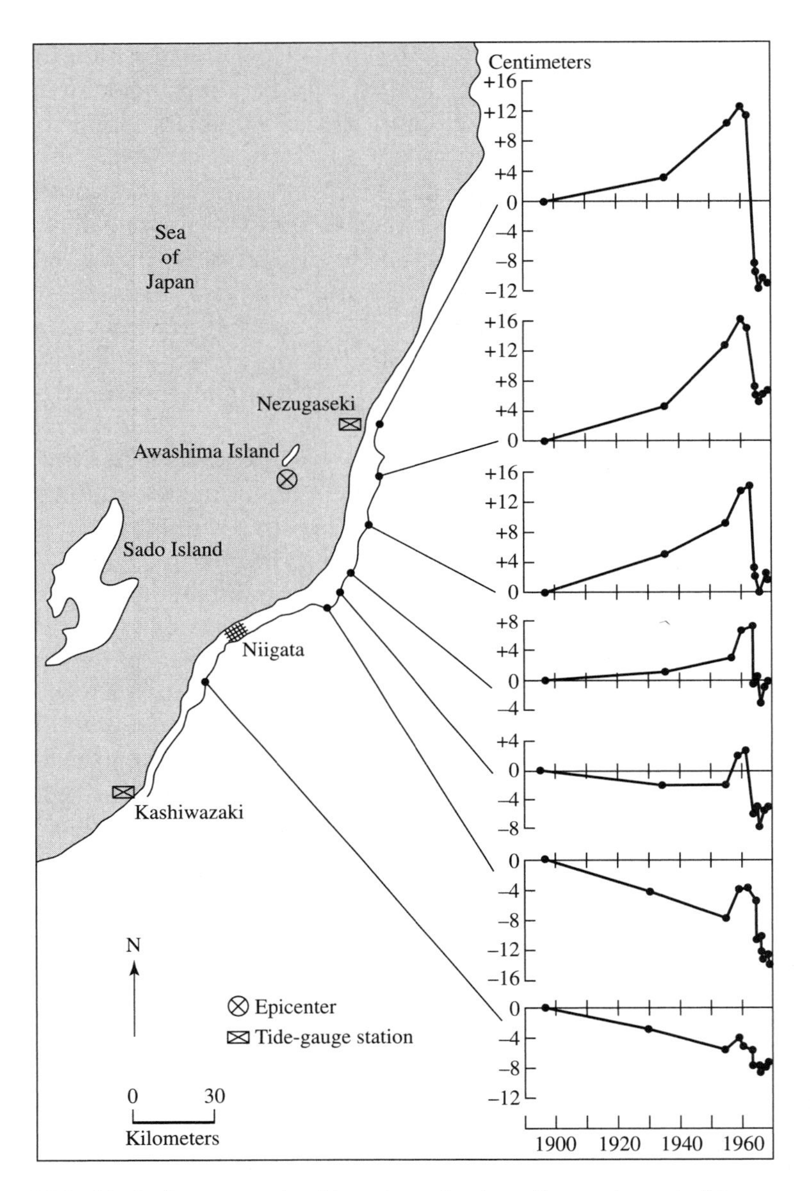

Figure 10.2 Vertical movements of bench marks along the west coast of Japan near the June 1964 Niigata earthquake, magnitude 7.5. Changes in level (in centimeters) before and after the earthquake are shown on the right. [From data of T. Dambara.]

quite complete lists for California, have been searched for such trigger effects without convincing results. For example, in 1974, a book for a general readership suggested that, about every 179 years, a near alignment of the planets takes place and the extra attraction would trigger great seismicity in California. The next such alignment was due in 1982. Because the San Andreas fault in *southern* California has not ruptured in a great earthquake since the large 1857 Fort Tejon shock, it was surmised that the San Andreas fault source might be particularly vulnerable to this planetary trigger mechanism in 1982. Fortunately for California, no such event happened. The book's argument had serious flaws. First, world seismicity catalogs show that the previous years in which this planetary alignment prevailed — 1803, 1624, and 1445 — were not ones of enhanced seismic activity. Second, the additional pull of the smaller or more distant planets is insignificant compared to that between the Earth and the sun. As a consequence, many more periodicities than 179 years would have to be considered, each corresponding to an alignment of the most significant planets.

A strong theoretical basis is required to make reliable predictions, such as is achieved in the prediction of the phases of the moon or the results of a chemical reaction. Unfortunately, at the present time there is not yet a precisely formulated theory of short-term processes in earthquake genesis. Nevertheless, limited though our present understanding of earthquake occurrence is, aspects of the elastic rebound theory allow us to make rough forecasts of when the next great rupture can be expected on a known active fault. In fact, after the 1906 California earthquake, H. F. Reid used it (as explained in Chapter 6) to argue that the next occurrence of a great shock near San Francisco would be about a century later.

In brief, Reid's argument was the following: survey measurements made across the San Andreas fault (see Figure 1.2) before the 1906 earthquake indicated that the relative displacement across the fault had reached 3.2 meters in 50 years. After the rebound on the fault on April 18, 1906, the maximum relative displacement along the fault was about 6.5 meters. If we do the arithmetic, we have $(6.5/3.2) \times 50 \approx 100$; therefore about 100 years would elapse before the next great earthquake. For this result we must make the somewhat tenuous assumptions that the regional strain continues uniformly and that the fault properties before the 1906 earthquake were not altered by the earthquake itself. Indeed, prudence requires that we do not rule out a series of more moderate earthquakes along the San Andreas fault in the next few centuries rather than another earthquake with a magnitude of 8.25.

Over the last 40 years, many prediction experiments have been conducted, and specific precursory symptoms (listed in the next section) have been tested. (The only "official" prediction in the United States up to 1999 is described at the end of this chapter in the section on the Parkfield experiment.) This overall effort, though substantial, has given little reason for optimism that practical prediction schemes will be realized in the near future in most areas of the world. Furthermore, those methods showing most promise require quite elaborate equipment and many workers.

Also, there is a fundamental dilemma inherent in earthquake forecasting. Suppose that seismological measurements hint that an earthquake of a certain magnitude will occur in a certain area during a certain period of time. Now presumably this area is a seismic one, or the study would not have been initiated in the first place. Therefore, it follows that by chance alone the odds are not zero that an earthquake will occur during the period suggested. Thus, if an earthquake occurs, it cannot be taken as decisive proof that the methods used to make the prediction are correct, and they may fail on future occasions. Of course, if a firm prediction is made and nothing happens, that must be taken as proof that the method is invalid.

In response to the significantly heightened activity in California on earthquake prediction, in 1975 a scientific panel was set up to advise the State Office of Emergency Services — and hence the state's governor — on the validity of predictions. The panel's important but limited role is to evaluate the data and claims of the person or group (normally a seismologist or seismologists conducting research in a government or university laboratory). By 1999, the panel had evaluated only three situations having a bearing on future California earthquakes.*

A valid prediction is defined by the California Panel as having four essential elements: (1) the period within which the event would occur, (2) the area of location, (3) the magnitude range, and (4) a statement of the odds that an earthquake of the predicted kind would occur by chance alone and without reference to any special evidence.

The social and economic consequences of earthquake forecasts are a subject of some controversy. As seismological research continues, numer-

Two national scientific panels to evaluate evidence on putative earthquakes deserve special mention. The first is the Japanese Earthquake Assessment Committee, of eminent seismologists, advisory to the Japan Meteorological Agency. Each member is alerted by a radio pocket beeper if anomalous signals are detected; a recommended alarm would, through the prime minister, shut down expressways, bullet trains, school, and so on, in a wide area. The second is the U.S. National Earthquake Prediction Evaluation Council, advisory to the U.S. Geological Survey.

Box 10-1

TYPES OF SOCIOECONOMIC IMPACTS AND ADJUSTMENTS TO AN EARTHQUAKE PREDICTION

EARTHQUAKE PREDICTION →

PROPERTY-RELATED CHANGES
- Decline in property values
- Decline in property tax revenues

FINANCIAL CHANGES
- Availability of insurance
- Reduced availability of mortgages
- Change in investment patterns

POPULATION MOVEMENT
- Temporary relocation (employers, employees, and citizens)
- Permanent relocation (employers, employees, and citizens)
- Availability of evacuation centers
- Evacuation urged in high-risk areas
- Avoidance of hazardous areas of the city

LEVEL OF BUSINESS ACTIVITY
- Cessation of work activity
- Decline in employment opportunities
- Decline in level of business activity

PREPAREDNESS ORGANIZATION AND EDUCATION
- Preparedness training and information
- Preparedness organizations
- Release of damage assessment maps

RESCHEDULING OF PUBLIC EVENTS

REDUCTION IN PUBLIC SERVICES

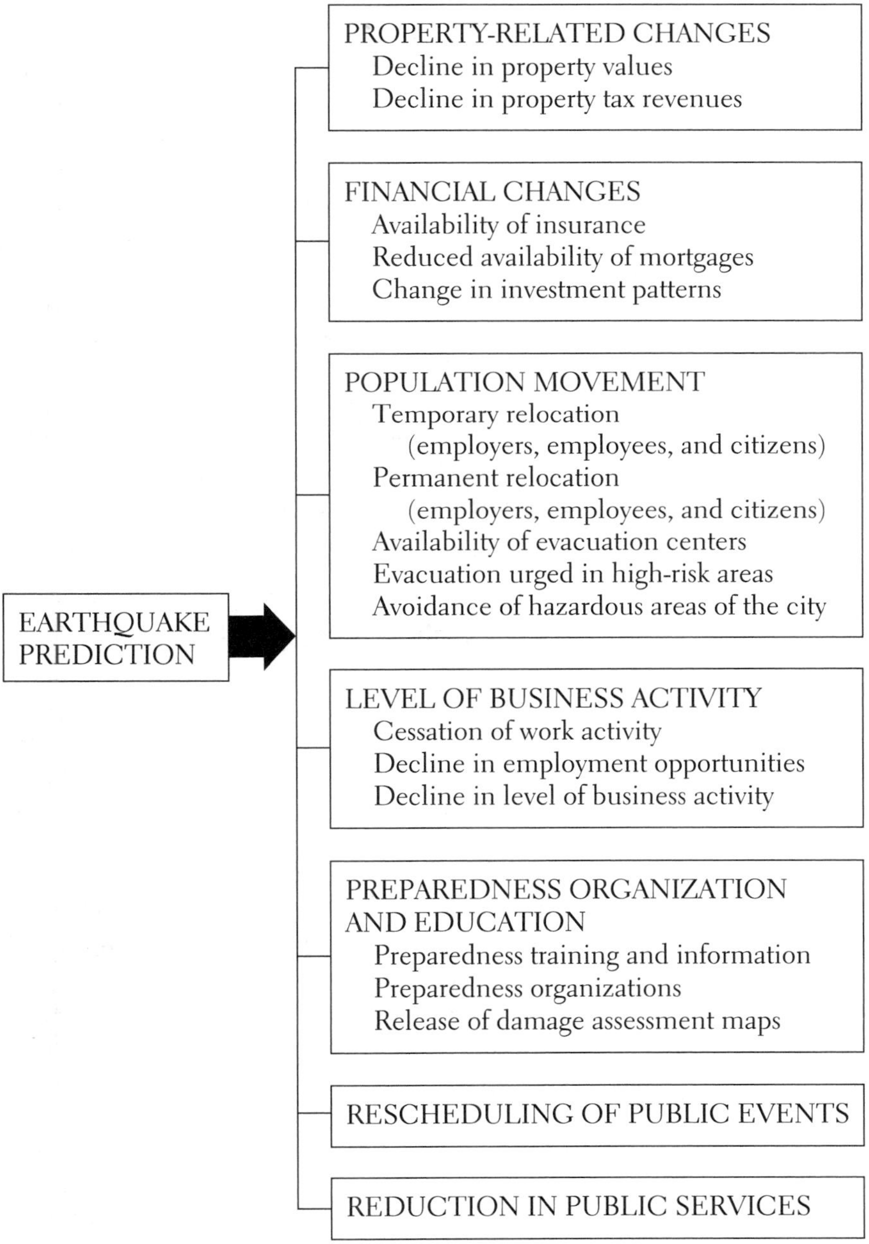

After J. E. Haas and D. S. Mileti. Socioeconomic Impact of Earthquake Prediction on Government, Business and Community. *[Institute of Behavioral Sciences, University of Colorado, 1976.]*

ous earthquake warnings from diverse sources will probably continue to be issued in various countries. For example, numerous forewarnings have been issued in China and will be discussed later in this chapter.

In western society, studies on the unfavorable as well as the propitious consequences of prediction have been made. For example, if the time of a large damaging earthquake in California were accurately predicted a year or so ahead of time and continuously updated, casualties and even property damage resulting from the earthquake might be much reduced; but the communities in the meizoseismal region might suffer social disruption and decline in the local economy. The major social and economic responses and adjustments that may occur are summarized in Box 10.1. Without an actual occurrence to draw upon, such assessments are, of course, highly tentative; the total reaction would be complex, because responses by the government, public, and private sectors could all vary. For example, if after the scientific prediction and official warning, massive public demand for earthquake insurance cuts off its availability, then temporary but drastic effects on property values, real estate sales, construction, investment, and employment might ensue.

Clues for Recognizing Impending Earthquakes

What could be harbingers of impending earthquakes? A multitude have been suggested,* but it is not clear which, if any, may become reliable. Certainly any operative scheme of practical prediction must be based on a combination of clues, so that decisions will be as firm as possible before warnings are issued.

Several of the more promising clues have already been discussed, such as the detection of strain in the rocks of the Earth's crust by geodetic surveys (see Chapter 6) and the identification of suspicious gaps in the regular occurrence of earthquakes in both time and space (Chapter 7). And a more precise but not foolproof tool is the observation of foreshocks, as in the 1975 Oroville sequence in California (Chapter 2) and, perhaps, in the 1975 Haicheng, China, earthquake.

To monitor such foreshocks, as well as to predict damaging local tsunamis, a specially designed set of seismographic stations was installed

One is an unnatural glow in the sky. There were reports, for example, of a luminous night sky to the north of the Friuli region in Italy before and after the May 6, 1976, earthquake. Objective measurements of such "earthquake lights" are needed. Currently, there is no satisfying explanation for such folksy claims.

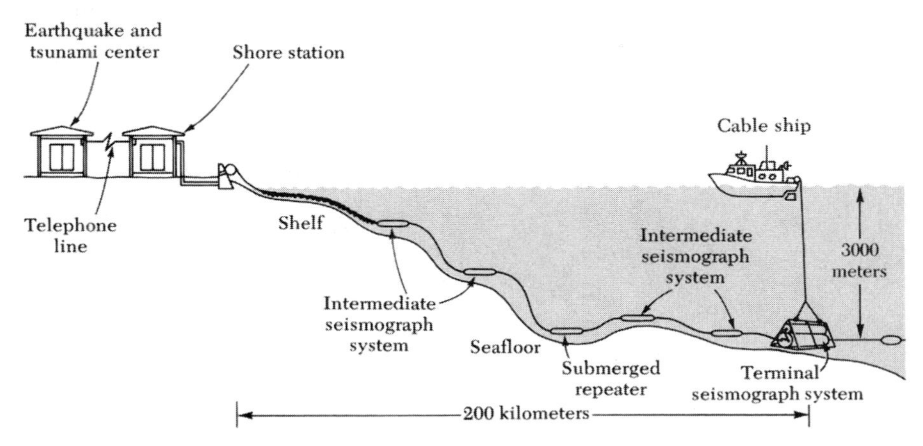

Figure 10.3 The network of seismographs on the ocean bottom to the east of Honshu, Japan.

more than a decade ago across the continental shelf and the ocean trench south of the Tokai special prediction area, Honshu, Japan. The scheme, which is part of the Japanese earthquake-prediction program, is illustrated in Figure 10.3.

In recent years, the major earthquake-prediction effort has been more precise measurements of fluctuations in physical parameters in crustal rocks of seismically active continental areas. Special sensing devices have been installed in order that long-term changes in the parameters might be observed very accurately. The number of such measurements is still limited, and results have thus far conflicted: in some, unusual behavior has been indicated before a local earthquake; in others, nothing significant has been seen or variations have occurred that are not associated with earthquakes. Five parameters thought to be auspicious are listed in the left column of Box 10.2: seismic P velocities, the uplift and tilt of the ground, the emission of radon gas from wells, electrical resistivity in the rocks, and the number of earthquakes in the region. How can each of these parameters be employed in a prediction scheme?

First, precursory changes in the seismic P velocities in a seismic area are of particular interest to seismologists because seismographic stations are designed especially to measure time very precisely. The idea behind the method is simple. If rock properties change before an earthquake, then the speed of seismic waves might also vary. Suppose, for example, the P velocity has changed by 10 percent through an area 20 kilometers across. Then the travel time of a P wave from one side to the other would change by

Box 10-2

Suggested Physical Clues for Earthquake Prediction

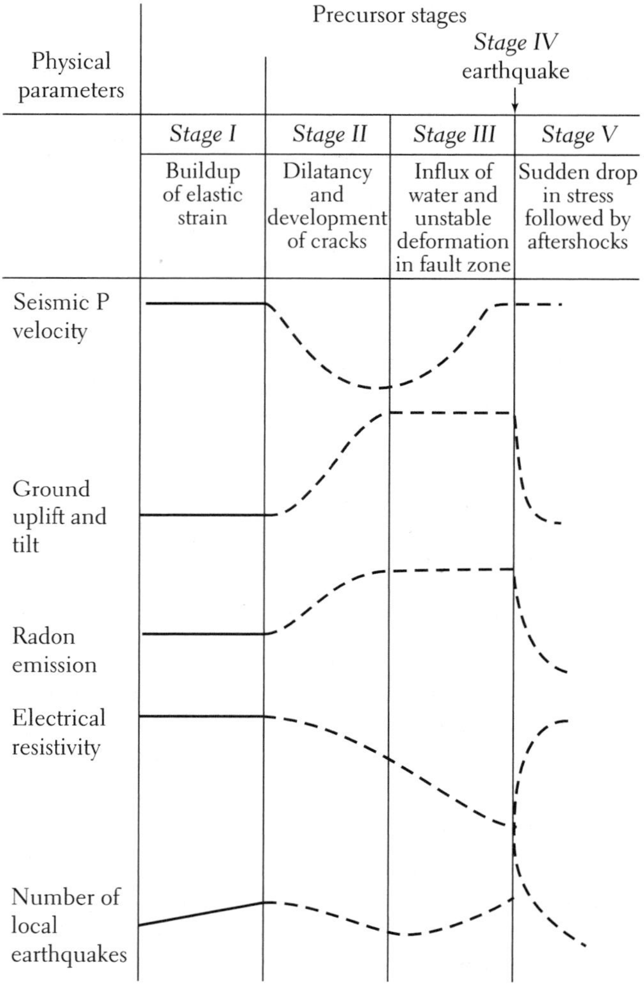

Precursor stages

Stage IV earthquake

Physical parameters	*Stage I*	*Stage II*	*Stage III*	*Stage V*
	Buildup of elastic strain	Dilatancy and development of cracks	Influx of water and unstable deformation in fault zone	Sudden drop in stress followed by aftershocks
Seismic P velocity				
Ground uplift and tilt				
Radon emission				
Electrical resistivity				
Number of local earthquakes				

After Predicting Earthquakes, *National Academy of Sciences, 1976.*

about 0.4 seconds. Such changes in time are easily measured with modern seismographs and chronometers. Some of the first information published on precursory changes in travel time of waves in moderate earthquakes was gathered as early as 1962 in Tadjikistan. Measurements there have suggested that P velocities change by about 10 to 15 percent before the occurrence of local earthquakes. More recent work has thrown doubt on the precision of these claims.

One difficulty is that when the source of the P waves is an earthquake, the earthquake focus has to be located precisely from the travel times themselves. Then small changes in the estimated focal depths of successive earthquakes, arising from migrations of foci along a fault zone (see Figure 2.5), are sufficient to explain some of the variations seen in the measured P travel times. A more promising procedure is to measure travel times from seismic sources with known positions and times of origin, such as chemical explosions or quarry blasts. So far, such measurements have not revealed convincing changes in the travel times before earthquakes.

The second parameter listed in Box 10.2 is precursory change in ground level, such as ground tilts in earthquake regions (see Figure 10.4). An initially hopeful study of this kind, but later discounted, made after the 1964 Niigata earthquake in Japan, has already been described in the first section of this chapter. Another celebrated claim of rapid ground uplift of a considerable area was around Palmdale in southern California, which apparently began around 1960 (see page 115). A subsequent critique of the surveying methods and errors threw doubt on the claim. To date no significant earthquakes have been produced in the area. The area is, of course, one in which major earthquakes have occurred in the past and are likely to occur again in the future.

The third parameter is the release of radon, a radioactive gas, into the atmosphere along active fault zones, particularly from deep wells. For example, it has been claimed that significantly increased concentrations of radon were detected just before earthquakes in some parts of the former Soviet Union. More recently, after the 1995 Kobe earthquake, research papers were published describing a tenfold increase in radon concentration 30 kilometers away from the epicenter, 9 days before the earthquake. Because so few measurements of radon concentration in various geological settings are available, however, it is currently difficult to determine whether increases are due to seismic or nonseismic causes.

The fourth parameter, to which attention is still given, is variation in the electrical conductivity of the rocks in an earthquake zone. It is known from laboratory experiments on rock samples that the electrical resistance of water-saturated rocks, such as granite, changes measurably just before the rocks fracture. A few field experiments to check this property in fault zones have been made in the former Soviet Union, the People's Republic of China, Japan, the United States, and elsewhere. From these studies, some workers reported decreases in electrical resistance before earthquakes, but the effect is often obscured by background human-made electrical systems.

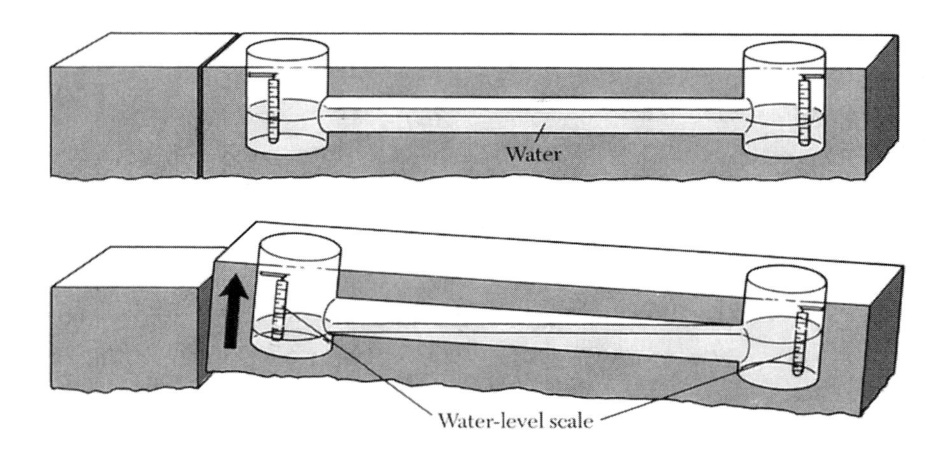

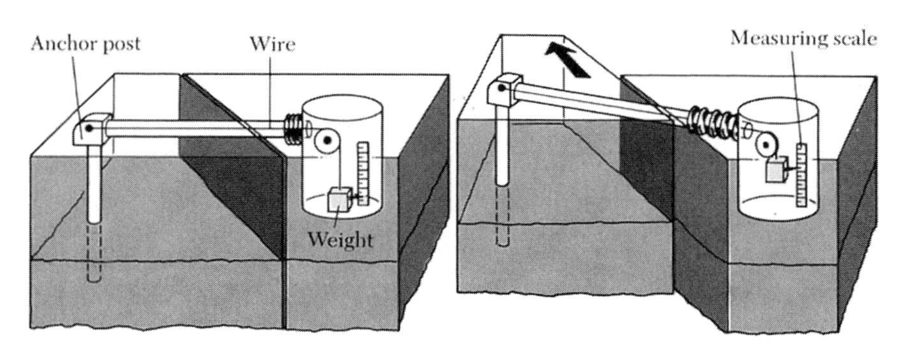

Figure 10.4 Diagrammatic models of a water-tube tiltmeter (top) and a simple wire-displacement meter (bottom). The arrows indicate ground movements or fault slip.

Variation in the seismicity rate is the fifth parameter. More information is available on this method than on the other four, but the present results are not definitive. In brief, a large change in the normal background earthquake occurrence is noted — for example, an increased rate of small earthquakes. Such changes were observed in 1975, before the Oroville earthquake (discussed in Chapter 2) and the Liaoning earthquake (to be discussed later in this chapter). Another forecast that seems to have been successful has been described by Italian seismologists. After the tragic principal shock of May 6, 1976, in the Friuli region (see Chapter 2), the aftershocks were monitored. In early September 1976, it was noticed that the number occurring per day in the region had increased significantly. On this basis, authorities issued a general warning that people living in buildings of dubious strength might be advised to sleep elsewhere, even in tents. On September 15, 1976, a major aftershock ($M_S = 6.0$) occurred at 5:15 p.m., collapsing many weakened buildings. Yet few were killed in this earthquake. The method of measuring the rate of occurrence using a *b* value has already been discussed (see Appendix G and the discussion of the Oroville earthquake sequence in Chapter 2). Statistics are not available on how many times such calculations have shown no significant change in the estimated *b* value and the result not published. Because we lack a specific theoretical basis, the method is at best of dubious standing.

Triggering of earthquakes by external forces has long fascinated the earthquake curious. "Earthquake weather," floods, and the atmospheric jet stream have all failed as candidate triggers; their forces are calculated to be trivial when matched against the resistance of rocks at depths of 5 kilometers or more. A more substantial suggested trigger is that the endless cycles of the gravitational pull of the sun and moon on the rocks of the Earth's crust might provide the necessary force. This attraction of gravity produces not only the 12- and 24-hourly ocean tides but also, not so widely known, a tidal rise and fall of the rocks of the Earth. The height of this tide of the solid Earth varies with latitude, but reaches several tens of centimeters (about a foot) near the equator. As a consequence, every day this gravitational deformation causes small variations in pressure across rock surfaces at depth. Could this be a possible trigger for sudden slip along strain-loaded earthquake faults?

Calculation shows, however, that such pressure increases and decreases are tiny, many orders of magnitude less than the ever-present gravitational compression caused by the weight of the overlying rocks at the focus of a normal tectonic earthquake. It follows that tidal triggering is highly unlikely, a conclusion confirmed in almost all studies that compare the time of origin of large earthquakes with the maximum total tidal pressures. The exact cause of the nucleation of an extensive fault rupture remains unclear. An exception is that small earthquake swarms near volcanoes sometimes do correlate in numbers with the peak total attractions.

Earthquakes and Prediction in China

On February 4, 1975, with the Cultural Revolution still churning, officials of the Manchurian province of Liaoning decided the evidence was

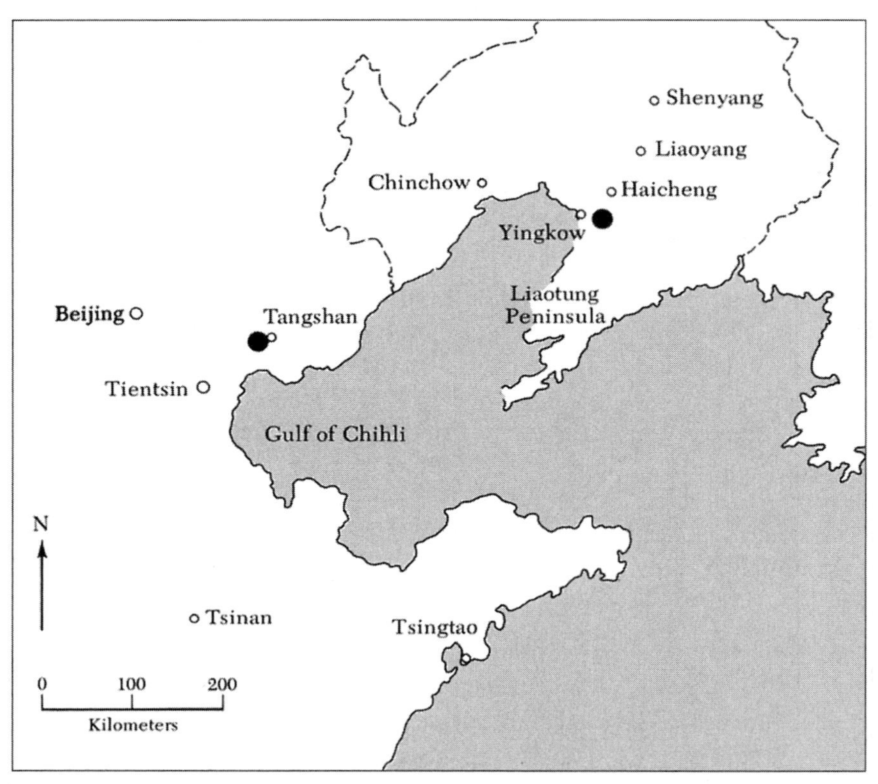

Figure 10.5 Map of northern China, showing the locations of the principal earthquakes near Haicheng (February 4, 1975) and Tangshan (July 27, 1976).

sufficient for them to issue an urgent warning to the populace to expect a strong earthquake within the next 2 days. Persons in the counties of Haicheng and Yingkow (see Figure 10.5) were urged to remain outdoors even though it was cold winter weather.

Then, at 7:36 p.m., a strong earthquake of magnitude 7.3 shook the Haicheng-Yingkow region. The event was described as follows in a later Chinese report:

> Most of the population had left their houses, big domestic animals had been moved out of their stables, trucks and cars did not remain in their garages, important objects were not in their warehouses. Therefore, despite the collapse of most of the houses and structures during the big shock, losses of human and animal lives were greatly reduced. Within the most destructive area, in some portions more than 90 percent of the

houses collapsed, but many agricultural production brigades did not suffer even a single casualty.

Western scientific observers, who visited the area subsequently, confirmed that damage had indeed been widespread and that large-scale reconstruction was taking place. In one Yingkow commune of 3470 people, most of the 800 dwelling houses were severely damaged, and 82 had completely collapsed. Yet, according to official Chinese reports, there were no casualties. In Haicheng, 90 percent of the structures were destroyed or seriously affected: entire buildings fell into the streets, and factories and machinery were damaged. In the countryside, dams, bridges, irrigation works, and houses were damaged. We would expect that a significant number of the 3 million people in this densely populated province would have died inside collapsed buildings. The exact number of dead is not known.

This historical event has been hailed as a milestone in seismology. The official claim is that effective forewarning of a destructive earthquake had prevented many deaths and injuries. It was immediately asked: Is routine earthquake prediction on a large scale imminent? Yet a quarter of a century has now passed with no similar case history.

In assessing the situation, it is only fair to cite the failures as well as the successes of the Chinese prediction program. Although there are claims that the time of several Chinese earthquakes has been forecast correctly — such as the 1975 Haicheng earthquake and a pair of earthquakes 97 minutes apart, magnitude 6.9, near the China-Burma border in western Yunnan on May 29, 1976 — other moderate to large ones have not (although specific statistics have not been published). Also, forecasts of impending earthquakes have been made in China that turned out to be false alarms. One was issued in August 1976 in Kwangtung province (normally not very seismically active) near Kwangchow (Canton) and Hong Kong. During the earthquake alert many people slept outdoors in tents for nearly 2 months and concern spread to Hong Kong for a time. No earthquake occurred.

The most publicized lack of forewarning was the tragic earthquake of July 27, 1976, which almost razed Tangshan (see Figures 10.5 and 10.6), an industrial city of 1 million people situated 150 kilometers east of Beijing. Official reports, confirmed as a minimum by various sources, estimated a death toll of about 250,000 in the meizoseismal area. About 100 persons were killed as far away as Beijing, where some mud walls and old brick houses collapsed. It is estimated that an additional 500,000 persons were injured. When the enormous industrial loss was added to this human

Figure 10.6 Horizontal and vertical offsets in a schoolyard of the fault that produced the 1976 Tangshan, China, earthquake. [Photo by Bruce A. Bolt.]

calamity, the economic aftermath for China was severe.* (The Tangshan earthquake also had political implications: a traditional Chinese view maintains that natural disasters are a mandate from heaven, and earthquakes have been claimed to mean trouble for governments as far back as the Sung dynasty.)

The Chinese prediction program began after a particularly lethal earthquake in 1966 about 300 kilometers southwest of Beijing in Hsingtai, Hopei province. Premier Chou En-lai called for an earthquake-prediction program "applying both indigenous and modern methods and relying on the broad masses of the people."

Such a narrow but specific seismological goal is perhaps particularly appropriate for this country with its over 1 billion people. Many structures are not earthquake resistant so that an effective short-term way of counteracting earthquakes is to evacuate people from their unsafe dwellings ahead of time. Second, the linking of seismological research to earthquake prediction enables scientists to work in a framework with approved, and indeed admirable, social ends. A third beneficial effect of

**On a revisit to Tangshan in 1996, I found a modern city with only a dedicated museum in the central park as a reminder of the 1976 earthquake.*

this approach is the widespread education of masses of people (if it is done in an objective way).

Let us now turn to the 1975 prediction claim. In Liaoning province, the subsequent site of the 1975 Haicheng shock, some observational programs of possible precursory phenomena were begun as early as 1970 by seismologists and nonprofessional but interested people. In late 1973 and 1974, a number of changes and variations in physical parameters were reported. For example, in the region of the Jinzhou fault in Liaoning province, it was claimed that the ground surface rose at 20 times its normal rate, an elevation increase of about 2.5 millimeters in 9 months.* Unusual fluctuations in the Earth's magnetic field were reported, as were changes in the elevations of the shorelines of Liaotung peninsula. Although similar phenomena, *not* associated with earthquakes, have been reported in many parts of the world, those in Liaoning were evidently considered symptomatic enough to warrant authorities to declare at least one abortive short-term emergency in 1974.

In the following year, more impressive clues came from seismographic stations near Haicheng, which reported that many small earthquakes had begun to be recorded. The specific prediction was mostly based on this increase in background seismicity (compare the Oroville earthquake story in Chapter 2). Throughout the region, people recounted incidents of peculiar animal behavior (see Figure 10.7). In addition, there were numerous reports, mainly from amateur observers, of changes in groundwater level. Tiltmeters showed that the ground had changed direction of tilting in some places but not others.

On February 3, the Shihpengyu seismographic station, east of Yingkow, suggested that the small earthquakes were the foreshocks of a large earthquake. As a result, on February 4, the party committee and the revolutionary committee of Liaoning province alerted the entire province. Special meetings were held at once to ensure that precautionary measures would be taken.

In retrospect, the efficacy of the mobilization speaks for itself. But the process of arriving at the decision to do so is less clear-cut. Some of the observations on which decisions were based were less than scientific; earthquake prediction was linked to the prevailing party ideological orthodoxy. There is no scientific evidence, for example, that animals can sense the

**A line of fresh* en echelon cracks, *extending into the shallow basement rock, was observed after the earthquake; the line is approximately parallel to the largest dimension (about 60 kilometers) of the aftershock zone but only about 5.5 kilometers long. The Jinzhou fault did not rupture.*

Figure 10.7 Cartoon in Chinese seismological textbook, suggesting that animals may give forewarnings of earthquakes. "Chickens fly up to trees and hogs stay quiet. Ducks go out of water and dogs bark wildly." [Courtesy of W. Lee and Francis Wu.]

advent of an earthquake.* Also, variations in the ground levels occur without being accompanied by earthquakes. Even the occurrence of the foreshocks is not an infallible harbinger, because there is no way of knowing that an earthquake is a foreshock of a large one until the large one occurs. In addition, some sizable earthquakes — like that in San Fernando, California, in 1971 — have not been immediately preceded by foreshocks. Also, swarms of earthquakes, unaccompanied by any principal shock, are common in earthquake country.

From a psychological perspective, the Liaoning foreshocks must have prepared much of the population for an official warning; it is known that many persons were already staying out of their weak adobe and stone houses before February 4 because of the repeated shaking. Thus, remarkable as the 1975 prediction was, there were other factors as well, such as continuous and widespread local and provincial government concern, telltale foreshocks, social discipline, political pressures — and a modicum of good luck.

**However, popular accounts of similar animal behavior before a major earthquake had previously been reported from various countries, e.g., before the 1906 San Francisco earthquake, before the 1923 Tokyo earthquake, and before the 1976 Friuli earthquake. Fine sensing of variations in magnetic and electric fields, radon gas concentrations, aerosol particles, and so on have been suggested as explanations, but no controlled experimental verification has been published.*

Fossil Earthquakes

Seismologists have for a long time sought ways of detecting the occurrence of great prehistoric earthquakes. Several methods have been tried. The first uses the uplift of seashores produced by sudden fault slip, which results in a change in the levels at which seashells and mollusca live in tidal regions. For example, after the massive 1964 Alaska earthquake (compare Chapter 1), tidal benches were dated from the fossil sea life, and the sequence of uplifts of the land surface associated with previous major earthquakes was determined (see Plate 1). Another scheme is the measurement of growth rings in large trees of great antiquity growing in earthquake country (see Plate 18). Severe shaking of the ground sometimes damages the root system of a tree, causing retarded growth in the following year. However, even after correlating the growth rates through a wide area, the range of uncertainty is large because of climatic variations.

More precise procedures have now been developed that can, under favorable circumstances, reliably track sequences of great earthquakes back in Holocene time. These procedures depend on the geological field studies of fault movement (see Figure 5.3). Detailed stratigraphic mapping has been successful, for example, along the San Andreas fault in California, where dated features are ascribable to specific large historical earthquakes. About 50 kilometers northeast of Los Angeles, the trace of the San Andreas fault transects a low-lying area that becomes a swamp from the waters of Pallett Creek during a rainy season. Trenches excavated by geologists across the fault in this area exposed a well-marked sequence of silt, sand, and peat (marsh-plant remains) layers (see Figure 10.8). The motivation for the work was that the displacement and liquefaction effects of great paleo-earthquakes might be preserved in such beds of sand and peat.

What happens is as follows: During the strong shaking of the ground, water-saturated sand layers at some depth below the surface become liquefied (see also Chapter 11). The overpressure of the rocks and soil above then causes the water and sand to rise to the surface, forming a layer of sand in the form of what are called *blows*, *boils*, or volcanoes (see Figure 10.9). As the cycle of wet and dry seasons continues, Pallett Creek and other neighboring streams carry down gravel and silt, which cover up the sand blows formed in the intense shaking; after the passage of time, another great earthquake occurs, producing further liquefaction and resulting sand blows at the surface; these are, in turn, also covered up. Thus, the sand blows, silt, and peat layers form a conformable sequence, with the younger layers lying over the older ones. The plants or other organic materials within each layer can be dated by radiocarbon methods.

Figure 10.8 Near-vertical slip plane of the San Andreas fault, Pallett Creek, California, exposed on the side of a trench. Dark bands are horizontal layers of sand and peat. [Courtesy of K. E. Sieh.]

At Pallett Creek, evidence was found for at least nine paleoearthquakes extending back more than 1400 years to 545 A.D. The dates, all but the first of them approximate, are as follows:

1857, 1745, 1470, 1245, 1190, 965, 860, 665, 545

The key year, 1857, marked the directly documented Fort Tejon earthquake of January 9 (see Appendix B), which is the last great earthquake produced by rupture of the southern section of the San Andreas fault. There is thus a direct check on the method.

Two conclusions follow from the work at Pallett Creek. First, large earthquakes along the southern reach of the San Andreas may break different segments of the fault at different times. Second, the average time between these past earthquakes is approximately 160 years, but there is a large variation. The greatest time interval was nearly 300 years and the smallest as short as 55 years.

Figure 10.9 Cemented sand-liquefaction column or "pipe". A system of cemented sand columns or pipes was exposed during trench excavations in the Tanlu fault zone in the province of Anhui, China. [Photo by Bruce A. Bolt.]

It should be noted that the Pallett Creek site gives information on earthquakes caused by rupture along only one large section of the San Andreas fault in southern California and not its whole extent even there. Also, damaging earthquakes may occur where groundwater conditions are not always appropriate to produce significant liquefaction by even a large earthquake in the vicinity (such as during a very dry season). Similar studies on liquefaction have since been carried out across other active faults in places such as China and Japan (see Figure 10.9). Difficulties arise in very wet and very dry climates; for example, long-lived organic material (such as from redwood trees) emplaced in the sand layers by water flow or animal action can produce spurious dates in the statistical analysis.

The Cascadia Subduction Zone in the Northwest United States

How well can forensic seismology and geology unravel the record of past great earthquakes? And further, how well can such paleoseismological evidence predict future earthquake hazard? The problems are well illustrated by recent geological studies in the Pacific northwest of North America.

In this region, a belt of volcanic mountains (see Figure 9.1), called the Cascade range, runs from California through Oregon and Washington state into British Columbia. In the terminology of the plate tectonic model, these volcanoes are the result of a subduction slab, which takes the Juan de Fuca plate and the Gorda plate down beneath the North America plate. The entire region, called the Cascadia subduction zone, stretches northward from Cape Mendocino, where the San Andreas fault swings westward into the Pacific, to north of Vancouver Island (see Figure 10.10). The mode of subduction is described in Chapter 7 (see Figure 7.3 and Plate 13): There is gradual sliding downward beneath the western margin of the continental plate at various rates. Sliding occurs in episodes of aseismic slip, interspersed with periods when the subduction slab is locked until the strain exceeds the strength of the rocks. A sudden slip then occurs, causing earthquakes. Such tectonic processes can spawn immense earthquakes such as the 1964 Good Friday Alaskan earthquake, described in Chapter 1.

The overall measured motions of the Pacific and North America plates indicate that, along the Cascadia subduction zone, the plates are converging (see Figure 7.2), with the Juan de Fuca plate slipping deep beneath the North America plate at an average rate of about 4 centimeters per year.

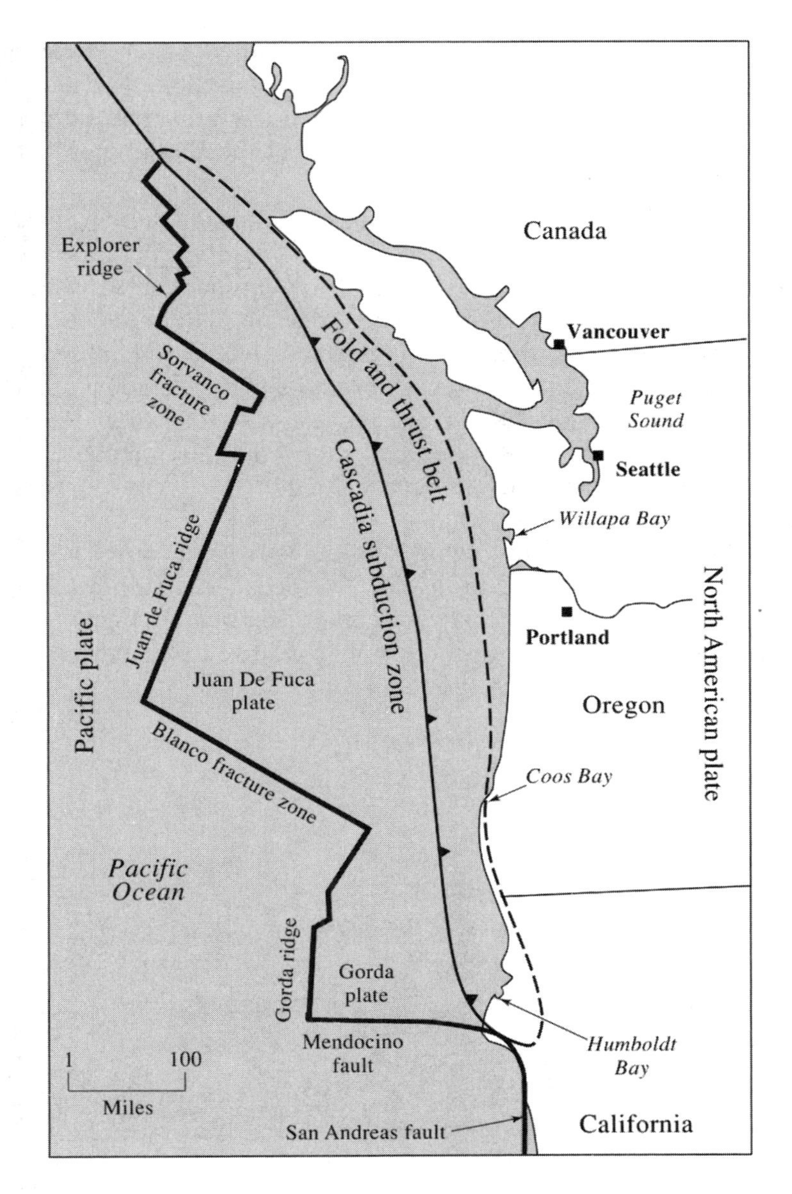

Figure 10.10 Diagram indicating the main generalized structure along the northern Pacific coast of the contiguous United States. Compare the structural triple junction near Humboldt Bay with the earthquake occurrence shown in Figure 2.6.

Molten rock in the form of magma rises to the surface, creating the Cascade range and its active volcanoes such as Mount Saint Helens. Along the continental margin itself, geodetic surveys confirm that the Cascadia range is being compressed and that portions of the Washington coast are being uplifted.

This tectonic model has changed attitudes toward seismic hazard in the northwest United States. The lack of significant historical high seismicity (see Plate 3) in most of the region had given rise to the orthodoxy that this zone was essentially aseismic. The absence of major earthquakes in western Oregon, for example, led to the inference that the Cascadia subduction zone was either no longer undergoing differential slip or, alternatively, the subduction process was occurring slowly without sudden fractures. Modern seismographs did not detect a Wadati-Benioff earthquake zone, such as occurs along the Chilean or Alaskan subduction zones.

A striking revision of this picture of quiescence and consequent lack of seismic hazard has occurred in the last two decades. The prevailing contrary argument is that, by analogy with other subduction zones, the downward movement of the North America plate relative to the offshore Pacific plates at Cascadia does sometimes jump suddenly.

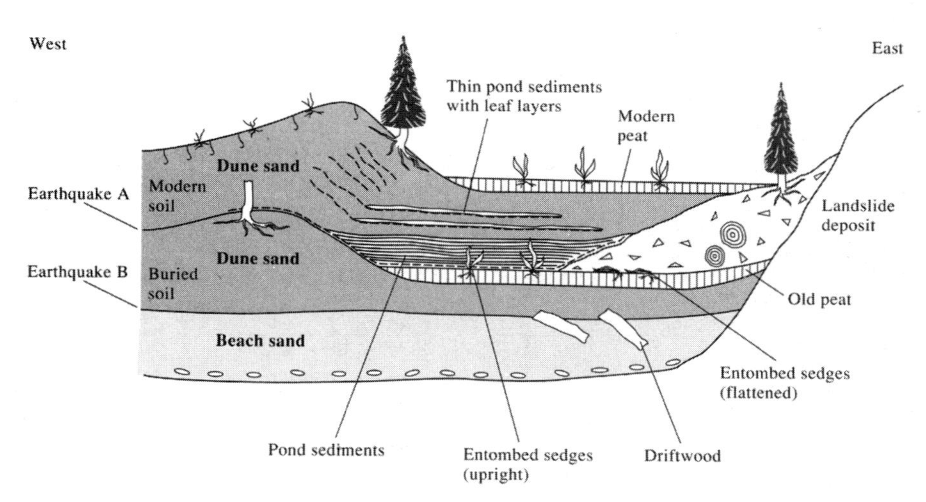

Figure 10.11 Diagram showing how depression of the land surface along a beach north of Cape Mendocino, California, due to sudden slip of the underlying subduction slab, leads to a series of peat layers. [After G. Carver.]

The evidence for the revision is as follows: First, deformation of water-saturated ocean-floor sediments of Holocene age (see Appendix D) has been mapped along the fold and thrust belt, shown in Figure 10.10. On land in the southern part of the Cascadia zone, many fold structures in the uppermost crust are of relatively young age, and the lateral shortening by folding is consistent with the continued pushing together of the two plates. Secondly, there is compelling evidence of cycles of subsidence of the ground surface along the coastline (see Figure 10.11). At a number of estuaries and tidal inlets, borings by metal cylinders into the water-saturated muds have been taken to depths of tens of meters. Long cylindrical mud cores result. These show lengths of soft, gray mud interspersed by thinner layers that contain fossil remains of tree stumps, driftwood, and the brownish, partly decayed plant matter called peat. These peaty layers indicate that at the time of formation, the land surface was above high-tide level, allowing vegetation to grow. Such marshes, covered with salt-tolerant plants, are prevalent along the Pacific coast today. Subsequently, such a marsh must have dropped beneath sea level and had fine sediments deposited over it for a considerable time. The organic material in these peaty beds can nowadays be dated with accuracies of a few tens of years using radioactive carbon techniques. In coastal areas along Washington state, such evidence of burial of extensive vegetative lowlands indicates at least six subsidence episodes of 0.5 to 2 meters in the past 7000 years. The most recent subsidence to be age-dated from a peaty layer occurred about 300 years ago.

Additional evidence comes from thin sandy sheets that mantle the peaty layers at some field sites. The argument is as follows. A slip of the Cascadia slab under the plate margin caused subsidence which buried the tidal marshes below sea level. The ocean-floor movement produced tsunamis which washed up the rivers and deposited layers of sand on the old marsh surfaces. Such effects are common in large subduction-zone earthquakes; in Alaska and Chile (see Chapters 1 and 7) for example, subsidence has occurred along the coastal lowlands and replaced fertile lowlands by water-covered estuaries. Over tens of thousands of years, we thus have a model of a cyclic rhythm between a drowning of fertile coastal areas, sometimes coincident with large thrust earthquakes, and the slow re-deposition of soils to re-form marshes.

Finally, there is remarkable detective work from historical tsunami records in a faraway place. On the night of January 27, 1700, a 2-meter-high tsunami was reported at three coastal sites in Japan. Computer calculations of Pacific Ocean waves show this could have been created by an extensive Cascadia zone seafloor slip, and not by a limited nearby one. This event then might correspond to the subsidence 300 years ago mentioned above.

The various arguments have led to new interpretations regarding earthquake occurrence along the Cascadia subduction zone. One extreme conclusion is that a mega-earthquake may be generated by slip of essentially the whole Cascadia subduction slab in one episode beneath the continent. A more temperate assessment is that, given the admitted slow convergence rate and the lack of present seismicity along the subduction slab, sudden slip on limited segments of the slab can be expected intermittently on slab segments, particularly adjacent to the Juan de Fuca plate. This would generate smaller but still substantial earthquakes.

The Gorda plate segment off northern California differs from the northerly section of the Cascadia zone, in that many earthquakes, including earthquakes up to magnitude 7.5, occur relatively frequently within it. This activity indicates that the ocean crust there contains many slipping faults on which elastic strain is released. Also in the Gorda segment, the thrust belt intersects the shoreline and extends inland down to an intersection near Cape Mendocino with the northwest-trending San Andreas fault. In this region, three quite different tectonic features merge; the San Andreas transform fault, the underwater Mendocino escarpment, and the Cascadia fold and thrust belt. The adjustment of stresses around this *triple junction* might be expected to modify the occurrence of large earthquakes on all three systems.

Slip on the Juan de Fuca subduction slab is not a matter of speculation. On April 25, 1992, a thrust of the subduction slab at shallow depth under Cape Mendocino (see Figure 1.2) near Petrolia produced a substantial earthquake (M_S = 6.9). Two damaging aftershocks followed the next day. The most notable damage in the area was to wood-frame houses, particularly in the town of Ferndale, that were thrown off their foundations: a serious but unnecessary result of large near-earthquakes (see Chapter 11).

The Cascadia case history is of particular scientific interest because it well illustrates the use of diverse geological and geophysical evidence to throw light on an earthquake hazard in a tectonic region where, unlike the San Andreas fault system in California, no direct historical evidence of major felt seismicity is available. At the moment, the Cascadia subduction zone studies do not allow any definite prediction in a span of a hundred years. They do mean, however, that there are valid reasons not to rule out significant seismic hazards in the northwest part of the United States and the Vancouver Island region of Canada.

The Parkfield, California, Earthquake Prediction: Failed but Postponed

As described above, there are two classes of credible earthquake predictions. The first involves general forecasting that gives probabilities over a considerable time span. The second type is more specific, stating the time interval, region, magnitude range, and probability. The only case history of such a definitive prophecy in the United States was that a moderate earthquake would likely occur near the town of Parkfield, California (see Figure 1.2), between 1987 and 1993.

In this rural region of rolling, open ranch land, remote from highly populated developments, the San Andreas fault trace is clearly visible, and its seismological properties along a 25-kilometer-long section are among the best understood in the world. The long-term seismographic record of the University of California observatories, operative since 1887, established that moderate-sized earthquakes (M_L 5.5 to 6) occurred near Parkfield in 1901, 1922, 1934, and 1966. There is also evidence from felt reports of similar earthquakes in 1857 and 1881. In addition, available seismograms have permitted quantitative estimates of size and location for the earthquakes of March 10, 1922; June 8, 1934; and June 28, 1966.

A simple subtraction between earthquake dates immediately suggests a cyclical pattern. There is an almost constant recurrence time of about 22 years. The exception is the 1934 fault slip (but note that 44 years elapsed between 1922 and 1966). If this cyclical pattern repeated, another Parkfield earthquake could have been expected about 1988. With an allowance for statistical variation, the window of occurrence was approximately 1987 to 1993.

More information is needed if reliance is to be placed on such correlation. Indeed, confirmatory evidence was marshaled. First, the last sizable earthquake in the area on June 27, 1966, was well recorded on seismographs in California and around the world. These seismographs yielded a magnitude 5.5 and indicated a rupture southeastward from a focus near Parkfield for a distance of approximately 16 kilometers. The fault rupture zone was confirmed by the presence of fresh cracks along the mapped fault trace south of the earthquake epicenter. Secondly, instrumental locations indicated that the earthquakes of 1922 and 1934 had an almost common epicenter with the 1966 earthquake northwest of the rupture zone mentioned above. Further, a cluster of foreshocks occurred in the 6 months before June 27, 1966, within about 3 kilometers of this epicenter. A similar foreshock pattern occurred in 1934. Such a foreshock cluster area has been called a *preparation zone*, on the theory that release of strain energy there occurs preparatory to a major rupture.

Finally, seismograms of the 1922, 1934, and 1966 Parkfield earthquakes show that similar amounts of energy were released. The elastic rebound theory (see Chapter 6) states that after a period of slow strain accumulation, the sudden fault slip is a consequence of the strength of the contacts between rock surfaces being exceeded. If the strain accumulates at a constant rate and the rock strength remains the same, it follows that there will be a constant time interval between the earthquakes. The 1934 Parkfield earthquake is an exception to such a constant pattern, but the other interoccurrence times are supportive.

As a consequence of the overall evidence and the favorable situation of the Parkfield area (low population, accessible fault zone, open countryside), a prediction experiment was undertaken there. A network of sensitive seismographs was placed around the postulated preparation zone. Surface-fault movements were monitored continuously by creep meters (see Figure 10.4) in the form of 10-meter-long devices across the San Andreas fault. Geodetic surveys began with special laser geodimeters that measure the distance across the fault between points that are 5 to 8 kilometers apart. Dilatation of the rocks began to be measured by balloonlike devices installed in boreholes; as the volume of the rock changes, the strain is measured as a change in pressure. The usefulness of watching for such precursors was suggested by reports of possible fault slip *before* the 1966 earthquake when an irrigation pipeline that crosses the San Andreas fault near Parkfield was found broken about 9 hours before the main earthquake.

Through 1999, none of these field measurements diagnosed abnormalities. Although small local shocks and measurement variations have triggered alerts, all were false alarms. It has been generally conceded now that the Parkfield prediction failed. Of course, some day a significant earthquake will be generated at Parkfield by slip of the San Andreas transform fault nearby. A new theory of initiation needs to be devised.

Calculating the Odds of an Earthquake

In Chapter 2, the patterns of recurrence of earthquakes around the globe were described. Both in broad regions and in local areas, individual earthquakes often appear to occur more or less randomly in time and space except that foci sometimes concentrate along active faults (see Figure 2.6).

Fort Wayne (Ind.) Journal-Gazette

Reprinted by permission of Dan Lynch

Usually, for hazard reduction purposes there is little concern about forecasting small earthquakes or remote earthquakes such as those along the ocean ridges. As with other nonpredictable natural hazards such as floods and windstorms, the best strategy is to state the odds that such an event will happen. In recent years, this approach to earthquake prediction has received considerable work. In particular, after the 1989 Loma Prieta earthquake publicity was given to probability statements concerning future damaging earthquakes in California. For example, a working group established by the U.S. Geological Survey calculated the chance of an earthquake of 7.0 (M_W) or greater recurring in the San Francisco Bay Area in the next 30 years. How are such calculations made?

Let us agree that probability is a numerical measure of the chance of occurrence of some event. It is generally accepted that the probability scale ranges between 0, which means no chance of the occurrence of an event, and 1, which means that the event is certain to occur. Numbers between these values give a measure of the relative probability of the event. For example, the odds of tossing a head in one throw of a coin is 50 percent and the odds of drawing one heart from a deck of playing cards is 25 percent. Of course, if the coin is unbalanced or the deck is irregular, these odds will vary.

From experience, most people have a reasonably correct idea of the probabilities involved in games of chance and in many common circumstances in life. Thus, few people would question that the chance of injury from driving on a crowded freeway is higher than from walking on a sidewalk. Similarly, there would be general agreement that the probability of injury from an earthquake is higher, in general, in the

Box 10-3

ODDS OF SEISMICITY FOR NORTHERN AND CENTRAL CALIFORNIA

The *Bulletin* of the Seismographic Stations of the University of California listed 3638 earthquakes ($3.0 \leq M_L \leq 6.9$), which occurred in the 280,000-square-kilometer area of northern and central California from 1949 through 1983. The cumulative number of earthquakes (N) expected with an assigned magnitude M_L is, on these data,

$$\log N = 4.23 - 0.815\, M_L,$$

normalized to earthquakes per year per 280,000 square kilometers.

The annual rate of seismicity ($r = 10^{\log N}$) in earthquake sequences per year and the percent probability of one or more earthquakes of magnitude M_L or larger occurring in one day, week, month, year, or decade is:

		Approximate percent probability in one				
$M_L \geq$	r (eq/yr)	day	week	month	year	decade
3.0	60.0	15.0	69.0	99.0	100.0	100.0
3.5	24.0	6.3	36.0	86.0	100.0	100.0
4.0	9.2	2.5	16.0	54.0	100.0	100.0
4.5	3.6	0.99	6.7	26.0	97.0	100.0
5.0	1.4	0.39	2.7	11.0	76.0	100.0
5.5	0.55	0.15	1.1	4.5	43.0	100.0
6.0	0.22	0.059	0.42	1.8	19.0	89.0
6.5	0.085	0.023	0.16	0.71	8.1	57.0
7.0	0.033	0.009	0.064	0.28	3.3	28.0

Los Angeles area than in Texas (see Plate 3). It would also be widely accepted that the chance of such injury would depend on whether a person were in an unreinforced brick building or in a wood-frame home bolted to its foundation.

The challenge is to give such beliefs the same definite numerical value as is achieved in calling the toss of a coin. Recently, progress has been made in reaching such a goal for earthquake occurrence. Of course, even in specially favorable geological circumstances, where regular repetitions might be expected, the reliability of the calculated odds does not approach that of the answer to the question, "heads or tails?"

Now let us turn to the ways of calculating the odds of an anticipated earthquake, or the sizes of earthquakes that will occur, in a certain area in a specified time interval. If the number and magnitudes of earthquakes in a region that occurred in 100 years is known, we might hope to calculate the average magnitude that is expected in the area or the odds that a specified magnitude will be exceeded every 10 to 20 years, say. In the San Francisco Bay Area, for example, between 1836 and 1991 (a period of 155 years) there have been five earthquakes with a magnitude of 6.75 or greater. We can then calculate that if these earthquakes occur randomly, another earthquake of the same magnitude or greater might be expected in the next 155/5 = 31 years with a high probability. Such calculation of odds of earthquake occurrence based on detailed historical seismicity is illustrated in Box 10.3.

A serious problem with this type of probability calculation is that earthquake occurrence in a given tectonic region is not exactly random but usually has systematic trends, such as gaps and clusters of earthquakes. An example of long-term gaps was given in Chapter 7 with regard to great Alaskan earthquakes. This variation in time and space renders the concept of the average odds of the occurrence of an earthquake above a given size not very useful for specific planning in the short term.

An alternative method of determining probabilities is based on the elastic rebound theory of the cause of earthquakes (see Chapter 6). The reader will recall that this theory explains earthquakes as the result of sudden slip on faults, segments of which rupture because they can no longer sustain the elastic strain built up in the neighboring rocks. As the strain increases, again after fault slip, the more probable it is that another earthquake will be generated. Thus geological or geodetic measurements should allow, in principle, the determination of which segments are most likely to slip in the future.

The first step in such an estimation is to define where fault segments begin and end; this is usually done by geological field work that maps bends or offsets in the fault or its intersections with other faults (see Chapter 2). The assumption is then made that the largest-magnitude earthquake that could be produced by any segment is that which involves the rupture of the whole segment. Thus, if sudden rupture occurs on a 40-kilometer segment of a fault like the San Andreas, as perhaps in the 1989 Loma Prieta earthquake, an earthquake of about magnitude 7 will occur. Smaller lengths will produce smaller-magnitude earthquakes and larger lengths will produce greater-magnitude earthquakes (see Appendix G).

The second step in probability assessment is to determine which fault segments along an active fault zone have slipped in the past and then to

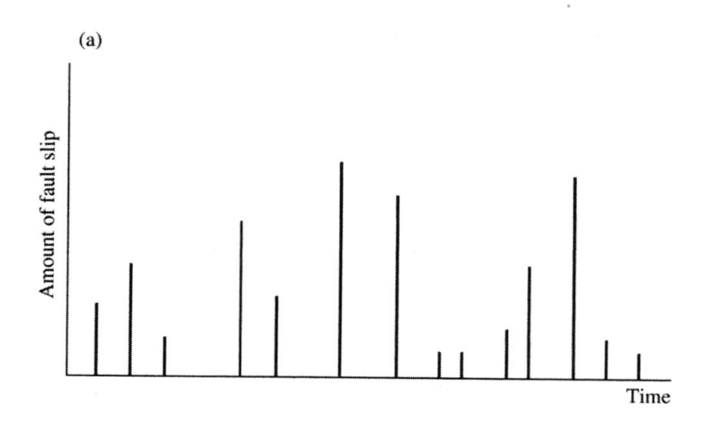

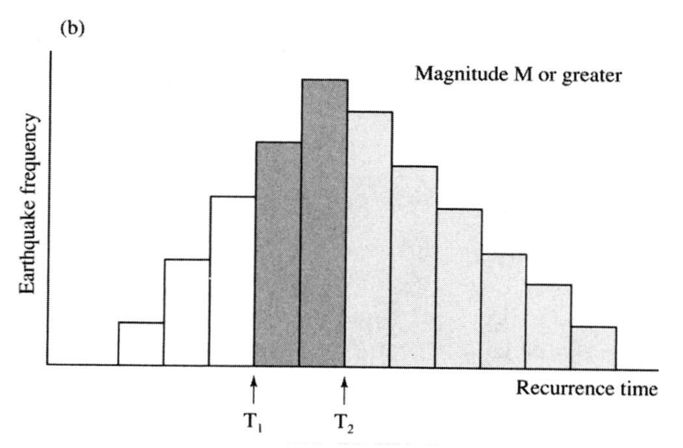

Figure 10.12 (a) Bar chart showing amount of surface slip of an active fault during successive earthquakes. (b) A histogram showing the number of earthquakes that occur with given times since the last earthquake.

calculate the rate of accumulation of strain for each segment. As an illustration, we may consider the information available in 1989 on the Santa Cruz Mountains section of the San Andreas fault which many assume ruptured in the 1989 Loma Prieta earthquake.* The average rate of slip across the

**But see the discussion when this earthquake is revisited in Chapter 12. That discussion provides a good example of one of the fascinating contemporary puzzles in seismology.*

San Andreas fault in this segment is about 1.5 centimeters per year. As a comparison, in the 1906 San Francisco earthquake, this segment of the fault is believed to have slipped about 1.6 meters, much less than the 5 meters of slip measured along other sections of the San Andreas fault to the north (see Figure 6.3). Thus from this slip rate, the probability of an earthquake rupturing the whole Santa Cruz Mountains segment (about 40 kilometers long) was, at least until the assumed rebound in the 1989 earthquake, higher than in the region to the north.

Next, consider the diagrams in Figure 10.12. The set of lines at the top indicates the fictitious geological record of slip of various amounts along a given fault segment. Each episode of slip can be related to the magnitude of the ensuing earthquake so that the intervals of time between earthquakes greater than a given magnitude can be counted. In this way, we determine the number of such earthquakes that occur with each specified recurrence time, for example, with a time between them of 50 years, 60 years, and so on. These numbers can be plotted in a histogram, such as the bottom diagram. The histogram shows the frequency that an earthquake above a given magnitude has for a specified recurrence time. From the histogram, we can calculate, for example, the most probable recurrence interval by finding the line that divides the area under the histogram into equal right and left areas.

Suppose, in the diagram, the time since the last earthquake in a specified magnitude range is T1. It follows that the recurrence time to the next such earthquake must exceed T1. Then, because we are assuming that there will be such an earthquake in the future, the probability of this earthquake occurring in time T1 to T2 years is the ratio of the area of the dark shading to the total area of the light and dark shading. As the longer recurrence time T2 increases, this ratio approaches unity, that is, the specified earthquake becomes certain. Such a calculation led to the statement issued in 1990 by a working group established by the U.S. Geological Survey that "the chance of one or more large earthquakes (M_L = 7 or larger) in the San Francisco Bay Area in the coming 30 years is about 67 percent."

This type of probability calculation, based on the amount of average slip that occurs on fault segments, is applicable only to seismic regions where the active faults are observable at the ground surface. The limitation is a serious one. One of the few seismic regions of the world where surface active faults are well mapped is the San Francisco Bay Area of California. A recently revised probability plot for this area is shown in Figure 10.13. It is clear that there are many assumptions underlying the published

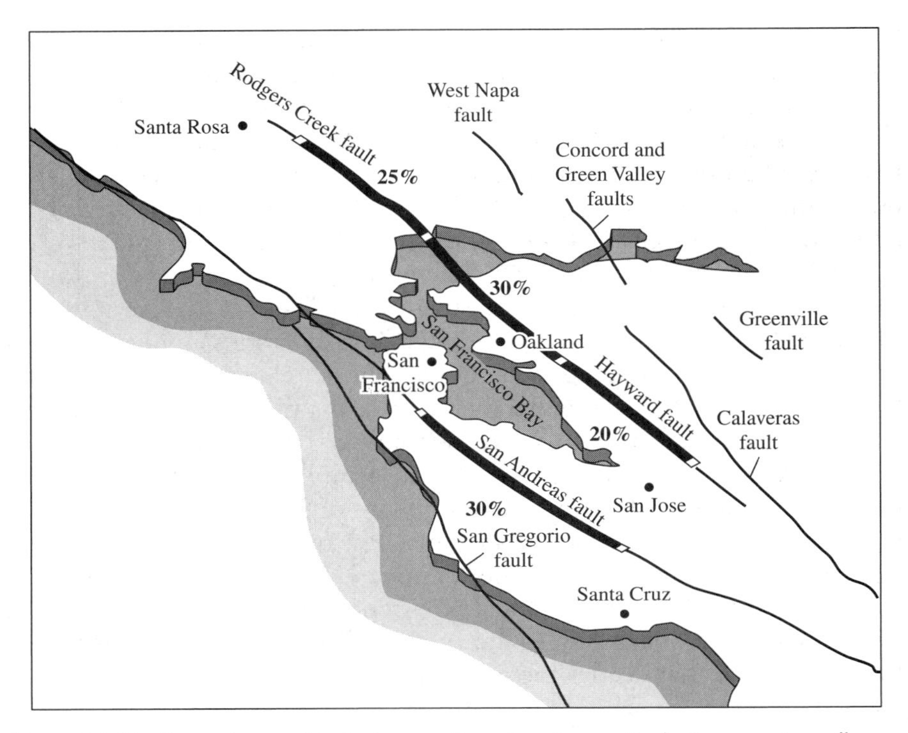

Figure 10.13 Percentage odds that an earthquake of magnitude 7 or greater will occur in 30 years on the three fault sections (heavy lines) in the San Francisco Bay Area, California. [From USGS, 1990.]

odds and that the numbers given have large uncertainties. Nevertheless, the hope is that the calculation will become more reliable as further geological and geodetic surveys are made. The probabilities of most interest are of the larger damaging earthquakes that can occur, but often large historical earthquakes in a histogram like that in Figure 10.12 are not well defined. Indeed, in 1996 through 1999 there were major changes in estimates of large historical earthquakes on the Hayward fault (see Figure 1.4).

The reader should beware of statements of probability that are not justified by the assessment method. An example of a *complete* probability statement would be: "In the selected region, or on the given fault segment, the probability of earthquakes with *magnitudes exceeding* 7 is between 10 and 50 percent in a given number of years." The key phrase is that the actual uncertainty is sign-posted by giving the percentage odds as a range, not a single value. If the magnitude range threshold in the quotation above were taken as 7.5 (say), the probability would decrease. Of course, adjacent fault segments along the same fault may rupture contemporaneously, such as in 1906 on the San Andreas fault, producing a larger magnitude (and moment) earthquake than contemplated in the assessment.

11 Dangers from Earthquakes

Interior damage over the altar in the church of Monte di Buia from the Friuli earthquake, Italy, on May 6, 1976. [Courtesy of James Stratta.]

I remember when (Britain) was shaken with an earthquake some years ago, there was an impudent montebank who sold pills which (as he told the country people) were very good against an earthquake.

—*Joseph Addison,* The Tatler

Every year there are killer earthquakes. Reports of dreadful loss of life from around the world continued through the last decades of this millennium. In the 1990s alone, over 100,000 people were killed by earthquakes (see Appendix A) with most loss of life being in Iran, India, Russia, and Japan. In addition, hundreds of thousands were injured, and the earthquakes produced enormous economic losses. Most casualties were directly caused by the collapse of weak houses and buildings.

The grim global statistics show that each year there are, on the average, over 150 earthquakes of magnitude 6 or greater, that is, about 1 potentially damaging earthquake every 3 days (see Appendix A). About 20 earthquakes with magnitudes of 7 or greater occur annually; this is about 1 severe earthquake every 3 weeks. Thus, the misfortune of the 1990s was not that more large earthquakes than normal occurred, but rather that many occurred by chance in susceptible highly populated regions.

In everyday speech the nouns "risk" and "hazard" are synonymous. By contrast, it is helpful in technical descriptions to give them distinct meanings. Thus, we define "hazard" as the event itself, that is, the earthquake ground shaking or the volcanic eruption; whereas "risk" is the danger the hazard presents to vulnerable buildings or persons. Earthquake risks must be balanced against other everyday risks. In the United States, the principal noncatastrophic risk to people in day-to-day life is posed by the automobile: about 300 of every million people are killed per year. Much farther down the scale come the main catastrophic risks such as fire (0.5 deaths per million per year) and tornadoes (0.4 deaths per million per year). The risk from earthquakes is lower still (but may take a particularly catastrophic form). It can be further reduced if we increase protective measures. Let us examine this statement by looking

Box 11-1

THE MAIN EARTHQUAKE HAZARDS

Ground shaking
Differential ground settlement
Land and mud slides
Soil liquefaction
Ground lurching
Avalanches

Ground displacement along a fault

Tsunamis and seiches
Floods from dam and levee failure and subsidence

Fires
Toxic contamination
Structural collapse
Falling objects

first at the main hazards involved with earthquakes, summarized in Box 11.1.

Types of Hazards

By far the most important hazard is the shaking of the ground (Plate 17). This in turn shakes buildings, causing objects to fall and structures to collapse partially or totally. A great deal can be learned about building safer structures by studying these effects as soon as possible on the spot, and many valuable studies of this kind have been published. Two such studies, already mentioned in earlier chapters, serve as models for seismologists today. They are the reports of Robert Mallet after the 1857 Italian earthquake and A. C. Lawson after the 1906 San Francisco earthquake.

Unfortunately, structural damage in historical earthquakes is usually not easy to evaluate. One intriguing debate centered on the biblical account of the falling of the walls of Jericho (Joshua 6:20). Some compilers of historical earthquake catalogs speculated that this event had been caused by an earthquake. A contrary opinion was voiced by the famous French seismologist, Montessus de Ballore. He argued that the walls should be the

Figure 11.1 Artist's impression of damage to Basel, Switzerland, after the October 1356 earthquake, shown on a woodcut from the "Basler Chronik" of Christian Wurstisen, 1580. [From *Basel und das Erdbeben von 1356*, (Basel: Rudolf Suter, 1956).]

strongest structures of the city, and yet Joshua's army had crossed the ruined walls "to burn the city with fire" — hardly necessary if strong shaking had already taken its toll.

For some major ancient historical earthquakes, the effects have been recorded in other ways. For example, the damage resulting from one that struck Basel, Switzerland, on October 8, 1356, is represented for posterity in a woodcut done 2 centuries later (see Figure 11.1). Collections of such earthquake "art" have now been achieved in specialist libraries (see the Further Reading section).

A few structures are vulnerable because they straddle an active fault that ruptures and causes displacements in the soil above the fault. This type of hazard can be minimized by taking care to construct buildings off the fault traces, as specified by geological mapping. To this end, special geological maps have now been drawn for various areas throughout the world. For example, a map of California published by the Division of Mines and Geology shows all known active faults (historical rupture, Quaternary displacement, etc.) in the state. Such broad hazard maps are, of course, not foolproof because some active faults may not have been detected at the time of publication (such as the 1971 San Fernando faulting, the 1975 faulting south of Oroville, California, and the 1983 Coalinga faulting);

some faults marked as active may not again be the source of large earthquakes; and some faults are "blind" thrusts (see Chapter 5).

The 1971 San Fernando earthquake north of Los Angeles provided firsthand observations of the effect of surface rupture on various types of structures. Flat-lying San Fernando is almost entirely built up with single-story wood-frame houses. The fault offset (up to 1 meter vertically and 1 meter laterally) itself produced no structural collapses, no deaths, and few serious injuries. Damage to houses along the fault scarp ranged from minor to that requiring expensive repairs, and a few homes were completely demolished. Water and gas pipes crossing the fault rupture were often compressed and ruptured, and concrete roadbeds were crushed.

Sometimes floods produced by earthquake shaking present a hazard. Along the ocean, tsunamis can cause more death and damage than the shaking itself; in lakes and reservoirs, water oscillation (seiches), the collapse of retaining walls, or landslides can pose a serious risk for people living downstream. As mentioned in Chapter 8, an earthquake (magnitude 7) on July 9, 1958, shook Lituya Bay, Alaska, and triggered a massive landslide into the bay that produced a water surge 60 meters high that ran up the bay sides over 500 meters.

Many other historical accounts of rock and debris avalanches and landslides triggered by nearby earthquakes can be cited (see Plates 19 and 20). Indeed, these gravitational effects are among the most common and hazardous results of earthquakes.

There is also the formidable threat of fire, such as resulted from the 1906 San Francisco, the 1923 Tokyo, and the 1995 Kobe earthquakes. Soon after the San Francisco earthquake, fires broke out in several places and spread for 3 days, burning 508 blocks of the city. The main problems were the highly combustible nature of many buildings, the lack of fire protective devices such as sprinklers, and the narrow streets. The ground shaking caused the city water-pipe system to break in hundreds of places so that, although there was ample water in the distribution reservoirs, little was available in the burning areas. In the 1923 Japanese earthquake, over 140,000 lives were lost in Tokyo, Yokohama, and other centers in Kwanto province, many in the fire storms in Tokyo which were fanned by high winds. After the 1995 earthquake in Kobe, many parts of the city burned.

The fire scourge is one hazard that can be sharply reduced by action and planning. Fire drills should be conducted in homes, schools, hospitals, and factories. Community fire-fighting services and regulations should be strong. In most cities the trend is encouraging. Certainly both Tokyo and San Francisco now have better equipment and water supply and fewer vulnerable buildings than at the time of their conflagrations. Nevertheless, fire

Figure 11.2 An example from the Niigata earthquake of June 16, 1964, in Japan, of liquefaction in the foundation sandy soil under apartment buildings. [Courtesy of Takeshi Minakami, University of Tokyo, Japan.]

is still a significant threat, as illustrated by the fires following the 1995 Kobe earthquake.

The ground shaking also damages the soils and foundation materials under structures, and much of the destruction in earthquakes is a consequence of this ground failure* (see Figure 11.2); these effects are particularly emphasized in the intensity descriptions given in Appendix C.

Liquefying Wet Sand

In historical descriptions of earthquakes, extraordinary pictures have been painted of spectacular changes to the ground surface in areas where water-saturated sands are common. In the great earthquakes of 1811 and 1812

**Considerable detail on these earthquake hazards is given in B. A. Bolt, W. L. Horn, G. A. Macdonald, and R. F. Scott,* Geological Hazards *(Berlin: Springer-Verlag, 1975).*

near New Madrid in the United States (see Figure 8.1), the accounts speak of high banks caving into the Mississippi River, sandbars of islands giving way and even disappearing. Large areas were covered with water which emerged from below through fissures and craterlets, and widespread landslides along the hillsides occurred. Many people were drowned when thrown into the river by collapsing banks. In all, an area of over 90,000 square kilometers was seriously affected by raised and sunken lands, fissures, soil sinks, sandblows, and large slides.

In more recent times, the reason for much of this ground failure has been shown to be the special behavior of fine-grained saturated soil when shaken. *Liquefaction* is the condition of soil losing its resistance to shearing. When, for example, wet sand is subjected to repetitious motions, it is found that the pressure of the water between the sand grains increases until eventually this pore pressure comes close to the external pressures on the soil. Sandy soil then takes on the characteristics of a dense liquid rather than those of a solid. The soils must have fine-enough sand grains that are not packed closely nor held together by cohesion produced by clay materials. After sufficiently intense or long duration of shaking, the strength of the soil is lost and the mix of sand and water flows. The pressure of the overburden forces this mix to the ground surface, producing the remarkable features called "sand volcanoes" and "boils" (see the illustration facing the opening page of Chapter 8).

Fine-grained sands that are water saturated are very widespread, particularly in low-lying areas, many of them used for agriculture. Certainly, in 1811 and 1812 in the Mississippi Valley the graphically described features were a prime example of liquefaction.

In the 1989 Loma Prieta earthquake in California, liquefaction of underground wet sand layers was widespread around the margins of San Francisco Bay and along the low-lying coastal areas. Indeed, considerable damage to apartment buildings in the Marina District of San Francisco has been traced to liquefaction of sand placed early in this century to fill the area.

A lasting impression for me, gathered in the field two days after the 1976 Romanian earthquake, was the liquefaction which occurred in farming areas along the Danube River. Although the source of this earthquake was over 400 kilometers from the affected area and the ground motion was only moderate, sand-filled fissures, sand volcanoes, and boils appeared throughout the area. In some water wells, wet sand had filled the excavation to the surface. In orchards, up to 20 centimeters of sand covered the soil, and on the concrete floors of farmers' houses, sand and water had squirted through cracks, leaving centimeters of sand and silt.

Liquefaction of sandy foundations has disastrous effects on structures. In the earthquake that affected Niigata in Japan in 1964, reinforced concrete buildings, otherwise structurally undamaged, tilted calamitously because of the liquefaction of the underlying soil (see Figure 11.2). Failures of walls along harbor facilities and bridge piers and embankments were also quite severe, as they were in Chile in 1985 and Kobe in 1995. The problem can often be somewhat mitigated. Engineers are able to test, by drilling or driving rods into the basement material, whether sand that is capable of liquefaction is present. Such site studies are commonplace for important structures such as large earthfill dams and bridges. Appropriate design can then be used in this construction — such as the use of deep pilings as structural supports. In other cases, the best procedure is to prevent, by zoning, the development of structures that would sustain costly damage due to liquefaction of the foundations.

Self-Protection in an Earthquake

The best protection advice for an individual is to be prepared and not to panic. Remember that earthquakes are strong vibrations of the ground that will greatly subside in less than a minute, often in less than 15 seconds. In this brief period of severe shaking, quickness of wit can prevent injury. People are often surprised by how calm they have been.

Box 11.2 lists suggestions for protection before, during, and after an earthquake. If you are in an open area or if you are in a car or other vehicle on the open road, you will have little to fear from even a high-intensity earthquake. If you are indoors when the shaking starts, get under the strongest structure in the room, perhaps an inside doorway or a strong table or chair. These help protect you from falling objects, such as light fittings and ceilings. Evacuate the building as soon as possible because it might be damaged, and aftershocks may collapse weakened buildings many hours after the mainshock. If you are in a city street during the shaking, move to the center of the street or into a doorway to avoid falling broken glass and building walls.

Householders and apartment dwellers should have a fire extinguisher easily accessible in the home. Then if a fire breaks out from, say, cooking oil on the stove, it can be dealt with. Even if water mains are ruptured, liquid can usually be found for minor fire-fighting purposes, first aid, and even drinking in toilet cisterns, water heaters, canned drinks, and elsewhere. A flashlight should always be on hand at night, because the power supply often fails immediately. A first-aid kit may be needed, particularly

Box 11-2

PERSONAL PROTECTION IN AN EARTHQUAKE

Before an Earthquake

At home

Have a battery-powered radio, flashlight, and first-aid kit in your home. Make sure everyone knows where they are stored. Keep batteries on hand.

Learn first aid.

Know the location of your electric fuse box and the gas and water shut-off valves (keep a wrench nearby). Make sure all responsible members of your family learn how to turn them off.

Don't keep heavy objects on high shelves.

Securely fasten heavy appliances to the floor, and anchor heavy furniture, such as cupboards and bookcases, to the wall.

Devise a plan for reuniting your family after an earthquake in the event that anyone is separated.

At school

Urge your school board and teachers to discuss earthquake safety in the classroom and secure heavy objects from falling. Have class drills.

At work

Find out if your office or plant has an emergency plan. Do you have emergency responsibilities? Are there special actions for you to take to make sure that your workplace is safe?

During an Earthquake

Stay calm. If you are indoors, stay indoors; if outdoors, stay outdoors. Many injuries occur as people enter or leave buildings.

PERSONAL PROTECTION IN AN EARTHQUAKE (Continued)

If you are indoors, stand against a wall near the center of the building, or get under a sturdy table. Stay away from windows and outside doors.

If you are outdoors, stay in the open. Keep away from overhead electric wires or anything that might fall (such as chimneys, parapets, and cornices on buildings).

Don't use candles, matches, or other open flames.

If you are in a moving car, stop away from overpasses and bridges and remain inside until the shaking is over.

At work

Get under a desk or sturdy furniture. Stay away from windows.

In a high-rise building, protect yourself under sturdy furniture or stand against a support column.

Evacuate if told to do so. Use stairs rather than elevators.

At school

Get under desks, facing away from windows.

If on the playground, stay away from the building.

If on a moving school bus, stay in your seat until the driver stops.

After an Earthquake

Check yourself and people nearby for injuries. Provide first aid if needed.

Check water, gas, and electric lines. If damaged, shut off valves.

Check for leaking gas by odor only (*never* use a match). If it is detected, open all windows and doors, shut off gas meter, leave immediately, and report to authorities.

Turn on the radio for emergency instructions. Do not use the telephone — it will be needed for high-priority messages.

Do not flush toilets until sewer lines are checked.

Stay out of damaged buildings.

Wear boots and gloves to protect against shattered glass and debris.

Approach chimneys with caution.

At school or work

Follow the emergency plan or instructions given by someone in charge.

Stay away from beaches and waterfront areas where tsunamis could strike, even long after the shaking has stopped.

Do not go into damaged areas unless authorized. Martial law against looters has been declared after a number of earthquakes.

Expect aftershocks: they may cause additional damage.

Figure 11.3 Damage to wood-frame structures, Melipilla, Chile, in the 1985 earthquake. [Photo by Bruce A. Bolt.]

for injuries due to broken glass. Leaking gas from broken connections to overturned gas heaters or broken gas mains can produce dangerous fires and explosions; therefore, open flames should be shut off until safety checks are completed. Finally, it is advisable to keep informed about the nature and extent of damage throughout the shaken area: keep a battery-powered radio in the house. The telephone may be dead, but even if it is not, it should be used only for emergency calls.

The odds are high (over 60 in 100) that a damaging earthquake will occur when most persons are at home. Where the standard of home construction is high or timber frame is used, this is to the good. But unfortunately, in many places, construction materials and methods are seismically hazardous. For example, in China, as well as regions of the Mediterranean, the Caucasus, South and Central America, and Asia, the types of housing almost guarantee heavy death tolls during an even moderate earthquake shaking (see Figure 11.3). As described in Chapter 5, this was what happened in the 1988 Armenian earthquake. The economic

Figure 11.4 Interior damage in a wood-frame house at Inangahua following the earthquake of May 23, 1968. Note that cupboard doors and drawers have come open, fixtures have moved from walls, and the electric range has fallen on the floor (becoming a possible fire hazard). [Courtesy of *New Zealand Weekly News* and R. D. Adams.]

resources to bring the present rural housing in most earthquake-prone countries up to adequate earthquake-resistant levels in a short time are just not available.

A disaster is less serious if most of the people in a damaged area survive, because the local labor force can immediately undertake reconstruction and repairs. This is one reason why so much effort is being poured into earthquake prediction in China, where a million workers may be in jeopardy from one great earthquake. However, as pointed out in Chapter 10, practical prediction activities, despite some successes, may, *on the average*, produce social and economic dislocations. Inexpensive modifications of rural and urban housing designs (such as the use of corrugated iron and simple wood and metal reinforcement) are the best long-term measures to prevent injury.

In contrast, the single- and two-story wood-frame houses typical of the United States and New Zealand, and the light wooden buildings of Japan are examples of places that are among the safest to be in an earthquake. These buildings can suffer damage, as shown in Figure 11.4, but it is minor in comparison with the total collapse that can and does occur otherwise (see Figure 11.3). But even in these countries the trend is to experiment with new materials and change the design of ordinary buildings, so that the increase in seismic risk may not be recognized until an earthquake occurs. For example, the 1971 San Fernando earthquake in California demonstrated that well-constructed concrete-block structures, unlike older weak masonry, have a high seismic resistance. However, some newly completed wood-frame houses of split-level design, presumably built according to code, collapsed. Unlike the older houses with quite small windows and a separate garage, there was insufficient shear bracing in the garage walls at ground level. Shaking collapsed the garage, causing the rooms above to drop, many on the family cars.

The 1983 Coalinga, California, Earthquake

In the United States, since the early 1970s, the tempo of federal and state government programs related to earthquakes has increased. In large measure, the stimulus was the February 9, 1971, San Fernando earthquake in southern California, which was the first of significant size to occur in a modern urban environment in the United States.

Starting in 1971, hearings were held and several bills related to seismic risk and hazard reduction were introduced into Congress. The culmination of these activities was the National Earthquake Hazards Reduction

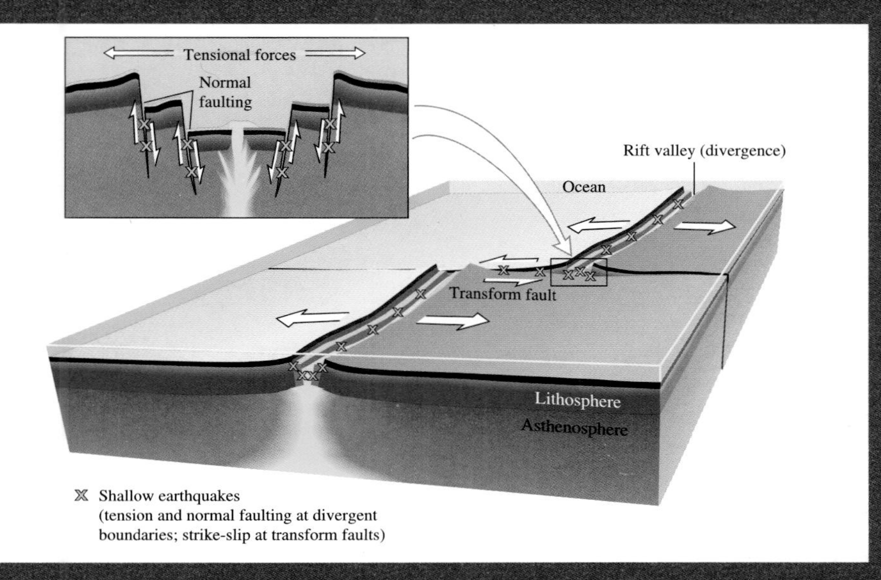

PLATE 12

Earthquake sources associated with two types of plate boundaries: divergent boundaries of ocean ridges and transform faults. [From Press and Siever, *Understanding Earth*, Second ed., New York: W. H. Freeman and Co.,1998.]

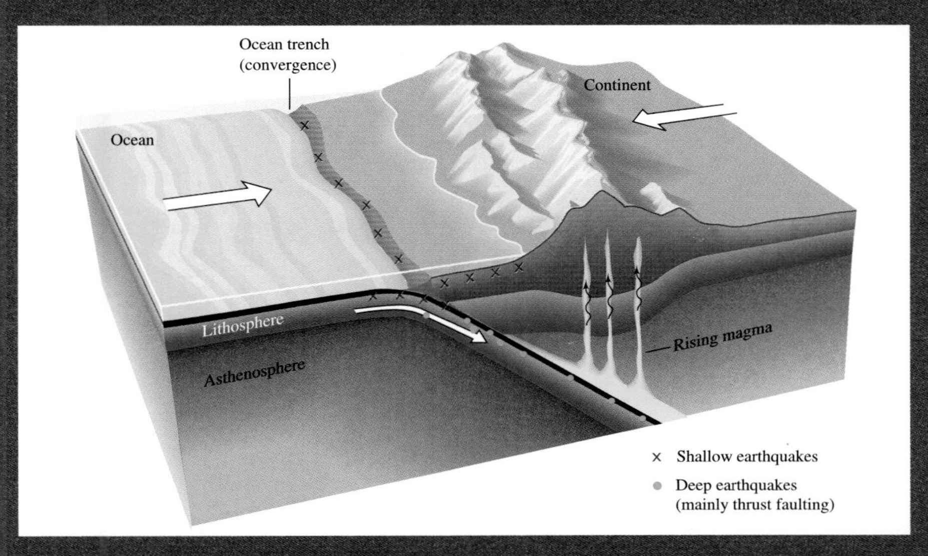

PLATE 13

Shallow- and deep-focus earthquakes along a convergent slab plunging underneath the upthrown mountain range. Magma is rising from the slab boundary at depth. [From Press and Siever, *Understanding Earth*, Second ed., New York: W. H. Freeman and Co., 1998.]

PLATE 14
A precarious rock in the Lovejoy Buttes, near Mojave, California, photographed in 1997 at a site only 16 kilometers from the seismically active San Andreas fault. The top boulder is about 2 meters across. No toppling puts a limit on the maximum earthquake intensity. [Courtesy of J. Brune.]

PLATE 15
Sea wave advancing on La Manzanilla in the southern end of Tenacatita Bay, Mexico, in the tsunami of October 19, 1995. By the time the lower photo was taken, the water had advanced more than 100 meters. The wave traveled about as fast as a person can run. The accompanying earthquake killed 40 people. [Courtesy of J. Martinez and C. Synolakis.]

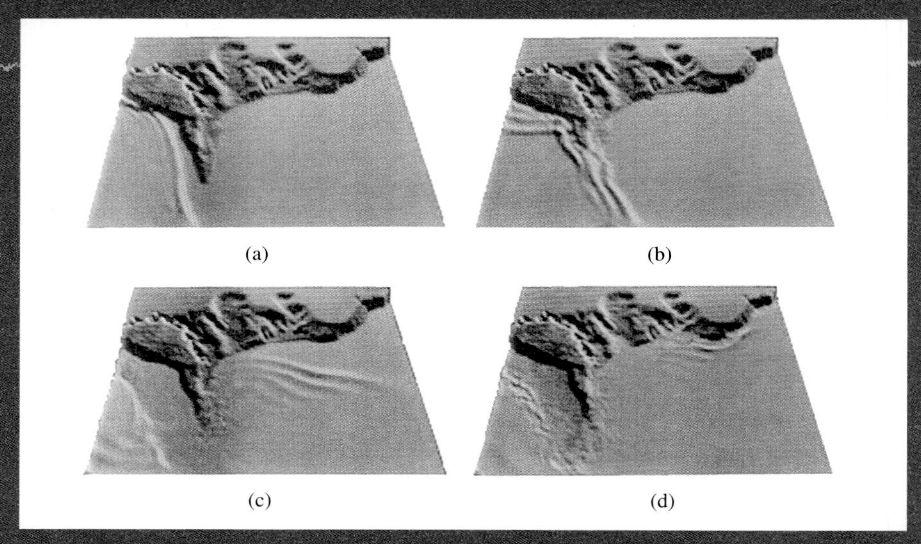

PLATE 16
Hokkaido, Japan, earthquake of July 12, 1993. Computer visualization of the generated sea wave as it (a) approaches, (b) overruns, (c) refracts around, and (d) passes by the Aonnae Cape on Okushiri Island. The time frames are (a) 5 minutes, (b) 6 minutes, (c) 8 minutes, and (d) 12 minutes after the earthquake. Note the tsunami run-up on the beach in (d). [Courtesy of V. Titov and C. Synolakis.]

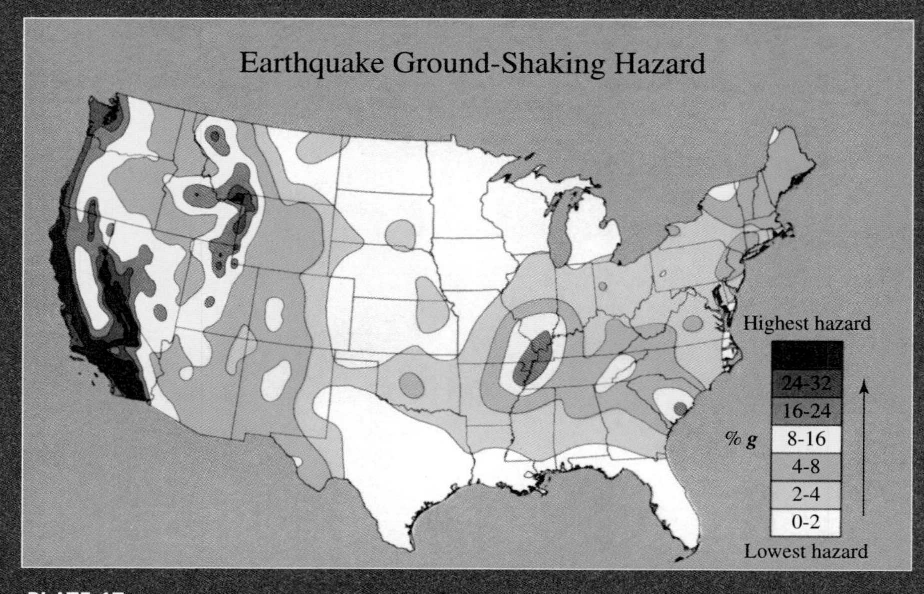

PLATE 17
Contours of the hazard of earthquake ground shaking in a horizontal direction as a percentage of gravity, having a 10 percent probability of being exceeded in 50 years, for firm soil or rock in the contiguous United States. [Courtesy of USGS, 1997.]

PLATE 18

This Jeffrey pine, called the Pool Tree, has been growing near the San Andreas fault in southern California for more than 360 years. The top was broken off in the large 1812 earthquake centered nearby. The root system was so damaged that, as its tree rings show, growth slowed for years afterwards. [Courtesy of G. C. Jacoby.]

PLATE 19
A view of Yungay looking east from Cemetery Hill, showing the town as it existed before the May 31, 1970, earthquake and the debris avalanche that it triggered.

PLATE 20
Same view of Yungay taken toward Mount Huascaran. Photo was taken after the devastating avalanche. [Courtesy of L. Cluff.]

PLATE 21
Photo showing a primitive kind of base isolation of a building against seismic ground shaking. The wooden columns of this Japanese temple, built in 1628 A.D., have feet that can slide freely on the rock platform. [Courtesy of B. C. Gerwick.]

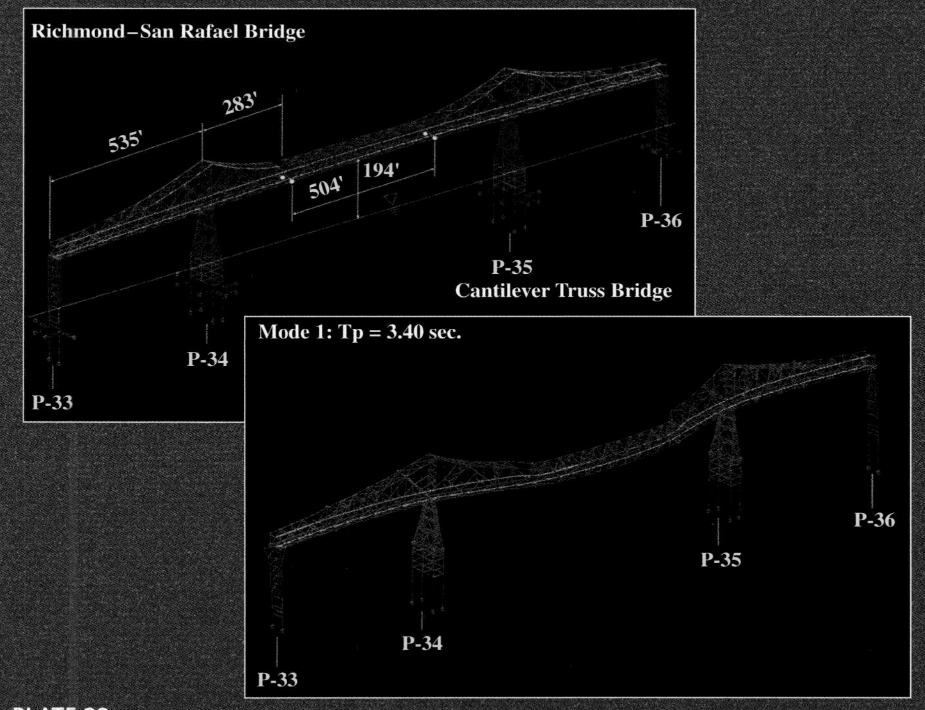

PLATE 22
Computer output model of a cantilever truss bridge (Richmond–San Rafael, California) demonstrating how modern dynamic analysis is made by engineers. The bridge model is subject to input earthquake motion and shakes like a vibrating string. Each mode of vibration is shown on the computer screen. (Top) Bridge model without input motion; (bottom) the first mode of vibration of the bridge. [Courtesy of E. Wilson.]

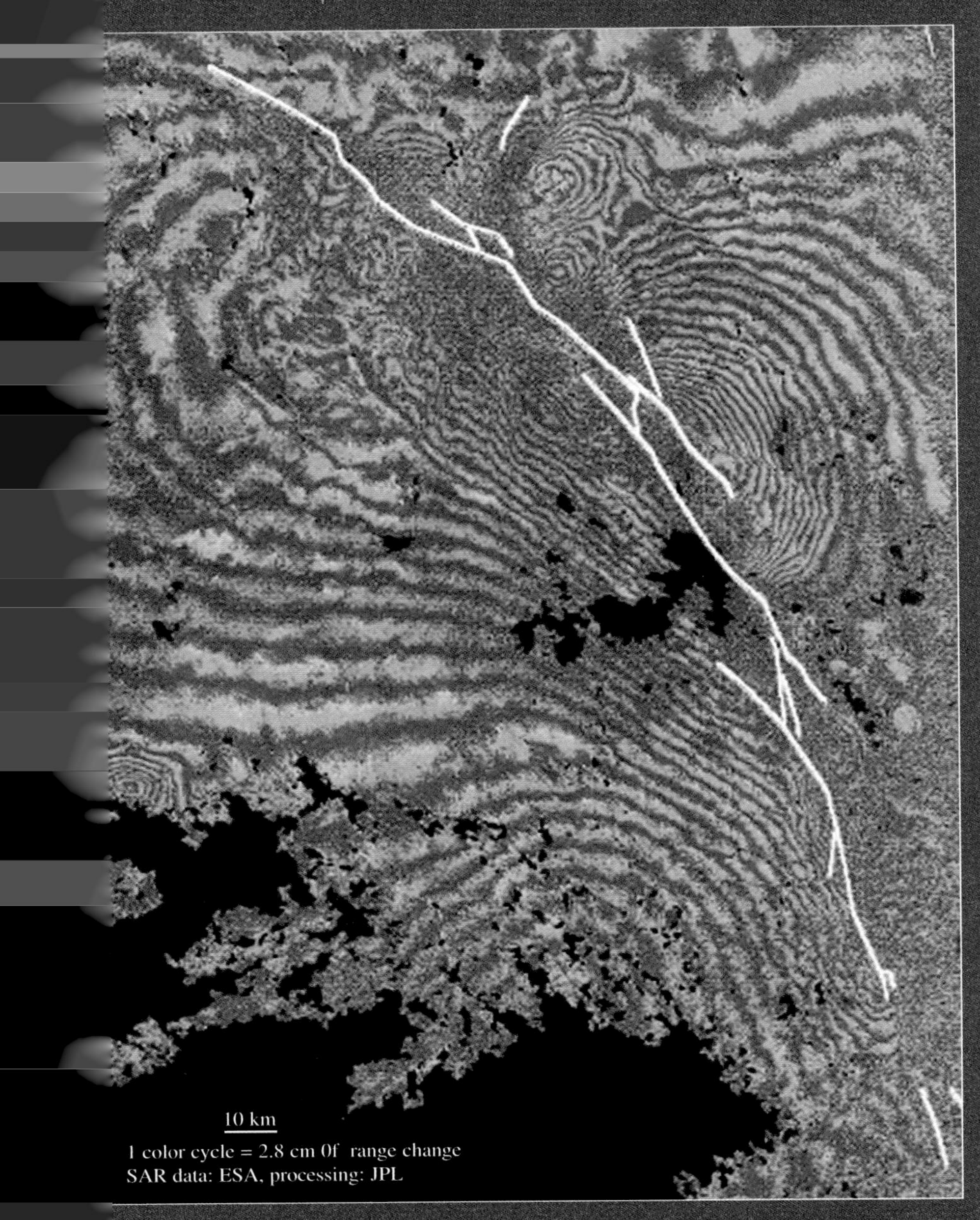

PLATE 23

Ground displacement fringes (similar to an interferometer in light studies) obtained by subtracting radar images from the ERS-1 satellite before and after the 1992 Landers, California, earthquake: one cycle of gray shading represents a difference of 28 millimeters between the two images. The irregular thin white line is fault rupture plotted in the figure at the beginning of Chapter 5. [Courtesy of D. Massonnet, M. Rossi, and F. Peltzer.]

Program, or NEHRP, a program based on the National Earthquake Hazards Reduction Act of 1977. In contrast to earlier versions, which stressed earthquake-prediction research, it has been very helpful because it focused on practical questions of mitigation of earthquake hazards. Responsibilities were assigned to various federal departments and agencies, with significant parts being played by state and local government, universities, private organizations, and individuals. In this way, the appropriated funds, continued in succeeding years, have had a far-reaching effect on every level of society touched by seismic dangers.

In California, government reaction to the 1971 San Fernando earthquake was quick. In 1972 the legislature created the Seismic Safety Commission* to be responsible for establishing state goals and priorities on earthquake-hazard reduction. The commission has broad independent powers of recommendation and coordination, including the proposal of needed legislation. The establishment of this commission is among the most significant accomplishments on earthquake-hazard mitigation in the United States, and specifically in California, since the study of the 1906 earthquake by the State Earthquake Investigation Commission (see Chapter 1).

One activity of the commission has been to study the aftermath of each damaging earthquake in California. One case, which had direct lessons for the general public, was the magnitude 6.7 earthquake that struck Coalinga, an isolated town in the San Joaquin Valley, on May 2, 1983 (see Figure 1.2). At that time, the population of the town was about 6500. The community was relatively young, yet it had a significant number of residents 65 years of age or older. It was not prosperous or particularly poor. Most residents lived in single-family, wood-frame residences of various ages.

The position of this earthquake is a reminder that damaging seismic shaking can occur in places not associated with major or even known active faults. Surface faulting was observed in a large aftershock but not in the mainshock. The latter fault mechanism was unusual: it was blind-thrust, not strike-slip faulting. The shallow dipping fault rupture in this case was concealed in the core of an anticlinal (upwardly curved) fold of the rock strata. The fold had been mapped by petroleum geologists who discovered oil there in 1898. The fold ridge rose 75 centimeters as a result of the thrust rupture below.

Such hidden or "blind" rupture present a special problem in estimating seismic hazard. The catalog of historical seismicity did show that

The author was a member of the Commission from 1980 through 1993.

moderate earthquakes were pervasive along the eastern front of the coast ranges of California running northwest through Coalinga; in 1982 and 1985, there had been sequences of felt earthquakes centered about 30 kilometers from the town. Nevertheless, Coalinga was essentially unprepared for an earthquake of even moderate magnitude. Nearly 200 people were

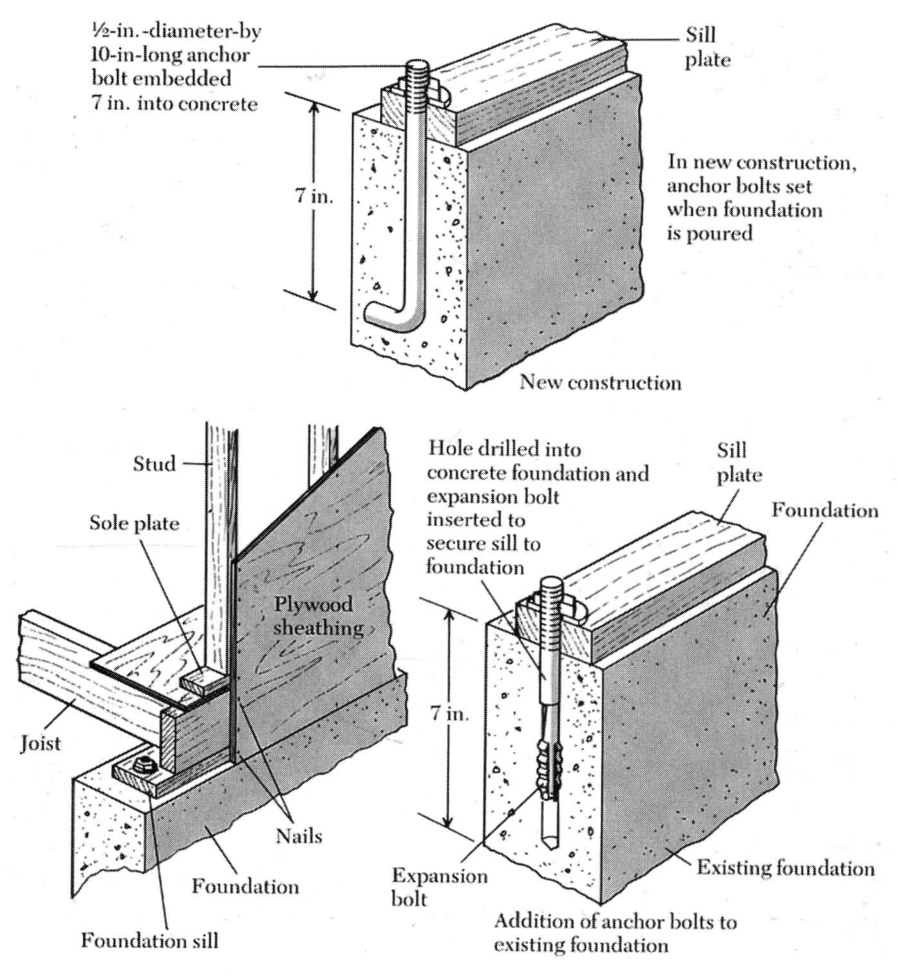

Figure 11.5 Detail of connections of existing wood-frame buildings to the foundation. Plywood sheathing should be securely nailed to sills and studs. Anchor bolts ($^1/_2$-inch in diameter) can be embedded into concrete or masonry foundation no more than 6 feet apart and within 12 inches of the end of each sill board.

injured and approximately 1000 were left homeless. The earthquake caused over $30 million in damage. Production in the surrounding oil fields was disrupted. The town was faced with replacing or repairing two-thirds of its housing and rebuilding essentially the entire 12-square-block business district.

The Seismic Safety Commission had hearings on the post-earthquake recovery and cataloged all the available information and studies. There were problems with recovery that may be common to many communities, including early demolition of damaged downtown structures that led to controversy and lawsuits. There were dilemmas related to financing the long-term economic recovery (as opposed to earthquake-damage repairs), and there were debates concerning the design of the new business district. The effect of the earthquake on sales and prices of houses was small, but it caused many rents to increase. The exodus of residents and businesses predicted after the earthquake failed to materialize, and employment eventually rose above 1983 levels. Some of the residents reported having earthquake-related problems, both emotional and financial, but most residents were generally satisfied with the manner in which various government agencies responded to their needs in the post-earthquake period.

One of the major lessons of the earthquake was the importance of bolting wood-frame homes to their foundations to prevent separation (see Figure 11.5). Surveys showed that about 15 percent of one-story wood-frame houses were damaged because they fell off their foundations. The maximum Modified Mercalli intensity in Coalinga was estimated to be VIII (see Appendix C), yet the large majority of houses with adequate foundation connections were hardly damaged; newer homes of all types showed no indication of significant structural damage, although some chimneys fell or cracked.

Steps to Reduce Hazards to Homes

Basic structural design aside, a householder can take certain steps to minimize earthquake vulnerability, such as the following:

1. Exterior sheathing of wood-frame houses should be waterproof plywood of 1 centimeter minimum thickness, adequately nailed. Because garage doors and large windows weaken the shearing strength, bracing such as plywood sheathing should be added.
2. Internal lighting fixtures and utility equipment (water heaters, refrigerators, wall stoves) should be fastened to structural elements

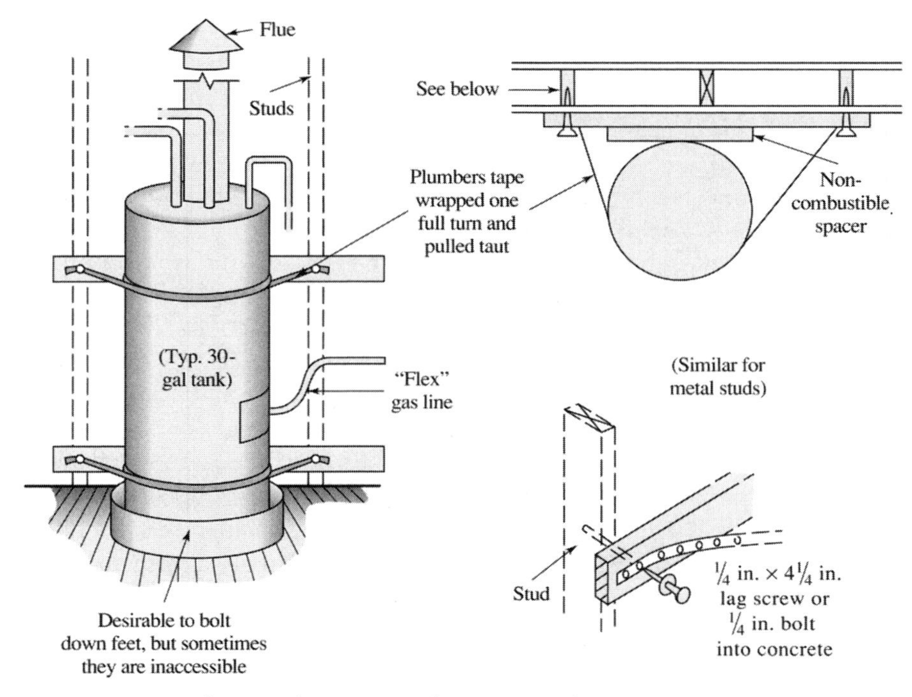

Figure 11.6 Attachment of a gas water heater to studs.

securely enough to withstand large ground acceleration (see Figure 11.6).

3. Brick chimneys should be adequately reinforced and braced to structural elements to prevent collapse into the living area; if they are not reinforced, the flues should be lightweight. Reinforcement with only four rods of vertical steel does not provide sufficient safety in high earthquake-risk zones.
4. The frame and sill plate should be inspected periodically to assure that the wood structure, built to resist lateral forces and *tied to concrete foundations,* has not been damaged by termites or fungus.
5. Since unreinforced brick and concrete-block walls often collapse during seismic shaking, all masonry walls should be reinforced and tied to adequate footings.
6. Roofs and ceilings should be of as light a construction as the climate allows.

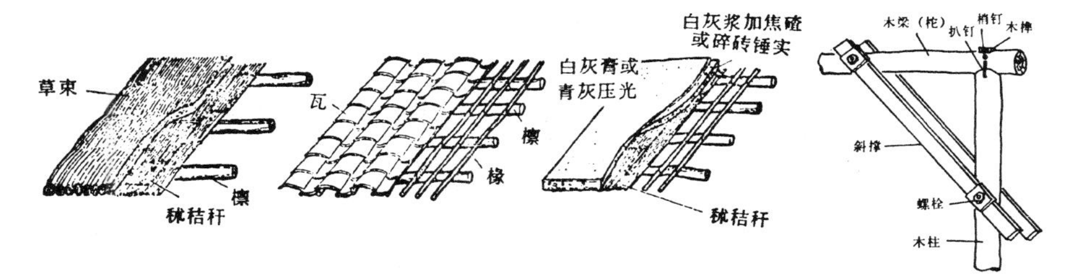

Figure 11.7 Diagrams showing how to build roof supports in Chinese peasant dwellings. [From a popular booklet on earthquakes distributed in the People's Republic of China.]

7. In high seismic-risk zones where foundation soils may move, flexible joints should be provided between the utility lines (particularly gas and water lines) and the outside mains.
8. Closets and heavy furniture should be fastened or strapped to the wall studs wherever these constitute a danger or contain valuable property.

Furthermore, every seismically active country could benefit from large-scale programs of instruction to the populace on how to make homes more earthquake resistant. For some decades, for example, in China the people have been taught how to strengthen ordinary rural dwellings (see Figure 11.7).

Even in small-magnitude earthquakes, such as the 1975 Oroville and the 1983 Coalinga earthquakes in California, the destruction and economic loss caused by falling objects can be high.* Attachments of light fixtures often need strengthening, and the catches on cupboard doors and the way items are stocked on the shelves should be effective (see Figure 11.8). Inexpensive restraining bars and supports should be placed on countertops and shelves, particularly in hospitals, where the drugs, chemicals, and equipment on shelves are vital and, if broken, can be lethal.† Also the

**So many bottles were broken in liquor stores in Oroville in the 1975 earthquake that the state legislature passed a special bill providing reimbursement for the owners from state funds!*

†Helpful ideas on preventing nonstructural damage can be found in R. Reithermann, Reducing the Risks of Non-Structural Earthquake Damage: A Practical Guide, *2nd ed. (Oakland, Calif.: Bay Area Regional Earthquake-Preparedness Project, 85-10, 1985).*

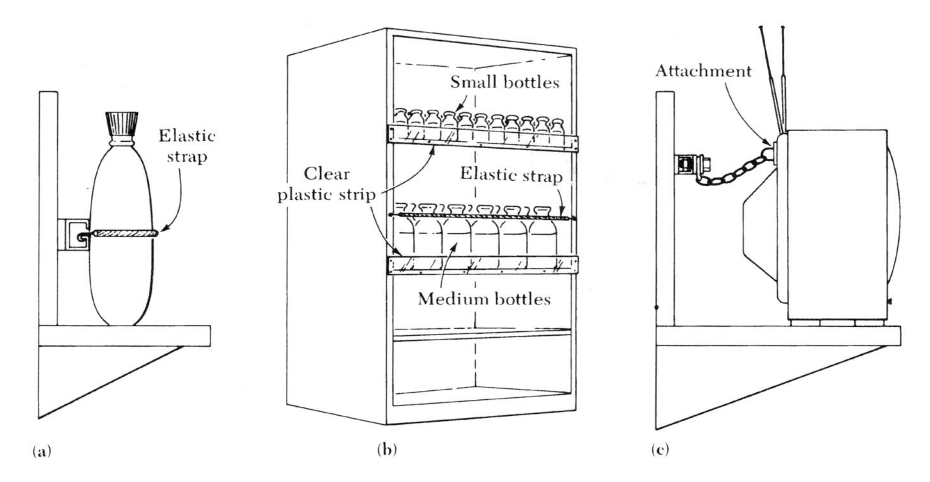

Figure 11.8 Methods of securing items on shelves. (a) Countertop items are secured by an elastic strap extending around the item. (b) Small items, such as those on the top shelf, can be affixed by a vertical strip, and the medium-sized bottles below by a strip and a horizontal elastic strap. (c) Large items on shelves can be attached at the rear by a chain that connects to a bolt or hook at the back of the shelf.

invaluable and irreplaceable art objects in museums and galleries should be secured to withstand lateral shaking. (Small hinges of strong transparent tape or loops of fishing line provide some protection.)

Despite the ample descriptions of the effects of earthquakes over the centuries on buildings and the Earth's surface around the world, there is surprisingly little information on the human reaction and social results. The folklore of earthquake countries such as Japan and China reflects the reactions of the human mind when under the influence of sudden and uncontrollable natural forces. Thus, the Japanese woodblock prints of catfish (see the illustration on page 104) contain conflicting themes; sometimes the catfish is the destroyer, attacked by the people who have suffered from the earthquake calamity. In others it is a benefactor who has, by damaging the homes of the wealthy, provided work for the artisan classes. Some namazu-e have political, humorous, or social content in the same spirit as modern cartoons.

After the San Fernando earthquake in 1971, I was impressed by the generally calm response of the population in the meizoseismal area.

During the subsequent balmy nights, many residents slept on lawns and in cars, as a sensible precaution against damaging aftershocks. Fortunately, unlike some other earthquake episodes, such as the Friuli sequence in Italy in 1976, aftershocks of the 1971 San Fernando and 1994 Northridge main shocks in California caused little additional damage. As time went on, however, there were disquieting reports of children showing emotional disturbances, particularly related to the unpredictability of the aftershocks. In modern society, after great natural disasters, the veneer of scientific and logical thought gives way to more primitive reactions and explanations that were connected in earlier times with superstitious beliefs and folklore. One cannot help wondering what the personal responses will be to future great earthquakes* in a society such as in California where the roots to folk legends and beliefs have been much attenuated.

Help from Earthquake Insurance

Availability of insurance coverage against earthquake damage varies a great deal from country to country. From an insurance company's point of view, earthquake insurance is not comparable to automobile insurance or life insurance but is rather like a stowaway that has crept into the business. The insurer is used to taking a calculated risk, but earthquake risk does not follow the usual rules. These are that the insured event must occur with some predictable regularity; the probability of sustained damage and its magnitude must be calculable; risk must be spread geographically; and the amount of damage must be limited.

Nevertheless in a number of countries, such as New Zealand, Japan, and the United States, insurance schemes covering earthquake risks have been instituted. In general, the cost of earthquake coverage is expensive but affordable in relation to the value of a dwelling and the protection provided, although there is frequently a substantial deductible sum that the insured must bear personally.

In the United States, earthquake coverage is available from private insurance companies. For example, in one typical 1985 scheme made available to homeowners, coverage for $100,000 costs annually $200, with 10 percent deductible allowance. There are also policies providing insurance

**In the damaging San Salvador earthquake of May 3, 1965, the handyman at the seismographic station of El Salvador became so disturbed by the violent shaking that he grabbed a pistol and fired it at the seismograph!*

Figure 11.9 Collapse of masonry walls of the California Hotel in the Santa Barbara, California, earthquake on June 29, 1925. No one was injured. Rebuilt and reinforced, the hotel stands today. [Photo by Putnam Studios.]

against tsunamis, landslides, and other geological hazards not necessarily related to earthquakes. It is common, however, for insurers to exclude secondary damage caused by earthquakes, such as fire, when policyholders do not secure an earthquake insurance rider.

In California, the insurance premium depends on the zone in which the building is located. For a typical modern, wood-frame dwelling, the coverage is usually written as an addition to a regular homeowner's policy, with a deductible amount calculated separately for the dwelling and contents. Earthquake insurance on poorly built dwellings may be quite expensive and is not common. In 1984, a state law required insurance companies to offer earthquake coverage as a special rider to homeowners' policies. The state has experimented with a California Earthquake Authority empowered to set insurance rates and provide coverage up to a limit. Nevertheless, 80 percent of homeowners do not have such insurance, perhaps because of the substantial deductible amount (see Figure 11.9).

The type of construction, the location of the dwelling, and the amount of damage that must be sustained before the policy becomes applicable are all factors that must be considered before a decision is made.*

Surprisingly, following the 1989 Loma Prieta earthquake in northern California, which caused an estimated $6 billion in property damage, earthquake insurance for homeowners became more widely available. In part, this was because insured losses (approximately $680 million) were not much more than most insurance companies would pay for damage from a widespread and severe summer hailstorm. However, because most private earthquake insurance is relatively expensive, with high deductibles, homeowners still turn away from earthquake insurance coverage. For a deductible of 10 percent of the insured value of the home, if the owner insured for $250,000, there would have had to be earthquake damage of $25,000 before the insurance policy helped. In addition, homeowners without earthquake coverage could still manage to collect some insurance because regular homeowner policies covered itemized possessions damage; if the home burned, costs would be fully recovered under the fire insurance policy. By such means, overall the 1994 Northridge earthquake near Los Angeles, California, led to the largest earthquake insurance loss in history, about $12.5 billion.

In Japan, where seismic risk is similar to that of California, local insurers divide the islands into earthquake zones according to earthquake frequency and degree of hazard. Zone 5, consisting of the prefectures of Tokyo, Chiba, Kanagawa, and Yokohama, is considered the highest risk area. Although earthquake coverage is available, it is granted only as an extension of fire coverage and is not often bought. For the individual homeowner, the maximum coverage available is capped at about 30 percent of the sum insured for fire. This is the main reason why, after the severe 1995 Kobe earthquake, insurance payments were only about $6 billion.

New Zealand has a government-backed scheme of indemnity for earthquake shaking and fire damage, as well as other geological hazards such as landslides. Under the government act, private insurance companies have the responsibility of collecting the premium. One-half of this premium

In 1999, a typical basic California Earthquake Authority policy in a metropolitan area, for a home insured for $194,000 (structure only), cost $729 per year. An available supplemental policy with 10 percent deductibility cost $281, covering $25,000 contents and $10,000 living allowance. Such Authority rates are based on the applicable earthquake-hazard zone.

may be claimed from a mortgagee, and property owners must bear 1 percent of the loss incurred from earthquakes. All insurance policies in New Zealand include this earthquake coverage, even those on automobiles, and the only way to avoid it is to be totally uninsured.

For Europe generally, the availability and cost of insurance reflects the generally low risk and low demand (see Plate 3). In the United Kingdom, where small-to-moderate earthquakes have occurred in historical times, coverage is readily available under comprehensive policies and is included within the overall rate. Earthquake coverage is not available in Germany or Holland, nor in the seismically quiet areas of Belgium, Denmark, Sweden, and Norway. In France, coverage is not usually available, although insurers may grant it on request as an extension to a fire policy.

Spain and Switzerland have compulsory earthquake coverage incorporated in general policies, and premiums are paid into a special government fund. In Spain, earthquake coverage is part of a package for catastrophe perils. The government must meet any claims, provided the earthquake damage results from intensities exceeding VII on the Modified Mercalli

scale. In Italy, with its belt of seismicity in the north and south, coverage can be added to a fire policy, but terms are subject to individual negotiation. In Greece, earthquake coverage is written separately from fire insurance. In Portugal, coverage is available as an extension to the fire policy. The country is zoned for rating purposes, the highest rates being in the south and west, and the lowest in the north and east.

In Canada, coverage is available as an extension of fire policies in all provinces, and the rates vary according to seismic zone. In Australia, earthquake insurance rates reflect the saying that the continent is "quiet but not silent." In Melbourne or Sydney, the rates for buildings are in the region of \$2 per \$1000 on building value, and \$5 per \$1000 on contents. The rate does not vary from one type of construction to another. The Newcastle earthquake of December 28, 1989 (see Chapter 12), produced insured losses of nearly \$750 million.

The capacity of private companies to handle large-scale earthquake loss is limited. One solution is for governments to take the initiative. This can be done in two ways: the institution of compulsory insurance like that in New Zealand; or the dispensation of emergency funds, as was done after earthquakes during the past few decades in California, Nicaragua, and Guatemala. It is interesting to note that some kind of ongoing insurance plan will be necessary if reliable earthquake prediction is ever attained (lest thousands of people suddenly rush out to purchase an earthquake policy in response to the first forewarning).

Finally, remember that any insurance scheme — no matter how well planned and executed — is at best short-term; it is no replacement for the reduction of earthquake risk by such preventive measures as applying current knowledge to the design and construction of new buildings and improving old buildings to make them more resistant (see the photo at the beginning of Chapter 7).

12 Reducing Earthquake Risk

Collapsed span of the San Francisco Bay Bridge after the 1989 Loma Prieta earthquake centered about 90 kilometers away in the Santa Cruz mountains of California. [Courtesy of Caltrans.]

Who builds his house on sands deserves a fool's cap.

—*Alexander Pope, An Essay on Man*

As was proposed in the last chapter, there are three key factors in earthquake danger: hazard, vulnerability, and risk. If structures were designed to be resistant to earthquake hazards, there would be no seismic risk. Structures in many earthquake countries are now being designed more safely, that is, they are less vulnerable. Not only have building owners become more conscious of the financial reasons for adopting earthquake-resistant design in construction and renovation but also public concern about the risk has increased.

This concern is due to the growing awareness of environmental issues and to the recognition that it is the public that will bear the cost of retrofit or reconstruction. Modern social developments are such that most industrial losses must be met by government programs, which are funded by taxes. As a consequence of these developments, regulatory agencies have been established by local, state, and central governments in many countries to protect individuals and general economic well-being.

Improvements in Planning and Zoning

The first step in a regulatory code is to draw up an appropriate set of rules. Since 1971, for example, cities and counties in California have been required by law to include a seismic-safety element in their general development plans. Technical studies are required that assess the consequences of the historical seismicity and activity of any local faults; the soil conditions and likelihoods of landslides, subsidence, and liquefaction in earthquakes; and otherwise. Land-use and emergency response plans are developed based on these judgments.

The observational bases for many of these special assessments are regional geological maps of various kinds. The fundamental maps show the geological structure, with emphasis on faults for which there is evidence of movement in Quaternary time (see Appendix D). They may be supplemented by maps of the type and thickness of surficial materials such as alluvium and filled areas. Seismic-intensity maps, showing reported intensities of historical earthquakes (see Figure 2.1), are also used if they are available.

From this basic geological and seismological information, seismic-zoning maps can be constructed on a variety of scales. A recent map outlining likely damage in the New Madrid zone,* the most active east of the Rocky Mountains in the United States (see Figure 8.1 and Plate 4), is reproduced in Figure 12.1. On the largest scale, these maps identify the regions of a country or province in which various intensities of ground shaking may have occurred or may be anticipated. If they are showing anticipated intensities, the probability of occurrence of a given intensity is implicit in the map. At present, there are zoning maps in America, Europe, the Balkans, the former Soviet Union, China, Japan, New Zealand, Australia, and elsewhere that are based on geological factors, earthquake- occurrence rate and magnitude, historical intensities, and subjective extrapolations from earthquakes in other parts of the world. In spite of their uncertainties, local seismic-hazard zoning maps are becoming common; and in a number of cities, such as Tokyo, they even show the blocks or streets.

On the maps, hazard is conceived as either *relative* or explicitly *probabilistic*. Most maps of relative hazards mark zones with an arbitrary numerical or alphabetical scale; for example, a now-superseded seismic-risk map for the United States had four zones ranging from no risk (zone 0) to most risk (zone 3). Maps of *probabilistic hazard* give an idea of the underlying statistical uncertainty, as is done in calculating insurance rates. These maps give, for example, the odds at which a specified earthquake intensity would be exceeded at a site of interest within a given time span (typically 50 or 100 years).

Changes continue to be made in the broad seismic zoning of the United States. New probabilistic maps have been developed as the basis of seismic-design provisions for building practice (see Plate 17). The most

**A nationally publicized "prediction" of a very severe earthquake in the New Madrid zone on December 3, 1990, led to much alarm (see page 223). Made by a nonseismologist, based on high tidal forces at that time, the traumatic event failed to occur anywhere in the world at the forecast time — let alone in Mississippi!*

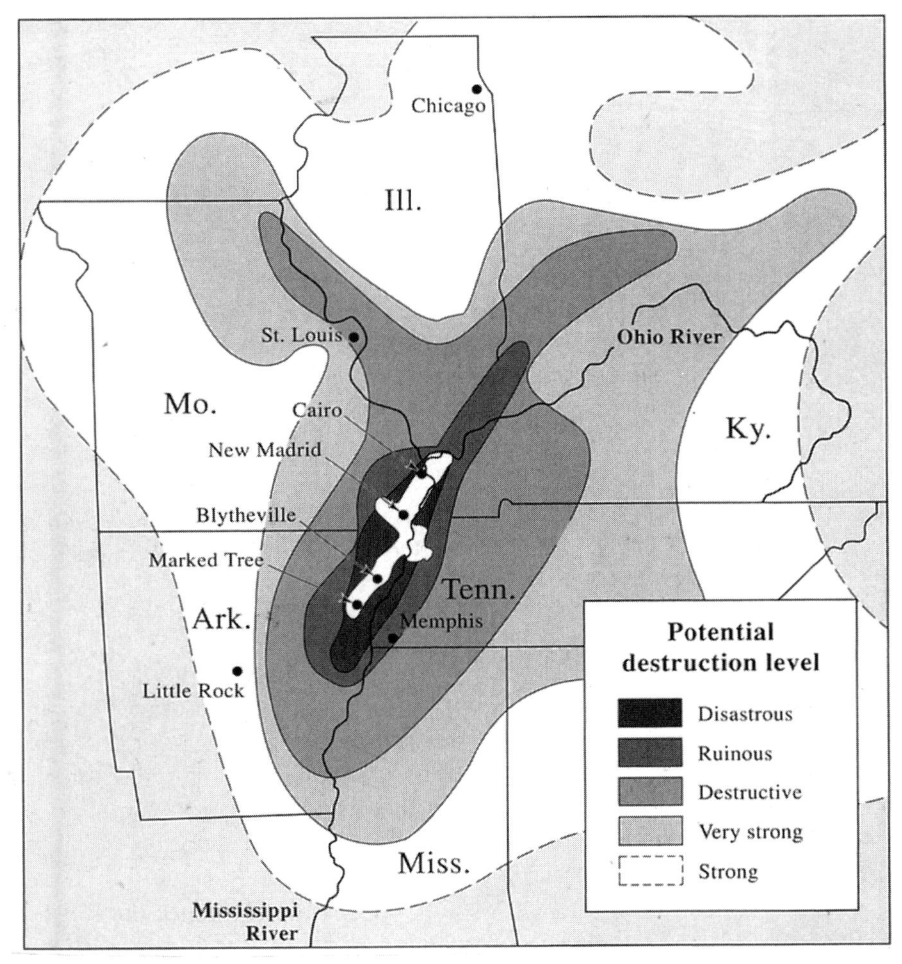

Figure 12.1 Zones forecasting likely destruction levels (to nonresistant structures) in the New Madrid earthquake region. [Courtesy of Federal Emergency Management Agency.]

usual give the expected intensity of ground shaking in terms of the *peak acceleration*. The peak acceleration can be thought of as the maximum acceleration in earthquakes on firm ground at the frequencies that affect sizable structures (high-rise buildings, factories, bridges, dams, and so on). Nowadays, also maximum ground velocity is often mapped. As explained in Chapter 8, ground acceleration and velocity (and, in some important cases, seismic ground displacement) can be correlated with seismic intensity in earthquake-engineering practice.

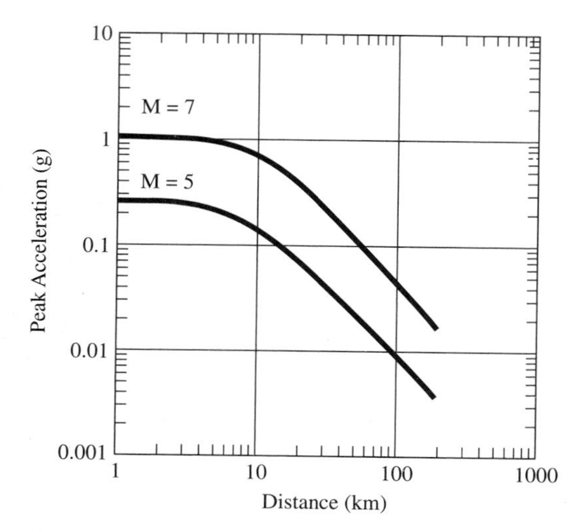

Figure 12.2 These curves show the average decrease with distance of the maximum ground acceleration in magnitude 5 and 7 earthquakes. The attenuation applies to intraplate earthquakes. Note that both scales are logarithmic.

The acceleration, velocity, or displacement values plotted as map contours have certain odds of occurrence, thus allowing for the caprice of earthquakes. A common value plotted has roughly a 90 percent probability of not being exceeded in 50 years. For example, the highest values are found in seismically active California, where near the contour marked 0.4 the chance is only 1 in 10 that active peak ground acceleration of 0.4 *g* or greater will occur within 50 years (see Plate 17).

Another factor that must be considered in the plotting of any seismic-hazard map is the *attenuation*, or decrease in average intensity of shaking with distance from the earthquake source, as illustrated in Figure 12.2. It is important because attenuation of seismic waves varies in different parts of the country (see Figure 12.3).

To avoid the drawbacks of previous zoning maps of the United States, the following general principles are followed in producing the new ones (such as that shown in Plate 17).

1. The map takes into account not only the size but also the frequency of earthquakes across the country.
2. The zoning pattern is based on the historical seismicity, major tectonic trends, intensity attenuation curves, and intensity reports.
3. Regionalization is defined by contours rather than numbers in zones so that the map shows the "hazard surface" for the whole

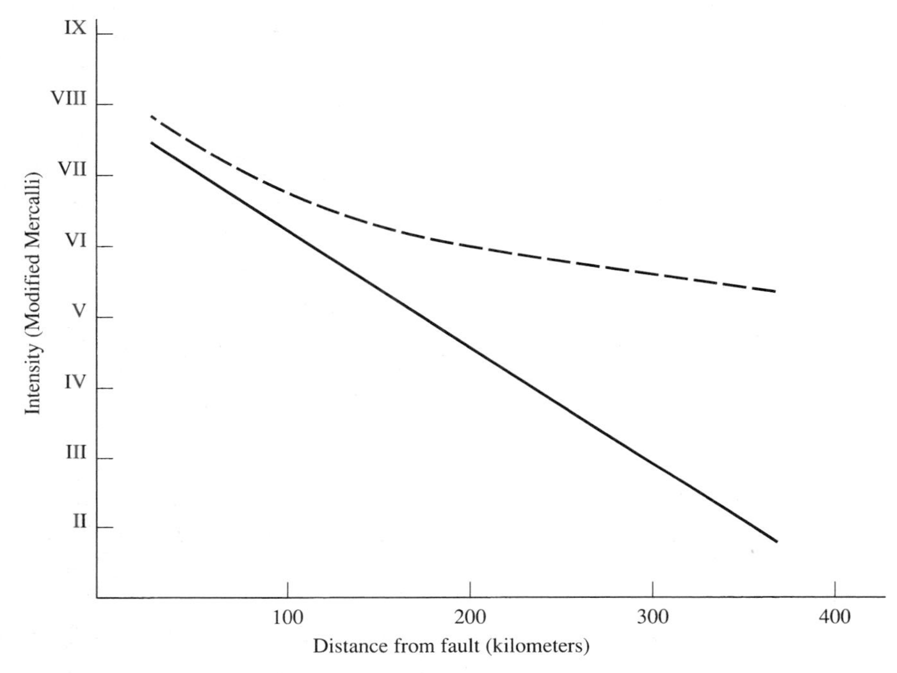

Figure 12.3 Graph showing how the average intensity of the strong ground shaking from seismic waves decreases with distance from the earthquake source. The intensity values refer to effects on firm ground. The solid line is the attenuation for California and the dashed line is that for the eastern United States.

country, with valleys of low seismic hazard and mountains of high seismic hazard.

4. The map is simple and does not attempt to subdivide the country into microzones. For this reason, only six highly smoothed contours have been selected.
5. The contours must be continuous.

The effective peak-acceleration map is intended for zoning, limited design, and site evaluation purposes; engineers can read expected accelerations by interpolation directly from the map. Thus it is hoped that, if precautions in construction are taken, the exposure to seismic hazard over the years will be approximately equal throughout the country.

Seismic-hazard maps are usually translated into building codes for application to construction. The seismic resistance of a structure is developed by engineers who follow the instructions in a building code and analyze the design specified by the code. All structures in a seismic region should conform to the local building code and, in general, more detailed analyses are applied only to the larger and more costly buildings. The codes are usually keyed to the hazard zone or a ground-shaking parameter such as the acceleration plotted in Figure 12.2.

Since the 1960s, a number of techniques have been developed to implement earthquake-resistant designs that are compatible with modern architecture. The more that seismologists understand earthquakes and engineers understand the response of buildings, the more effective the protective measures will be. As we have seen in the previous chapters, scientific understanding of earthquakes has advanced to a stage at which the causes of earthquakes and the types of ground shaking are reasonably well explained. As more geological mapping and critical observations from strong-motion seismographs become available, more light is thrown on the intensity of seismic shaking under various circumstances. Thus, the studies of the 1971 San Fernando earthquake, the 1985 Chile earthquake, the 1985 Mexico earthquake, the 1994 Northridge earthquake, the 1995 Kobe earthquake, and others worldwide have all helped to improve earthquake-resistant design.

Engineers working on earthquake problems have also improved their abilities to analyze building motions, often using high-speed computers. Already, some dynamical analyses not only are supported by a strong theoretical basis but have also been verified in actual earthquakes such as the 1989 Loma Prieta earthquake in the San Francisco area. A number of high-rise buildings designed according to earthquake-engineering codes were found to have shaken during the earthquake almost the way the designer had predicted. Sometimes, structural designs are proven inadequate, particularly if the architectural form is unusual or the materials untried (see, for example, the failure at Olive View Hospital in the 1971 San Fernando earthquake shown in Figure 12.10). A surprise to many engineers after the 1994 Northridge, California, earthquake was the cracking of some welded joints in steel-frame high-rise buildings.

Seismic risk to critical structures is often especially reduced these days.* The methods used to predict what the ground might do in a large earthquake vary in detail from site to site and from country to country. Obviously, the amount of work and expense invested in making geological-hazard studies for any site in order to minimize risk depends very much on the type of facility involved. Some large facilities hardly affect the population directly, and the main task is to minimize the cost of structural

Of course, this was always a goal. Plate 21 shows an early example of base isolation, *using large rubber bearings, for retrofit of historic and special structures.*

damage. But others, such as large hospitals, utilities, and key bridges must be functional through the aftermath of a damaging earthquake in the region. Consequently, in most geotechnical studies, hard decisions have to be made; and consulting seismologists, in offering advice to design engineers, usually wish they had much more observational material.

Collapse of Unreinforced Masonry: Newcastle, Australia, 1989

Whenever earthquake hazards are faced, one vulnerability dominates: unreinforced buildings made of brick, stone, concrete block, and similar material. The various earthquake case histories discussed in this book—such as San Francisco, 1906; Chile, 1985; Armenia, 1988; Loma Prieta, 1989; and Kobe, 1995—provide graphic evidence for the danger. The widespread nature of the threat can be illustrated from an unusual earthquake in Australia.

As in the eastern United States, there is little general public concern for earthquake risk in Australia. Yet intraplate earthquakes do occur in both these large continental areas. In the eastern United States, no surface fault rupture has ever been observed associated contemporaneously with an earthquake.

Similarly in eastern Australia (see Figure 12.4), the seismicity has occurred to the present without surface faulting, while the larger earthquakes in the west have often been produced by faults that rupture the ground surface. The most recent Australian earthquakes unequivocally associated with surface faulting were three events that occurred near Tennant Creek in the Northern Territory on January 22, 1988. They had magnitudes of M_S 6.3, 6.4, and 6.7 and were associated with surface faulting along two discrete arc thrusts for a distance of approximately 35 kilometers. The maximum slip was 1.2 meters. Previously on March 30, 1986, near Marryat Creek in central Australia, a magnitude M_S 5.8 earthquake was generated by a fault slip that appeared at the surface in the shape of a boomerang about 13 kilometers long with a maximum displacement of 0.8 meters. As is usual for intraplate earthquakes, the faulting was a combination of thrust and strike-slip produced from pressure across the region. Earlier earthquakes with surface scarps were Meckering in 1968 ($M_S = 6.8$), Calingri in 1970 ($M_S = 6.0$), and Cadoux in 1979 ($M_S = 6.2$), all nearby in western Australia. In each case, scarps of length 37 kilometers, 3 kilometers, and 15 kilometers, respectively, were formed at the surface, with fault mechanisms of almost pure thrust. The largest

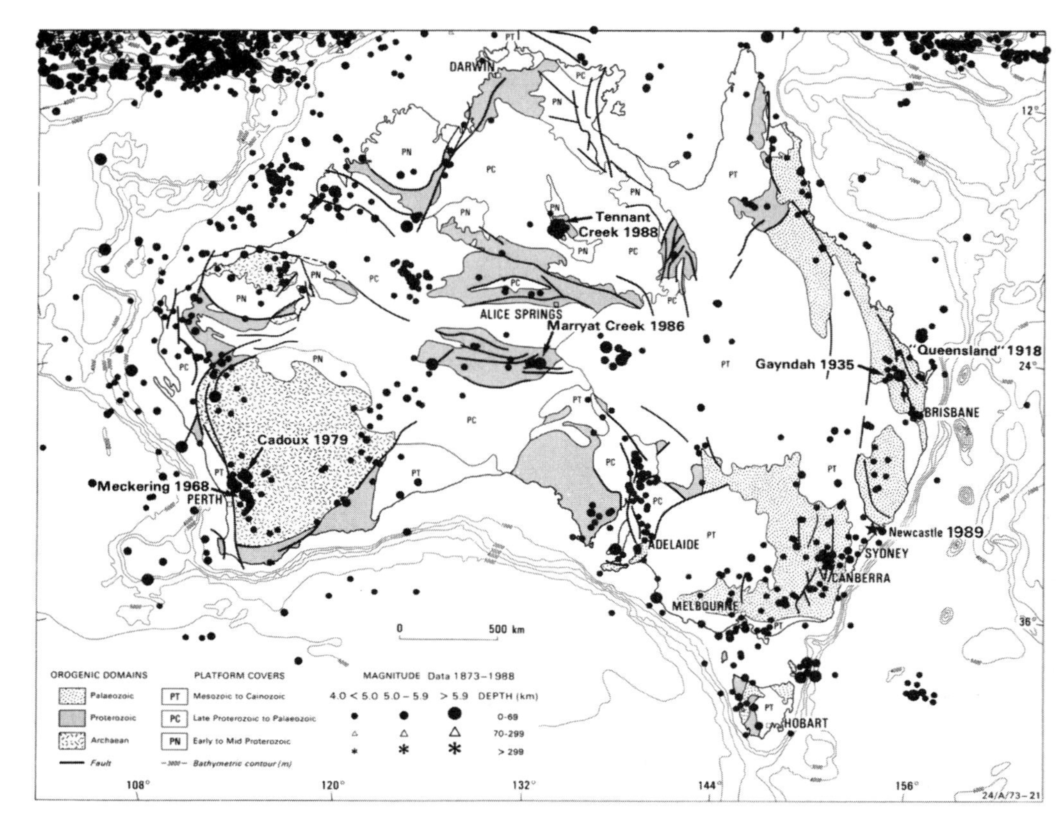

Figure 12.4 Seismicity of Australia from 1873 to 1988 plotted on a tectonic map. [Courtesy of D. Denham, Australian Bureau of Mineral Resources, Geology and Geophysics.]

known Australian onshore earthquake occurred in 1941 near Meeberrie about 500 kilometers north of Cadoux ($M_S = 7.0$). It can be associated with a recently discovered fault scarp near the epicenter.

In the eastern United States, moderate-size earthquakes relatively frequently cause damage to unreinforced buildings (see Figure 12.2). Particularly notable is the South Carolina earthquake of August 31, 1886, which shook Charleston, doing $5,500,000 worth of damage and killing 60 people in a population of about 50,000 (see Figure 7.1). This region of South Carolina had been free of damaging earthquakes from 1680 to 1886 and has been since. The fault slip that produced the strong Charleston shaking remains enigmatic to the present day. An unusual feature was the tremendous area of perceptibility, probably about 8 million square kilometers, for an earthquake that is estimated to have had a magnitude of

7.2. By contrast, the Loma Prieta earthquake of October 1989 in California of similar magnitude was felt over only 1 million square kilometers.

Given the low seismicity of eastern Australia, it was a great surprise when on December 28, 1989, a magnitude M_S 5.6 earthquake struck Newcastle, causing major damage and killing 13 people. From a geological point of view, no fault maps were available predicting seismic activity near Newcastle. There was no surface fault rupture. But by the method of remote sensing outlined in Chapter 4, the source of the Newcastle earthquake was determined seismologically as slip along a thrust fault on the margin of what geologists have called the Sydney basin. This sedimentary basin contains extensive seams of coal that have been mined since early in the nineteenth century.

A study of historical records after December 28, 1989, indicated that the Newcastle earthquake was not the first to have damaged towns around the Sydney basin. Earthquakes had occurred in June 1868 and December 1925 near Newcastle but had caused negligible damage, and memory of them had dimmed.

Even with its relatively low rate of occurrence of intraplate earthquakes, Australian engineers and seismologists had developed a seismic-hazard map to guide the design of earthquake-resistant buildings. Because of the lack of clear geological evidence, the 1979 version did not include Newcastle in a zone where specific seismic design was required, although some lateral force resistance was provided by the requirements for resistance to strong winds. As it turned out, most of the damage occurred to unreinforced masonry buildings that had been constructed in the early decades of the twentieth century. Most were situated on the recent alluvium and fill in the plain of the Hunter River.

On a field trip to the damaged area 5 months after the earthquake, I was struck by the extent of damage to old brick structures. One- and two-story residences in the city typically have two independent walls, or "wythes," of brick separated by a 1-inch cavity. These separate brick wythes are tied together by galvanized iron wire. Unfortunately, in many cases early construction practices led to inadequate use of the ties; even when in place, they were corroded from the salt-air conditions in this seaside city. Moreover, older masonry buildings that failed often had poor-quality mortar.

Over 3000 residences were damaged throughout the city, with walls, parapets, eaves, and chimneys often crashing to the ground. Wood-frame houses, which are usually supported on piers of concrete or brick and were not generally bolted to the foundations in Newcastle, were not noticeably damaged, perhaps indicating relatively small ground motion. The large steel mills were not severely damaged, and most utility lines were not significantly affected. Modern structures were generally not damaged; among

Figure 12.5 The partially demolished ruins of the Newcastle Workers' Club where 10 people were killed. [Courtesy of A. Page, University of Newcastle.]

the exceptions was the Newcastle Workers' Club, where a section built in 1972 collapsed, killing 10 people (see Figure 12.5).

In the Newcastle earthquake, no strong-motion seismograph was present to record the ground motion for a magnitude 5.5 earthquake, with only about 2 or 3 seconds of high acceleration. A larger earthquake would have produced much more widespread damage. An unusual feature is that almost the entire city is built over a grid of now-water-filled tunnels from early coal mining.

The severe hazard from unreinforced masonry structures is of course well known. Indeed, in California it is probably the greatest threat to economic viability and human lives from large-to-moderate earthquakes. In the Loma Prieta earthquake of October 17, 1989, at least eight deaths and hundreds of injuries were caused by the collapse of unreinforced masonry buildings. Surveys indicate that over 840 such buildings were damaged, of which 374 were vacated and 40 were demolished. Most reinforced brick buildings, which resisted this earthquake, were concentrated in the San

Francisco–Oakland area, a distance of over 80 kilometers from the source of the earthquake; at that substantial distance, the overall ground shaking was much reduced, as the waves had died down (see Figure 12.2). A fault source nearer the Bay Area will result in much more extensive damage to these nonresistant buildings.

To meet the problem of unreinforced masonry buildings in California, a law was passed by the legislature in 1986. Since that time, 95 percent of the affected jurisdictions have taken some action to reduce the hazard to those types of buildings. The law requires cities and counties within the highest seismic-hazard zones in California to do two things: first, list all unreinforced masonry buildings in the jurisdictions, and second, set up programs to mitigate the earthquake hazards in these buildings. The programs must include notification to the building owners that the buildings are a definite earthquake hazard. In addition, local governments must establish standards for seismic retrofits of the listed structures and enact measures to reduce the number of occupants in these buildings. From 1986 to 1990, about 25,000 unreinforced masonry buildings have been placed on the inventory in 57 counties. The cost of carrying out adequate retrofit for such structures is estimated to be about $4 billion, a substantial sum, which, however, pales in comparison with a $60 billion loss anticipated in a single major earthquake near the urban areas.

The Mexico Earthquake of September 19, 1985

In comparison with California, which has had 5 earthquakes with magnitude greater than 7 in this century, Mexico has had over 40, many of which have brought great toll in human lives. The tragic earthquake of September 19, 1985, with a magnitude of 8.1, had its source in the subduction slab under the Pacific coast (see Figure 7.2). It occurred in a seismicity gap that had been pointed out by seismologists for over a decade. It is enlightening to compare the cause and effects of the Mexico earthquake with the less-damaging Chile earthquake (see Chapter 1) of comparable size in the same year, which was also produced in a subduction zone.

Fortunately, the Mexico earthquake struck in the early morning, at 7:17 a.m. local time, when businesses and schools were not occupied, because many such structures were severely damaged in Mexico City. Even so, casualties in Mexico City, over 350 kilometers away from the focus, amounted to over 9000 deaths, with 30,000 injured and about 50,000 homeless. Severe damage or destruction occurred to about 500 buildings (see Figure 12.6), with an estimated $4 billion worth of total damage.

Figure 12.6 Removal of debris from a collapsed reinforced concrete building in the Lake Texcoco zone of Mexico City, September 1985. [Photo by V. V. Bertero.]

From one point of view, because Mexico City has a population of over 18 million people and about 800,000 buildings, these statistics indicate that the shaking severely affected only a small fraction of the city.* The damage along the coast nearer the source was significant but much more limited, partly because of the more favorable types of buildings and geological conditions there.

Extensive measurements of the strong ground motion of this earthquake were provided by a network of accelerometers of the National Autonomous University of Mexico (UNAM). Under cooperative arrangements with the University of California at San Diego, strong-motion instruments had also been placed along the Pacific coastline most affected by the shaking.

The intervening distance (about 350 kilometers, see Figure 12.2) between the earthquake source near the coastal region and the Valley of

**By chance, a commissioner of the California Seismic Safety Commission, Mr. L. Cluff, was staying overnight in Mexico City and experienced only moderate shaking; he went to breakfast in his hotel and became aware of major destruction in the city only at about 10 a.m. that morning.*

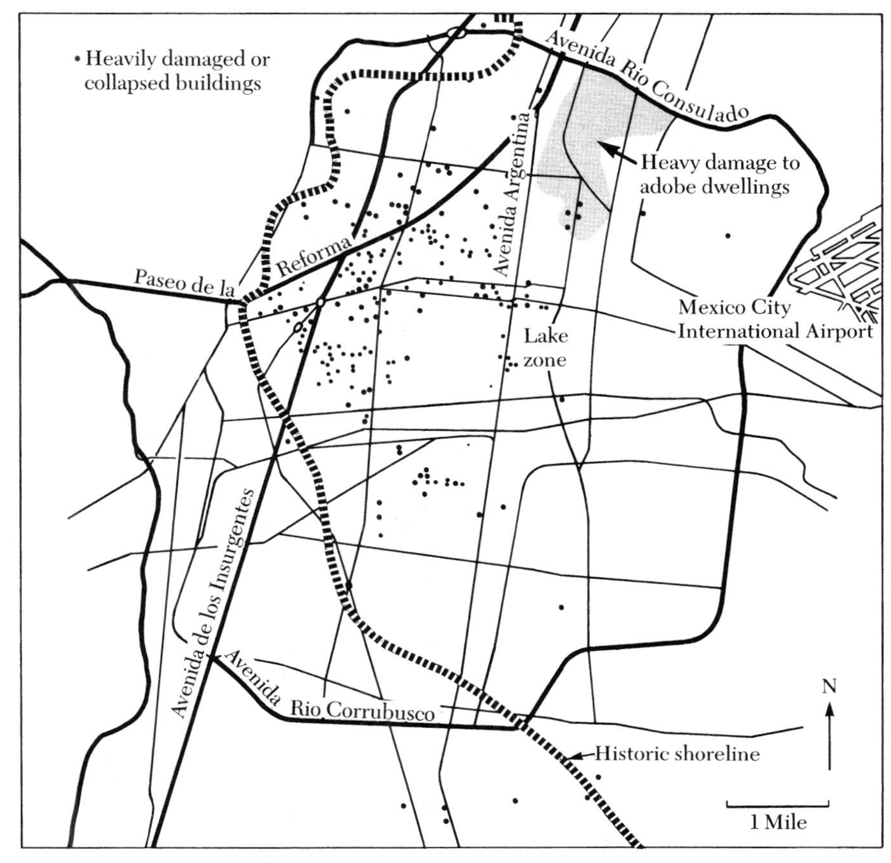

Figure 12.7 Map showing the concentration of heavily damaged and collapsed major structures in the old lake zone of Mexico City after the 1985 earthquake.

Mexico greatly reduced the amplitude of seismic waves in Mexico City, so that few structures built on firm soil and rock suffered damage. However, the near-surface geology in part of Mexico City created unfavorable conditions. Over recent geological time, rains had carried gravel, sands, and clays into the valley and had deposited them in Lake Texcoco. This lake was drained by the Spanish after the conquest of the Aztecs to allow development of the city. Modern Mexico City is to a large extent built on the higher ground surrounding the old lake bed, but near the city center there are parts underlain by a thick deposit of very soft, high-water-content sands and clays. This zone contained most of the buildings that collapsed during the September 19 earthquake. The map in Figure 12.7 shows the striking concentration of structural damage. (A similar pattern of damage occurred previously in Mexico City during a 1957 earthquake.)

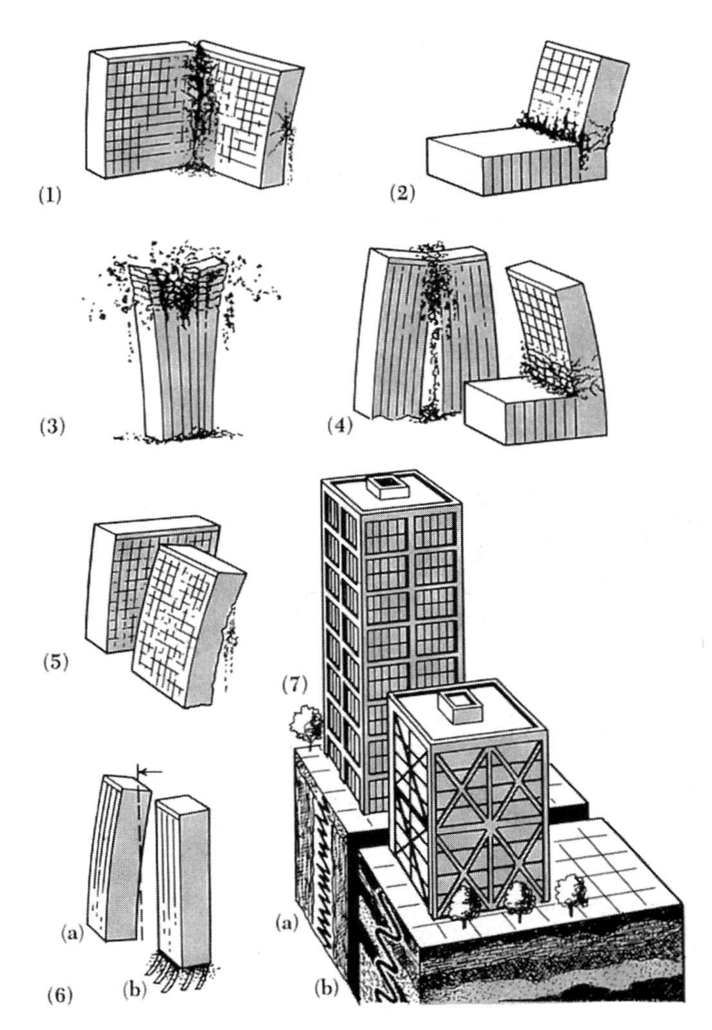

Figure 12.8 Diagram showing the effect of strong ground shaking on high buildings with differing shapes. (1) Two wings at right angles with different response causing damage at the connection points. (2) Building of varying height producing different resonance frequencies. (3) As waves move up the building, shaking is amplified at the top. (4) Pounding between adjacent buildings because of differently phased motion. (5) Enhanced swaying because of alignment of structures relative to the direction of incoming waves. (6) Flexible high-rise buildings with different architecture: design (a) remains elastic, while "soft" ground floor in (b) has no shear resistance. (7) Pair of buildings on different soils: (a) on rock which transmits higher-frequency waves; (b) on softer soil layers which set up wave resonance requiring special bracing of the structure.

What is the explanation? Let us start with the radiation of the seismic waves from the sudden movement along a fault in the subduction zone of the west coast in the Michoacán area. Locally, seismic waves radiated upward to the surface through distances of 20 kilometers or more. Accelerometers along this part of the coast showed moderate ground shaking amounting to 16 percent of gravity in a horizontal direction. As the waves traveled outward through the crustal rocks toward Mexico City, they spread out in time and space, and their amplitudes decreased overall. Those that shook the firm surface materials in the higher parts of Mexico City, such as at the National University of Mexico, were reduced to horizontal accelerations of only 4 percent of gravity and no damage resulted.

In the lake zone, however, surface waves of about 2-second period were preferentially amplified by the clay layers. Moreover, these surface waves had become extended over 10 cycles in the dispersed wave train (see Figure 3.8). In this area, strong-motion instruments on the surface measured peak horizontal accelerations of up to 40 percent of gravity. This resonance was again amplified by the shaking properties of buildings between 10 and 14 stories high, which have characteristic vibrational periods of between 1 and 2 seconds. Such buildings, when their foundations are pushed sideways, sway back and forth like inverted pendulums, and resonance leads to large displacements and a variety of structural failures (see Figure 12.8). Yet, even in the most heavily shaken zone, most buildings in Mexico City were generally not structurally damaged. These included most shorter buildings and higher "skyscrapers," such as the Latin American Tower of 37 stories which was constructed in the 1950s. Its extreme height produced a vibration period of 3.7 seconds — above the period of the most intense seismic surface waves.

Major Engineered Structures and Earthquake Risk

Simplified regional seismic-hazard maps (such as that shown in Figure 12.1) are adequate for designing a majority of structures and for zoning and planning purposes. However, in earthquake country, much more specific seismic site evaluation should be carried out for such structures as large dams, bridges, freeways, offshore oil-drilling platforms, high-rise buildings, and nuclear reactors (Figure 12.9). The costs of erecting them, and their importance to the surrounding community and nation, are too great to permit reliance solely on broad regional hazard maps. In the recent decade, numerous special studies for these structures have been made in the United States, Japan, the North Sea, and elsewhere by teams of geologists,

Figure 12.9 An example of the collapse of a major facility in a large earthquake. Failure of the earthfill Sheffield Dam due to the Santa Barbara, California, earthquake of June 29, 1925. [Photo by Putnam Studios.]

seismologists, soil engineers, and others. Consequently a great deal has been learned about earthquake occurrence and planning to mitigate earthquake hazard. The main points of such studies are listed in Box 12.1. In most projects, the study begins with analysis of the geological history of the region. The last steps are to calculate numerical values for the maximum accelerations (or velocities) and durations of the predicted seismic waves appropriate for the design and to provide appropriate earthquake ground motions.*

**Since the 1989 Loma Prieta and 1994 Northridge earthquakes in California, I have been advising on the estimation of such predicted ground motions for use by engineering designers of several major bridges in California. The new eastern section of the San Francisco Bay Bridge will be designed to withstand maximum credible earthquakes from the San Andreas fault. The new bridge will cost about $1.3 billion.*

Box 12-1

SEISMICITY RISK STUDIES FOR A PARTICULAR SITE

Geological studies

- Regional tectonics and patterns of deformation
- Mapping of significant capable faults within 100 kilometers
- Determination of fault types (strike-slip, dip-slip, and so on)
- Evidence for and against recent displacements along faults
- Field location of any landslide, ground settlement, or water inundation problems

Soil engineering studies

- Field report on foundation soils and their shaking susceptibility
- Special treatment of slope instability and subsidence when necessary
- Modification of strong-motion parameters when necessary

Seismological studies

- Study of local historical earthquake records
- Mapping of earthquake epicenters
- Determination of earthquake intensity and magnitude recurrence relations over time in the region
- Study of all historical intensity information near site
- Correlation of earthquake locations with mapped faults
- Estimation of future seismic intensities (acceleration, velocity, duration) near the site with stated odds of recurrence
- Selection of strong-motion records from past earthquakes that best represent the probable intensities and appropriate wave patterns

Structural design for new construction or building retrofit using modern methods of dynamic engineering analysis follows (Plate 22).

The design of electric power plants, including nuclear reactors, in particular, must ensure that ground shaking would not be enough to prevent the continued operation of the plant without undue risk to the health and safety of the operators and the public.

Earthquakes can also affect energy production in other ways. In some parts of the world, such as Alaska, California, and the North Sea, huge oil-drilling platforms may be subject to large-amplitude seismic waves. The design of drilling rigs therefore requires checking to test effects of earthquake shaking. The consequences of earthquake damage to an oil rig are, of course, not likely to be as widespread as, for example, to a nuclear reactor. Yet seismic-risk evaluation is still prudent and economically desirable for protection of workers' lives, capital investments, and the environment.

Oil pipelines too may be subject to significant earthquakes. Thus the 1260-kilometer trans-Alaska pipeline had to be designed to withstand seismic hazards. This pipeline crosses a number of earthquake-prone areas on its way from the Arctic Sea to Valdez port in Prince William Sound. The pipes are designed to resist large horizontal ground offsets at fault crossings, liquefaction of the ground supporting them, and various levels of shaking intensity appropriate to the section traversed. My involvement in 1972, in assessing the appropriate levels of ground motion for the design of this pipeline, opened my eyes to the seismological challenges in such work and the need for more seismographs to measure strong ground motion (see Chapter 3).

Safe Operation of Hospitals and Schools

One grave result of the February 9, 1971, San Fernando earthquake in southern California was the damage to hospitals. Some wings of the old Olive View Hospital of masonry construction collapsed but, fortunately, were unoccupied at the time. More disturbing were structural failures at the new Olive View Hospital built of reinforced concrete: the first floor of the psychiatric ward caved in completely, and the main hospital facility was heavily damaged (see Figure 12.10) and later demolished (although only two persons were killed at Olive View). The committee of the Structural Engineers Association of Southern California gave the following opinion:

> The lateral force design of the (new) Olive View hospital structures generally complied with the building codes in existence at the time. Failures in both units occurred in columns due to increases in vertical loads resulting from vertical seismic accelerations of the ground together with high lateral accelerations.

We now know that they were subject also to a large velocity and displacement wave pulse, which I called at that time the "fault fling" effect, causing severe shear and bending stresses. This energetic S-wave pulse (particulary potent in the direction of travel of the fault rupture) was confirmed by records of later earthquakes, such as the 1994 Northridge earthquake. It has become of critical importance in predicting ground motion for engineering design of large structures near seismically active faults.

To the southeast of Olive View, in the northeastern San Fernando Valley, also in the meizoseismal zone of the 1971 shock, was a Veterans Administration hospital. The facility opened in 1926, and in February 1971

Figure 12.10 Collapsed tower at the Olive View Community Hospital after the 1971 San Fernando, California, earthquake. Note the damage to the reinforced concrete columns at the first floor of the main building (later demolished). [Photo by Bruce A. Bolt.]

consisted of a 456-bed general medical complex. The buildings were located within 5 kilometers of the fault rupture observed in the 1971 earthquake. Engineering surveys at the hospital indicated that 26 buildings and additions constructed before 1933 suffered the greatest structural damage. Four of these buildings totally failed during the shaking, killing 38 persons. Buildings that were constructed after 1933 and had masonry or reinforced concrete shear-resisting walls generally did not collapse. In 1972 it was decided to abandon the site, and most buildings have since been demolished.

In all, four major hospitals were seriously damaged in the 1971 earthquake. As a result, an urgent drive was launched to ensure that more adequate seismic resistance was incorporated into hospital structures in the United States. One program was the seismic-risk evaluation and strengthening of all Veterans Administration hospitals in the country.* Geotechnical consultants made geological and seismological evaluation of

**The author was a member of the four-person advisory board.*

sites, both in use and proposed, following procedures similar to those listed in Box 12.1. The purpose of these studies was to locate any special geological hazards at the site as well as establish seismic parameters describing the strong ground motion that the site might experience. These seismic parameters were used as the basis for engineering analysis.

A special requirement for hospitals is that they — perhaps more than any other public facility — must remain operational after an earthquake for the treatment of injured and sick persons. Often not much attention is given to certain mechanical, electrical, and architectural elements, which — although not part of the structural frame — are crucial to the maintenance of a working system: these include facades, stairways, power systems, switching gear, elevators, alarms, sprinklers, medical equipment, and boilers. Also, all heavy suspended items such as ceilings must be braced to prevent falling. In short, although a hospital need not be completely functional in all aspects after the shaking, adequate post-earthquake emergency services, utilities, and access facilities must be available.

The safety of certain critical smaller items in medical facilities is too often overlooked. Medical supplies, for example, should be protected from breakage during the earthquake shaking. Much of the same advice mentioned in the previous chapter for family homes applies to hospitals. Above all, because significant earthquakes, even in seismically active areas, are infrequent, safety in hospitals and similar public service facilities requires constant training and vigilance by staff and continued revision of emergency plans.

Finally, the problem of schools has to be mentioned. Chance escape from tragedy has played too great a role in recent years in many seismic countries. Earthquakes have caused school buildings to collapse, but they have often occurred on weekends or at night (see Figure 12.11). One of the most famous examples of such an occurrence was the magnitude 6.3 earthquake near Long Beach and Compton in southern California on March 10, 1933. The local time of the shock was 5:54 p.m. Fortunately, by this hour most of the schools and colleges were vacant, for many of them suffered extreme damage and several collapsed. Only about 120 fatalities are listed for this earthquake, but 3 were teachers killed in the high school at Huntington Beach.

This event provoked such public resentment against shoddy construction that the California legislature passed an act the same year to control the construction of new public schools thereafter. The Field Act set firm standards for mitigating the earthquake risk in the public schools in California. It applied primarily to new public schools. Since then, regulatory standards for other categories of buildings, such as private schools,

Figure 12.11 Wrecked modern school building in Government Hill slide area, Anchorage, Alaska (1964). The building withstood the shaking but not the landslide. The earthquake occurred during a school holiday. [Courtesy of USGS.]

places of public assembly, colleges, and public service buildings, have also been set in various ways, but generally without the strict supervisory control or penalties that pertain to the public schools.

The Field Act requires supervision and enforcement by the Office of the State Architect of the State of California. The efficacy of the Field Act has been tested several times, and generally the results have been most satisfactory. The test with the 1971 San Fernando earthquake was particularly gratifying because this shock was of the same magnitude as the 1933 Long Beach earthquake. A special study in the San Fernando Valley area after the earthquake showed that of some 568 older school buildings that did not satisfy the requirements of the act, at least 50 were so badly damaged that they had to be demolished. But almost all the 500 or so school buildings in the district that met seismic-resistance requirements suffered no structural damage. The odds of children suffering injury if these schools had been in session would have been very small.

Throughout the years in California, several steps have been taken to reduce hazards posed by older school buildings not dealt with in the Field Act. The favorable publicity derived from the confirmatory statistics in the 1971 San Fernando earthquake further encouraged the legislature and school districts to push ahead with the urgent task of replacing substandard school buildings. The California law is now such that nearly all unsafe public school buildings in the state have been closed, demolished, or repaired.*

The same cannot be said for some other earthquake countries. As mentioned in Chapter 2, many school buildings were heavily damaged in the great 1985 earthquake in central Chile, fortunately again not occurring during school hours. Most knowledgeable observers feel that, had the earthquake occurred on a weekday, the number of injured students and teachers would have been high. The positive aspect is that many other school buildings, particularly of recent vintage, suffered little damage beyond a few broken windows, showing that the necessary engineering knowledge is available.

Worldwide, it is generally accepted that if a government requires people, especially children, to congregate in certain buildings during certain hours, it has the responsibility to ensure that the buildings are resistant to geological hazards such as earthquakes. It has been demonstrated in California and elsewhere that the application of earthquake regulatory codes to school buildings and other public buildings is effective when supervision is vested in a capable organization with clear guidelines and the power to impose penalties for violations.

Sakhalin, Russia, Disaster, 1995

There can be little question that we will continue to see many horrendous disasters from earthquakes around the world. Despite the growing knowledge of the causes and variation of earthquake waves through actual rock and soil structures and the undoubted advances in earthquake-resistant design of buildings, a majority of habitations in earthquake countries still remain prone to severe damage in seismic shaking.

There is no better illustration than in the remote large island of Sakhalin on the south side of the Sea of Okhotsk north of Japan in the far northwest Pacific (see Figure 7.2). On May 28, 1995, a fault trending northeast ruptured over many kilometers, within a few kilometers of the oil town of Neftegorsk. The resulting earthquake of magnitude 7.6 essentially leveled the town of 3000 inhabitants and killed or severely injured as

**But such is not the case in 1999 for all publicly used state-owned buildings. Many structurally deficient buildings remain, for example, on my own University of California campus at Berkeley. In 1998, there was another political push to amend or even abolish the Field Act.*

many as 2000 people. Sakhalin does not have high historical seismicity. Previously the largest earthquake known had a magnitude 6.8 in 1924 in the southwestern part; earthquakes greater than magnitude 5.5 are a rarity, although the complete seismicity record is known for less than a century. It is not surprising, therefore, that government estimates of earthquake hazard on the island were low, with earthquake elements of building codes not required until 1971.

Neftegorsk grew rapidly from a village during the boom years of oil development in the early 1960s. Although the city had some high-rise buildings, the housing was largely of single- and two-story buildings of wood construction, and three-story buildings of prefabricated concrete with little or no steel reinforcement. These apartment buildings are known in Russia as "khrushchoby." Many such apartments were constructed at the time throughout the former Soviet Union, including in Armenia.

Seventeen khrushchoby in the center of Neftegorsk collapsed completely into a pile of rubble, trapping the inhabitants as they slept, in the same way as the failure of prefabricated concrete houses in the 1988 Armenian earthquake (see Chapter 5). Rescue operations in the 1995 northern tragedy were hampered by the cold, fog, and inaccessibility. The nighttime temperatures were below freezing, and efforts to bring in heavy equipment, supplies, and food were frustrated because of damage to railroads and roads.

This heartbreaking story has lessons for many moderately seismic regions of the world where little or no attention has been given to earthquake hazard because of evidence from only a short-term geological record and lack of funds. For example, in 1997, I attended a special seminar in Alma Alta in Kazakhstan that considered the seismic risk in the eastern republics: Kazakhstan, Kyrgyzgstan, Uzbekistan, Tajikistan, and in the Caucasus region. In all these countries, it is not improbable that earthquakes of magnitude 6.5 and greater can occur near their most populated areas. Their present economic condition is such that adequate resources to increase the seismic resistance of many of the multiple-story unreinforced buildings are not likely to be available for decades.

The Loss of Art: Assisi, September and October 1997

Like the percussion sounds of an orchestra, earthquakes appear throughout the evolution of western civilization. From the Minoan world and the empires of Greece, Rome, and Byzantium, the effects of earthquakes in the Mediterranean region repeat over and over again. As scrutiny of the plate tectonic map shows (see Figure 7.2), the eastern Mediterranean region is dominated by subduction of the Africa plate northward, producing a trench south of the Aegean Sea between Greece and Turkey. But to the west, the Italian peninsula is also active seismically and volcanically, with many famous earthquake and volcanic eruption episodes.

A classical case history of earthquakes in Italy has been the Calabrian sequence that occurred in February and March of 1783. The rank of these earthquakes comes not only from their destructive power, which devastated over 180 towns and villages and killed more than 30,000 people, but also because the sequence was studied and described extensively. Notably, they led to the appointment of the first scientific commission to investigate a great earthquake. Five scientific members and several artists were appointed by the Neapolitan Academy of Science and Fine Letters. The commission made a field trip to the heavily shaken region of the Aspromonte Massif that occupies the toe of the boot of the Italian peninsula. It reported in detail on the intensity of the continuing earthquakes.

Charles Lyell later highlighted these earthquakes for the geological world in his *Principles of Geology* (see page 134). He states, "[Calabria] affords the first example of a region visited both during and after the convulsions, by men possessing sufficient leisure, zeal and scientific information to enable them to collect and describe with accuracy such physical facts as to throw light on geological questions." One of these geologists was a young French aristocrat, Deodat de Dolomieu, who visited the area in February and March of 1784. His descriptions contributed much to the knowledge of the geological setting of earthquake sources and the types of landscapes that are produced by heavy ground shaking and displacements.

Of course, these early scholars of earthquakes did not understand that sudden rupture of earthquake faults produced such earthquakes (see Chapter 6), and they had no unified geological theory to guide them. In fact, the juxtaposition of the plates and their movements in the Mediterranean is not simply explained by plate tectonic arguments. Because Italy is not strictly a plate margin, the earthquakes that extend along its central spine, the Apennines Mountains, are not exactly intraplate earthquakes. Although until late Tertiary time, thrusting and folding were taking place to form the mountain range, through Quaternary geological time to the present (Appendix D), the main deformation of the Italian peninsula appears to be extensional. The latter conclusion comes from the principal contemporary earthquakes along the Apennines where, particularly to the north, normal-type fault slip now dominates.

The continuing high seismicity of the Mediterranean has had a calamitous effect in both the short and long term on many of western civilization's greatest treasures. We have already mentioned the damage caused by the 1787 Calabrian earthquakes and the 1976 Friuli earthquake in northeast Italy (Chapter 10). Indeed, few years pass without reports of earthquakes that destroy precious heirlooms from these cradles of art.

An illustration is the loss of artistic paintings in the sequence of earthquakes in the Umbria-Marche provinces of the central Apennines, Italy, 1997. This sequence started on September 26, 1997, with a magnitude 5.7 earthquake, followed by a larger one 8 hours later. The sequence continued with several more earthquakes of about the same size, culminating in a magnitude 5.7 earthquake on October 14. Surface ruptures were reported from the epicentral area, which may be surface manifestation of the local faults whose slip caused the earthquakes.

By good fortune, in this 1997 seismic sequence the damaging earthquakes were preceded by smaller ones. In the early hours of September 26 the citizens were thus awakened and went outdoors, reducing deaths to 10 and injuries to approximately 500. But aesthetic tragedy resulted from irreparable damage to monuments of the Middle Ages throughout the area. The greatest regret is the damage at the Romano-Gothic Sacred Convent of Saint Francis in the town of Assisi, the birthplace of Saint Francis, standing on a hill above fertile valleys. The Basilica of Saint Francis consists of upper and lower churches whose construction began immediately after the canonization of Saint Francis in 1228. Both churches have frescoes by some of the most gifted artists of the Middle Ages, such as Giovanni Cimabue, Giotto, and Pietro Cavallini. While the damage to these frescoes in the lower church where Saint Francis is buried was slight, there was collapse of the vault roof of the upper church and significant cracking of the facades. Part of a Cimabue fresco crashed to the floor, killing four people.

It is interesting that these and other churches, monasteries, and related monumental structures around Assisi had been damaged in previous earthquakes. Some show various attempts to "strengthen" them against earthquakes in their 800-and-more year lifetimes, with varied success and degradation of aesthetic form. These observations make clear that the extent of retrofit of historical treasures in all countries deserves more thought than hitherto afforded this subject. Retrofit of such structures using the latest earthquake-resistant techniques should never be inconsistent with the architectural styles of the historical monuments. This principle is supported not only by the desire to preserve the older artforms but also by the economic benefits such historical buildings bring to the people of the region.

The 1989 Loma Prieta and 1994 Northridge Earthquakes Revisited

The California earthquakes mentioned previously in this chapter focused public attention on earthquake safety more than any other events in recent decades. The moment magnitudes* of the 1989 Loma Prieta and 1994 Northridge earthquakes were 7.0 and 6.9, respectively, making the energy release over 30 times less than that of the 1906 San Francisco earthquake. Nevertheless, as was stressed in Chapters 1 and 6, they left an indelible mark on the economy of California and serious, although often intangible, social effects. The first was the largest earthquake to occur in the San Francisco Bay Area since the great earthquake of 1906. A Modified Mercalli intensity VIII was rated over an area 50 kilometers long and 25 kilometers wide, including the cities of Los Gatos, Watsonville, and Santa Cruz (see Figure 12.12). In pockets of San Francisco and Oakland, Modified Mercalli intensities of IX were judged appropriate. Strong-motion seismographs gave the highest peak horizontal acceleration of over 60 percent of gravity close to the source and as high as 26 percent of gravity at places around the San Francisco Bay Area, over 70 kilometers away.

The extent of the faulting in the 1989 Loma Prieta source can be inferred from the map of the foci of aftershocks in Figures 1.4 and 2.5. The aftershock maps define in a general way the part of the fault that ruptured in the main earthquake, about 40 kilometers in length and 20 kilometers in depth. The simplest explanation on this assumption is that the rupture spread north and south in a bilateral sense. Because the rupture speed is about 2.5 kilometers per second, the rupture would have been completed in about 8 seconds. If the slip had begun at one end of the zone, it would have ruptured over an 80-kilometer-long fault segment and hence the seismic waves would have been generated for up to 16 seconds. In other words, the duration of strong shaking in the Loma Prieta earthquake was perhaps only half as long as may occur in a similar magnitude 7.0 earthquake. The actual distribution of accelerations measured in central California in this earthquake at some 150 widely scattered points provided a rich harvest of

Nothing invites debate more than earthquake magnitude. Several articles ran in local newspapers announcing "corrected" Loma Prieta magnitudes. Often these involved using another scale (see Chapter 8 and Appendix G). The local Richter magnitude *as estimated at the University of California at Berkeley seismographic station was 6.7.*

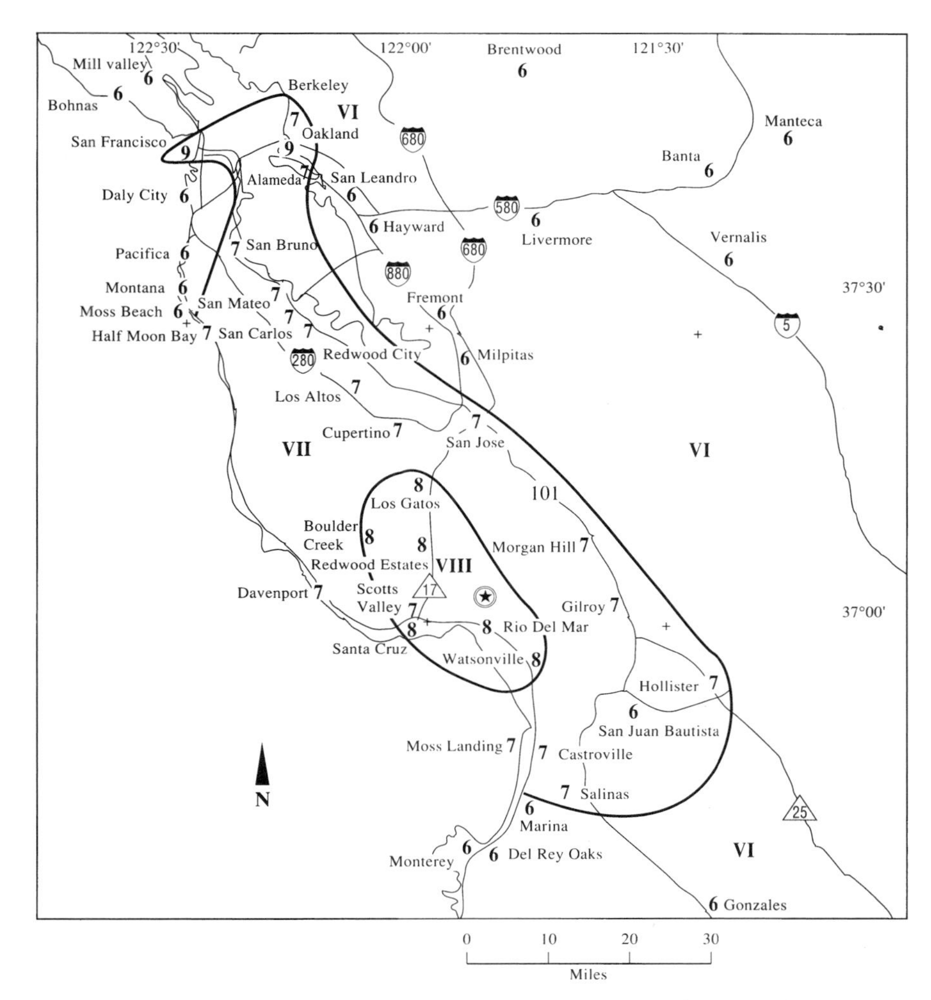

Figure 12.12 Isoseismal map showing the distribution of Modified Mercalli intensities for the Loma Prieta earthquake. [From Plafker and Galloway, 1989.]

information concerning the strong ground shaking for different rock and soil conditions.*

Extensive geological field work after both the 1989 and 1994 earthquakes failed to find any fresh surface fault break related to the slip that produced the mainshock. The size and aftershock foci of each earthquake,

**My involvement with the 1994 "home" earthquake was multifold, but much time was taken up as a member of the State Seismic Safety Commission, which had numerous hearings on all of its aspects.*

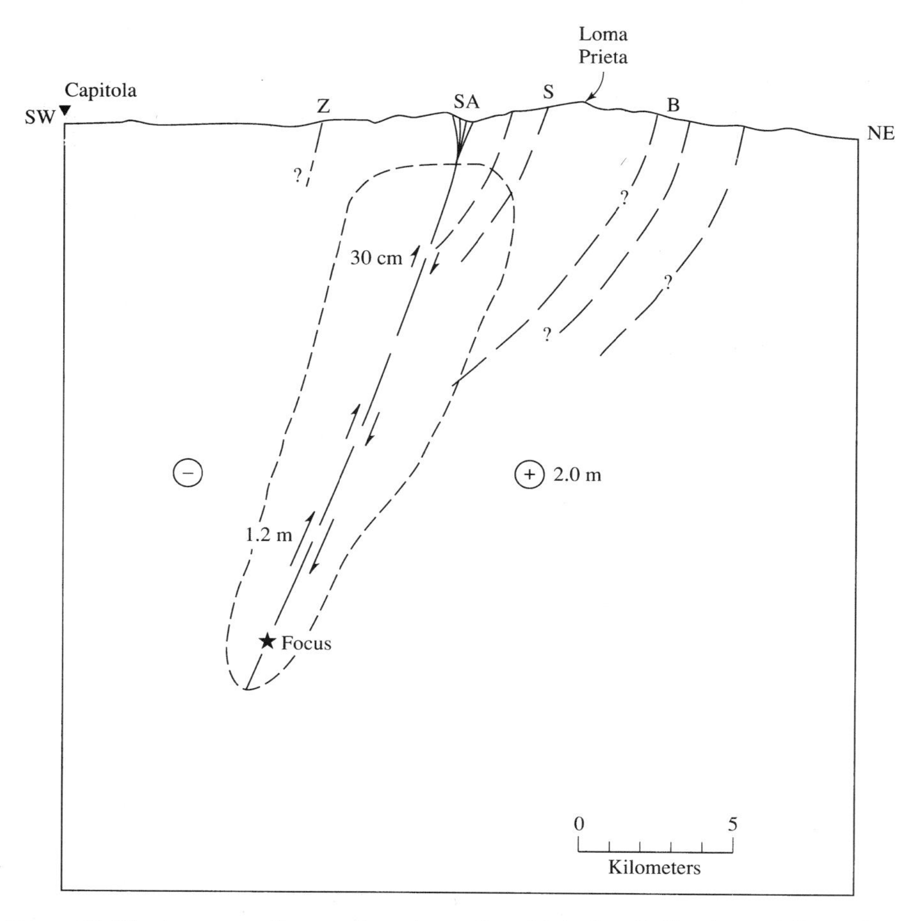

Figure 12.13 A cross section southwest–northeast into the California crust through the mountain called Loma Prieta. The dashed line encloses the area of aftershocks of the 1989 sequence (see Figure 1.4). ★ is the focus; arrows show the fault slip. Z is the Zayante fault; SA is the San Andreas fault; S is the Sargent fault; and B is the Berrocal fault. The numbers show the maximum vertical and horizontal fault displacements.

however, indicate that blind-thrust faulting must have occurred deep in the crust (see Figures 12.13 and 12.14). In the Loma Prieta case, the west side of the Santa Cruz Mountains was elevated a distance of about 1 meter. It remains a mystery why, although the San Andreas fault is a clear ground surface rift through the Santa Cruz Mountains, the fault rupture did not break the surface in this case as it has many times over hundreds of thousands of years. Another puzzle is the movement upward of the western side of the Santa Cruz Mountains in this fault rupture. Because Loma Prieta,

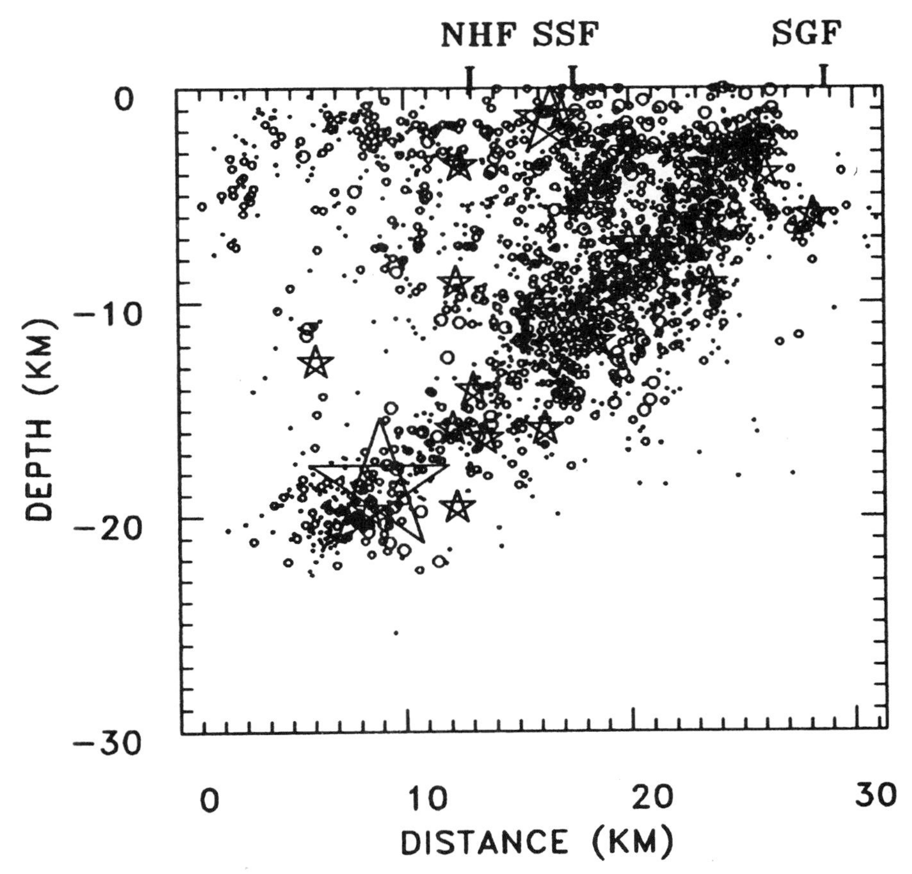

Figure 12.14 Side view of the Earth's crust showing the mainshock focus (large star) and aftershock hypocenters of the 1994 Northridge earthquake. The dense zone maps the dipping blind-thrust fault that rebounded. NHF, Northridge Hills fault; SSF, Santa Susanna fault; SGF, San Gabriel fault. [Courtesy of California Institute of Technology/USGS.]

the highest peak, is on the east side, we would expect the east side to rise. If slip like that in 1989 continued, the highest mountains should be to the west! Such problems may be solved in future earthquakes by the use of satellite geodesy to measure regional uplift in detail (see Plate 23 and Chapter 6).

The 1989 earthquake caused notable liquefaction of sandy soil in recent sediments along the Pacific coastline, near the earthquake source, and around San Francisco Bay (see Chapter 9). Theoretical models showing

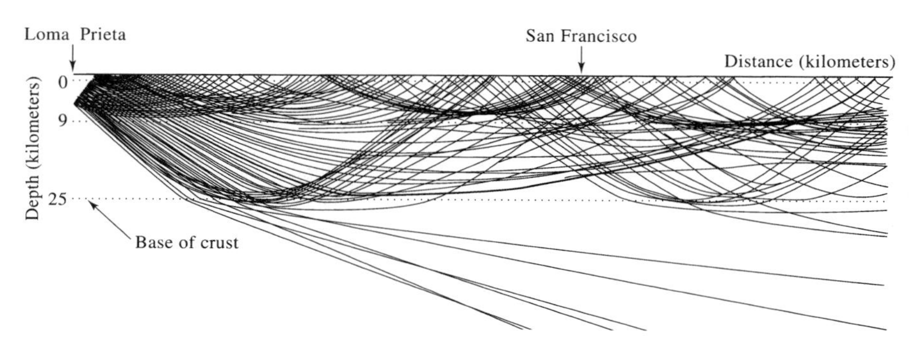

Figure 12.15 Rays showing the likely paths of seismic shear (S) waves from the focus of the 1989 Loma Prieta earthquake through the crust northward under San Francisco Bay (see Figure 1.4). [Courtesy of A. Lomax.]

plausible paths of seismic rays (see Chapter 4) from the fault slip in the Santa Cruz Mountains to the San Francisco Bay (see Figure 12.15) suggest an uneven distribution of intensity of ground shaking. In contrast, the 1994 Northridge earthquake, which shook the relatively arid and dry alluvium and soil of the San Fernando Valley/Northridge area, led to few significant liquefaction and ground failure effects.

In both earthquakes the significant collapse of road transportation systems occurred. The January 17 Northridge shaking led to the collapse of seven bridges of the Los Angeles freeways: out of some 2000 state highway bridges in the shaken area, five had been identified in 1993 for retrofit and two as not high risk. Two collapsed bridges on the busy Santa Monica freeway were removed, new spans constructed, and normal traffic flow established by May 20, 1994.

Many questions were raised among the public. For example, how safe are California's freeways and bridges? Boards of Inquiry were set up by the Governor of California to investigate this question. The main conclusion was that although most California freeways were seismically safe, key ones were not.

The report of the 1989 Board invoked an immediate response from the governor, who issued an executive order of singular importance. The order established formally, for the first time in California, that it is the policy of the state that seismic safety be given priority consideration in the allocation of resources for transportation, construction projects, and the new design and construction of all state buildings, including transportation structures and public buildings. The executive order also required the University of California and the California State University system to give priority to seismic safety in the allocation of resources available to construction projects. Because of this action, the Loma Prieta and Northridge earthquakes may prove to be of lasting benefit to all citizens of California and other earthquake-prone lands.

Emergency Warning Systems

An encouraging development for reduction of earthquake hazard is the connection of digital seismographs with emergency centers. These communication systems allow preparation and quick response before and after many of the larger earthquakes, tsunamis, and volcanic eruptions. Already, within 24 hours, most earthquakes in the contiguous United States over magnitude 3.5 (and in Hawaii, the Aleutians, and most foreign earthquakes of magnitude over 5.0) can be analyzed and their earthquake parameters sent electronically by pager, satellite, Internet, or telephone to state, federal, and international civil defense agencies. Also, such digitally based information can be transmitted to groups operating critical public facilities, such as dams, power plants, and railroads, and to national and international news media, to scientific institutions, and even to private citizens who request information. (See information on World Wide Web resources at the end of the book.)

In some seismic urban regions with dedicated seismographic systems, even more detailed information is becoming available. The first remarkable developments are that experiments in *early-warning systems* are being tried. These alert government officials, police and fire departments, and operators of vital facilities that earthquake waves are on the way. The scheme is based simply on the different speeds of travel of the S and P waves that make up the main parts of earthquakes (see Chapter 1 and Appendix I).

There are two prime examples. One is the already-tested warning system for Mexico City, which can provide up to 50 to 70 seconds of advance warning using the onset times of the two types of waves on seismographs. This interval represents the time delay between the P and S motions from earthquakes generated along the Pacific plate margin of Mexico, about 300 to 350 kilometers from Mexico City. The second example is from Bucharest, Romania. There the critical earthquakes are generated under the Carpathian Mountains at a depth of about 100 kilometers. In this case the S minus P warning time is 12 to 20 seconds. Such schemes offer considerable challenges, including extremely fast and reliable detection and computer identification, coupled with automatic response of the lifelines and transportation control centers.

The second example, perhaps more widely useful than the first, involves specification of actual intensity of the shaking that has *just occurred.* The necessary connected strong-motion seismographs have been put into place already in some cities in Japan and Taiwan and in Los Angeles, California. These strong-motion accelerometers (see Chapter 4) provide interactive computers with measurements of maximum ground motions at various seismographic stations throughout the shaken area. These recordings of the wave amplitudes are then correlated with seismic intensity such as Modified Mercalli values (see Appendix C) and contours automatically drawn between the points. The resulting intensity maps (see Figure 4.3, for example) are then transmitted within seconds to emergency response centers.

What Is Acceptable Risk?

All statements of risk contain elements of chance. Yet acceptance of estimates of the probability of earthquakes or structural vulnerability, both for engineering design and for public policy decisions, has been slow in coming. This public reluctance stems partly from a perception of differing opinions of experts, and partly from the observation that publicized odds of natural hazards and building resistance are often in error. On the positive side, the hesitancy is accompanied by some appreciation of the major simplifications that are needed to evaluate complex systems — such as the response of a building to intermittent earthquake shaking of uncertain size.

Numerical statements of the odds of risk are also difficult to interpret unless they are compared with odds for other hazards. Thus, the risk in the United States of death per year to an individual from a motor vehicle accident can be stated simply as about 1 in 4000; from earthquakes in the most exposed metropolitan areas, the risk is perhaps 1 in 50,000. But clearly much more is needed than such simple propositions before a practical local decision can be made. Personal risk clearly varies with individual situations and may be different from the collective or societal risk.

The first widely discussed estimate of the odds of a major earthquake in California was 50-50 in 10 years, as given in 1979 by the late Professor R. Jahns and myself. Criticism of this announcement was twofold: first, that such statements were not specific enough, and second (erroneously), that an even chance was not much of a risk! On balance, however, the reaction to this early probability description was favorable; the major benefit was a clearer awareness that the risk was immediate (10 years) and not indefinite ("we can expect 'a big one' sometime"). There have been refinements to our numbers since 1979, such as those discussed in Chapter 10, due to the much more extensive geological information; but a major remaining problem is that the import of many probability statements is not given in a way that nonspecialists can understand.

Education is needed on this and many other aspects of risk application if probability assessments are to be adopted widely as a basis for risk reduction; caution and care are needed in formulating the statistical statements. Among the explanations required are: What is the range of earthquake size involved rather than the specification of a particular magnitude? What are the overall uncertainties in the calculations? Are the statements meant as predictions — or only summary accounts of past events? Acceptance by a wider public depends on replacing ill-defined probability statements, which lead to needless anxieties, by statements in which the extent of reliability is clear.

Even when the calculations of earthquake odds are explained, the difficulty remains that there is often lack of agreement on the major goals of hazard abatement. Unquestionably, the approach in recent years in many countries has been to maximize safety rather than economic loss. For example, a recent Uniform Building Code in California (used as the guiding basis for major construction) specifically stated that "the purpose of this code is to provide minimum standards to safeguard life or limb, health, property and public welfare while regulating and controlling design and construction." Not only may these aims be incompatible but, when minimal building standards apply, damage to structures can be significant even though casualty loss is low.

Older structures, not built under recent building codes, present the greatest risk. The trade-off between safety and reconstruction costs is well illustrated by recent studies of the seismic resistance of state-owned buildings in California. It is estimated that over $20 billion of state properties are involved, and much of this property is vulnerable to damage. One of the lessons after the 1989 Loma Prieta and 1994 Northridge shaking was the seismic fragility of many crucial facilities even in a developed urban and industrial society. The point was driven home that failure of "lifelines" — electrical power, water, sewage, communication, and transportation — can prostrate the economy (see Figure 12.16). Severance of the San Francisco Bay Bridge on October 17, 1989, and the widespread power failure in San Francisco, 70 kilometers from the Loma Prieta seismic source, prove this point. The lesson is that, in decision making on risk reduction, the failure to allow for the functioning of key facilities, in addition to life safety, can have the gravest consequences.

Figure 12.16 Punching of a support column through the roadbed of a connector bridge on Freeway I-5 following the 1994 Northridge earthquake. [Photo by F. Seible.]

The same danger has long concerned authorities in Japan. Soaring real estate values in cities have encouraged the filling of coastal land tracts, and these have become heavily populated industrial and commercial zones. The moderate 1995 Kobe earthquake became Japan's worst natural disaster since the great 1923 Kwanto earthquake; into the 3-kilometer-wide plain between mountains and Osaka Bay was squeezed the densely developed urban area, which suffered the strongest shaking. Liquefaction of soils and harbor fill contributed to damage to bridges and gantry cranes. Was such economic loss regarded as acceptable, perhaps because of lack of widespread informed debate among the public?

In the mustering of broad political supports to remove earthquake threats to broadly acceptable levels, it is paradoxical that the practical aspects of earthquake hazards both contribute to and inhibit achievement of the ultimate safety goals. Although the benefits of research and its application would appear to be obvious, in fact, both are subject to deadlines, feasibility questions, and tendentious conflicts of interest that dampen enthusiasm and public demand. Space scientists have been successful in

FRANK AND ERNEST

[Reprinted by permission of Newspaper Enterprise Association.]

obtaining funding for space vehicles worth many tens of millions of dollars. In terms of national welfare, it might be expected that the risk involved in earthquakes would give special force to the claims for resources to enhance seismic safety. Seismological history tells otherwise. Risk reduction is characterized by bursts of activity and political support *after damaging earthquakes* and by decay curves that have a half-life of a year or so before public effort recedes.

The earthquake safety programs of the 1990s coincided with the International Decade of Natural Disaster Reduction. This initiative was agreed to by the United Nations as a major effort to reduce significantly, before the year 2000, the risk from earthquakes, volcanoes, floods, and other natural hazards. It was not well supported financially, but some progress was made. I have described in this book how major earthquakes near huge metropolitan areas at seismic risk such as Mexico City, Istanbul, Los Angeles, and Tokyo will have deep economic effects, not only regionally, but nationally. Industries and institutions will not be able to operate effectively for a considerable time, reducing the living standards of the whole country. Yet despite the remaining ground motion prediction difficulties in seismology and technical gaps in engineering, there are really no insurmountable scientific reasons why earthquake risks to both the individual and society cannot be reduced in the first decades of the new millennium to levels comparable with those of more familiar dangers.

Appendix A

World Earthquakes and Seismicity Rates

NOTABLE WORLD EARTHQUAKES AND SEISMICITY

Year	Date (UT)	Region	Deaths	Magnitude (M_s)	Comments
856	December	Greece, Corinth	45,000		
1038	January 9	China, Shensi	23,000		
1057		China, Chihli	25,000		
1268		Asia Minor, Silicia	60,000		
1290	September 27	China, Chihli	100,000		
1293	May 20	Japan, Kamakura	30,000		
1531	January 26	Portugal, Lisbon	30,000		
1556	January 23	China, Shensi	830,000		
1663	February 5	Canada, St. Lawrence River			Maximum intensity X; chimneys broken in Massachusetts
1667	November	Caucasia, Shemakha	80,000		
1693	January 11	Italy, Catania	60,000		
1737	October 11	India, Calcutta	300,000		
1755	June 7	Northern Persia	40,000		
1755	November 1	Portugal, Lisbon	70,000		Great tsunami
1783	February 4	Italy, Calabria	50,000		
1797	February 4	Ecuador, Quito	40,000		
1811	December 16	Missouri, New Madrid	Several		Intensity XI; also January 23, February 7, 1812

1812	December 21	California, offshore Santa Barbara	Several injuries		Maximum intensity X; reported tsunami uncertain
1819	June 16	India, Kutch	1543		
1822	September 5	Asia Minor, Aleppo	22,000		
1828	December 18	Japan, Echigo	30,000		
1857	January 9	California, Fort Tejon			San Andreas fault rupture; intensity X–XI
1868	August 13	Peru and Bolivia	25,000		
1868	August 16	Ecuador and Colombia	Ecuador 40,000 Colombia 30,000		
1872	March 26	California, Owens Valley	About 50		Large-scale faulting
1886	August 31	South Carolina, Charleston-Summerville	About 60		
1891	October 28	Japan, Mino-Owari	7000		
1896	June 15	Japan, Sanriku	22,000		Tsunami
1897	June 12	India, Assam	1500	8.7	
1899	September 3 and 10	Alaska, Yakutat Bay		7.8 and 8.6	
1906	April 18	California, San Francisco	700	8.25	San Francisco fire

NOTABLE WORLD EARTHQUAKES AND SEISMICITY *(continued)*

Year	Date (UT)	Region	Deaths	Magnitude (M_s)	Comments
1908	December 28	Italy, Messina	120,000	7.5	
1915	January 13	Italy, Avezzano	30,000	7	
1920	December 16	China, Kansu	180,000	8.5	
1923	September 1	Japan, Kwanto	143,000	8.2	Greak Tokyo fire
1932	December 26	China, Kansu		7.6	Much damage
1935	May 31	India, Quetta	60,000	7.5	
1939	January 24	Chile, Chillàn	30,000	7.75	
1939	December 27	Turkey, Erzincan	23,000	8.0	
1948	June 28	Japan, Fukui	5131		
1949	August 5	Ecuador, Pelileo	6000	6.9	
1950	August 15	India, Assam	1526	8.6	Surface faulting
1960	February 29	Morocco, Agadir	14,000	5.9	
1960	May 22	Southern Chile	Over 3000	8.5	
1962	September 1	Northwest Iran	14,000	7.3	
1963	July 26	Yugoslavia, Skopje	1200	6.0	
1964	March 28	Alaska	131	8.6	Damaging tsunami
1968	August 31	Iran	11,600	7.4	Surface faulting
1970	May 31	Peru	66,000	7.8	$530 million damage; great rock slide

1971	February 9	California, San Fernando	65	6.5	$550 million damage
1972	December 23	Nicaragua, Managua	5000	6.2	
1975	February 4	China, Liaoning Province	Few	7.4	Predicted
1976	February 4	Guatemala	22,000	7.9	200-kilometer rupture — Motagua fault
1976	May 6	Italy, Friuli (Gemona)	965	6.5	Extensive damage; no surface faulting
1976	July 27	China, Tangshan	Over 250,000	7.6	Great economic damage, also perhaps 500,000 injured; not predicted
1977	March 4	Romania, Vrancea	2000	7.2	Damage in Bucharest
1977	August 19	Indonesia, South of Sumbawa Island	100	8.0	Tsunami at Sumbawa Island and northern Australia
1979	December 12	Near coast of Ecuador	600	7.7	20,000 injured
1980	October 10	Algeria, El Asnam	3500	7.7	Extensive damage
1980	November 23	Southern Italy	3000	7.2	About 2000 missing; 7800 injured
1981	June 11	Southern Iran	3000	6.9	
1981	July 28	Southern Iran	1500	7.3	
1982	December 13	Yemen	2800	6.0	About 300 villages badly damaged
1983	May 26	Japan, Oga Peninsula	107	7.7	Tsunami caused extensive damage
1983	October 30	Turkey	1342	6.9	
1985	March 3	Chile, Valparaíso	177	7.8	Extensive damage; 2575 injured
1985	September 19	Mexico, Michoacán	9500	7.9	More than $4 billion damage; 30,000 injured; small tsunami

NOTABLE WORLD EARTHQUAKES AND SEISMICITY *(continued)*

Year	Date (UT)	Region	Deaths	Magnitude (M_s)	Comments
1986	October 10	El Salvador, San Salvador	1000	5.4	10,000 injured and 200,000 homeless
1987	March 6	Colombia-Ecuador border	1000	7.0	4000 missing; 200,000 homeless and extensive damage
1988	August 20	Nepal-India border	1450	6.6	Thousands injured in northern Bihar, India and eastern Nepal
1988	November 6	Burma-China border	730	7.0	4808 injured and severe damage About 3.2 million people affected
1988	December 7	Spitak, Armenia	25,000	7.0	13,000 injured; 500,000 homeless and severe damage
1989	August 1	West Irian, Kurima District	90	5.8	15 injured by landslides which buried 2 villages
1989	October 17	California, Santa Cruz Mountains	63	7.0	3757 injured; $5.6 billion damage
1989	December 28	Newcastle, Australia	13	5.6	First known death from Australian earthquake
1990	June 20	Iran, Caspian Sea region	Above 40,000	7.7	Surface faulting; 400,000 homeless; extensive landslides
1990	July 16	Philippines, Luzon	1700	7.8	Major rupture on Digdig fault

1992	June 28	Landers, California	1	7.5	Surface fault rupture over 70 kilometers long
1993	September 29	Latur, India	10,000	6.4	Surface faulting; many villages destroyed
1994	January 17	Northridge, California	56	6.9	No surface faulting
1995	January 16	Kobe, Japan	5400	6.9	Over 100,000 buildings destroyed or damaged and 27,000 injured. Large fire areas
1995	May 27	Sakhalin, Russia	2000	7.5	Apartment-building collapses
1996	February 3	Lijiang, China	309	6.4	Over 100,000 people homeless
1998	February 4	Afghanistan	Above 2500	6.1	Many homeless
1999	January 25	Armenia, Colombia	700	5.9	Widespread damage

Source: U.S. National Oceanic and Atmospheric Administration and U.S. Geological Survey.

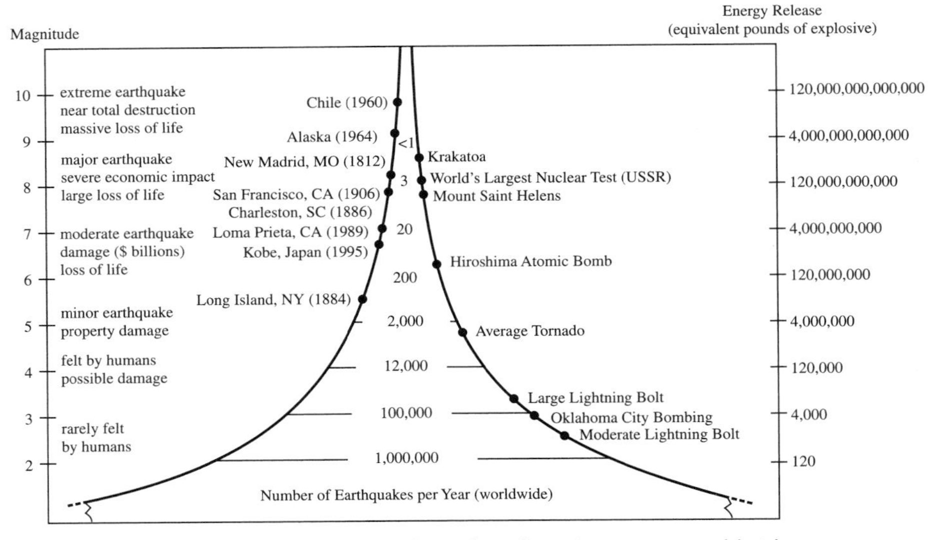

Curves showing the approximate number of earthquakes a year worldwide as a function of magnitude and energy release. The energy in the Krakatoa and Mount Saint Helens eruptions is the total released in the eruption, *not* that in the accompanying earthquakes. [after IRIS]

Appendix B

Important Earthquakes of the United States and Canada

IMPORTANT EARTHQUAKES OF THE UNITED STATES AND CANADA

Year	Date (local time)	Place	MM Intensity	Remarks
1638	June 11	Massachusetts, Plymouth	IX	Many stone chimneys down; chimneys down in shocks in 1658 and probably other years
1663	February 5	Canada, Three Rivers, lower St. Lawrence River	X	Chimneys broken in Massachusetts Bay area
1732	September 16	Canada, Ontario	IX?	7 killed at Montreal
1755	November 18	Massachusetts, near Cambridge	VIII	Many chimneys down, brick buildings damaged, stone fences generally wrecked; sand emitted from ground cracks; felt from Chesapeake Bay to Nova Scotia
1769	July 28	California, San Pedro Channel area	X	Major disturbances with many aftershocks
1790	?	California, Owens Valley	X	Major shock with appearance of fault scarps
1811	December 16	Missouri,	XI	Three principal earthquakes; New Madrid destroyed,
1812	January 23	New Madrid		extensive changes in configuration of ground and rivers,
1812	February 7			including the Mississippi River; chimneys down in Cincinnati and Richmond; felt in Boston; several killed; the three shocks had Richter magnitudes of about 7.5, 7.3, and 7.8
1812	December 8	California, San Juan Capistrano	IX	Church collapsed, killing 40

1812	December 21	California, near Lompoc	X	Churches and other buildings wrecked in several towns including Santa Barbara
1836	June 10	California, San Francisco Bay area	X	Ground breakage along Hayward fault from Mission San Jose to San Pablo
1838	June	California, San Francisco	X	Fault rupture phenomena along San Andreas rift; this earthquake has been compared with the earthquake of April 18, 1906
1857	January 9	California, Fort Tejon	X–XI	One of the greatest historical Pacific coast shocks; originated on San Andreas fault in northwest corner of Los Angeles County; buildings and large trees thrown down
1868	April 2	Hawaii	IX–X	The largest historical Hawaiian earthquake; occurred off the southern tip of the Big Island and was felt for over 350 miles; generated a tsunami, maximum height 65 feet; 148 lives lost
1868	October 21	California, Hayward	X	Many buildings wrecked and damaged in Hayward and East Bay; severe damage at San Leandro and San Francisco; 30 killed; rupture of Hayward fault
1870	October 20	Canada, Montreal to Quebec	IX	Widespread; minor damage on coast of Maine

IMPORTANT EARTHQUAKES OF THE UNITED STATES AND CANADA *(continued)*

Year	Date (local time)	Place	MM Intensity	Remarks
1872	March 26	California, Owens Valley	X–XI	One of the greatest earthquakes in Pacific coast area; 7-meter scarp formed; 27 killed at Lone Pine out of 300 population; adobe houses wrecked
1886	August 31	South Carolina, Charleston	X	Greatest historical earthquake in eastern states; 102 buildings destroyed, 90 percent damaged, nearly all chimneys down; $5.5 million damage; about 60 killed; felt in Boston, Chicago, and St. Louis
1887	May 3	Mexico, Sonora	XI	Widespread in border states; chimneys down in several towns, including El Paso and Albuquerque
1895	October 31	Missouri, near Charleston	IX	Felt in Canada, Virginia, Louisiana, and South Dakota; acres of ground sank and lake formed; many chimneys demolished
1899	September 10	Alaska, Yakutat Bay	XI	Great earthquake; widely felt; slight damage because area uninhabited; shoreline rose 15 meters
1899	December 25	California, San Jacinto	IX	Nearly all brick buildings badly damaged in San Jacinto and Hemet; chimneys down in Riverside; 6 killed; another severe shock in 1918
1900	October 9	Alaska, Kenai Peninsula	VII–VIII	Felt from Yakutat to Kodiak; severe damage in Kodiak

1906	April 18	California, San Francisco	XI	Great earthquake and fire; about 80 percent of estimated $400 million damage due to fire; over 1000 killed; greatest destruction in San Francisco, Santa Rosa; horizontal slipping along San Andreas fault, 6.5 meters; greatest damage on poorly filled land
1909	May 26	Illinois, Aurora	VIII	Many chimneys down; felt over wide area
1915	June 22	California, Imperial Valley	VIII	Nearly $1 million damage; 6 killed; well-constructed buildings were cracked
1915	October 2	Nevada, Pleasant Valley	X	Widespread; adobe houses and water tank towers wrecked; fault break 35 kilometers with 3.5 meters vertical throw in one place
1925	February 28	Canada, Murray Bay	VIII	Felt in many eastern and central states; damage less than $100,000
1925	June 27	Montana, Manhattan	IX	Landslide blocked entrance to railroad tunnel; some buildings wrecked and many chimneys fell; $300,000 damage
1925	June 29	California, Santa Barbara	IX	$6 million damage; 13 killed; 70 buildings condemned
1927	November 4	California, offshore Lompoc	IX	North of Point Arguello; tsunami
1929	August 12	New York, Attica	IX	250 chimneys toppled
1929	November 18	Canada, Grand Banks, off Newfoundland	X	Submarine shock broke 12 transatlantic cables, some breaks 240 kilometers apart; some deaths by tsunami along Burin Peninsula; some chimneys in Canada toppled

IMPORTANT EARTHQUAKES OF THE UNITED STATES AND CANADA *(continued)*

Year	Date (local time)	Place	MM Intensity	Remarks
1931	April 20	New York, Lake George	VIII	Chimneys fell
1931	August 16	Texas, near Valentine	VIII	All buildings damaged, many chimneys fell
1932	December 20	Nevada, Cedar Mountain	X	In sparsely settled region
1933	March 10	California, Long Beach	IX	$41 million damage, 120 killed; fire damage insignificant
1934	Mach 12	Utah, Kosmo	VIII	Marked changes in terrain north of Great Salt Lake; 2 killed
1935	October 18 October 31	Montana, Helena (strong aftershock)	VIII	$3.5 million damage, 4 killed, less than 50 injured; more than half of buildings damaged, from 2.5 to 100 percent; second shock strongest of many aftershocks
1935	November 1	Canada, Timiskaming	IX	Widespread; landslide near origin
1940	May 18	California, Imperial Valley	X	$6 million damage, 8 killed, 20 seriously injured 65-kilometer fault appeared with maximum horizontal displacement of 4.5 meters
1941	June 30	California, Santa Barbara	VIII	$100,000 damage
1941	November 14	California, Torrance, Gardena	VIII	About $1 million damage; 50 buildings severely damaged
1944	September 5	Canada-New York, Cornwall and Massena	IX	On St. Lawrence River, $1.5 million damage reported; 90 percent of chimneys in Massena destroyed or damaged

1946	April 1	Alaska, Aleutian Islands	?	Great earthquake; tsunami destroyed a light station and caused severe damage in Hawaii; estimated damage $25 million
1947	October 16	Alaska, Nenana	VIII	Rock slides and damage to Alaska railroad
1949	April 13	Washington, Puget Sound	VIII	$25 million damage; 8 killed, directly and indirectly; damage confined mostly to marshy, alluvial, or filled ground; many chimneys, parapet walls, and cornices toppled
1952	July 20	California, Kern County	X	$60 million damage, 12 killed, 18 seriously injured; railroad tunnels collapsed and rails bent in S-shape; surface faulting with about 0.5 meter of vertical, as well as lateral, displacement
1952	August 22	California, Bakersfield	VIII	2 killed, 35 injured; damage $10 million
1954	July 6	Nevada, Fallon	IX	Extensive damage to irrigation canals; several injured
1954	August 23	Nevada, Fallon	VIII	Surface ruptures; damage more than $91,000
1954	December 16	Nevada, Dixie Valley	X	Surface ruptures along 88 kilometers linear distance and up to 4.5 meters vertical throw in sparsely populated desert
1957	March 22	California, San Francisco	VIII	Damage in Westlake and Daly City area
1958	April 7	Central Alaska	VIII	Severe breakage of river and lake ice, pressure ridges and mud flows

IMPORTANT EARTHQUAKES OF THE UNITED STATES AND CANADA *(continued)*

Year	Date (local time)	Place	MM Intensity	Remarks
1958	July 9	Alaska, Lituya Bay	XI	Major earthquake, landslide created water wave that denuded mountain side as high as 540 meters; long fault break; cables severed; 5 killed by drowning
1959	August 17	Montana, Hebgen Lake	X	Huge landslide dammed river and created lake; fault scarps with 4.5-meter throw; maximum vertical displacement 6.5 meters; 28 killed; $11 million damage to roads alone
1964	March 27	Alaska, Prince William Sound	X–XI	Great (Good Friday) earthquake; damage to public property $235 million, real property $77 million; in Anchorage, extensive damage to moderately tall structures (45 meters or less) and to poorly constructed low buildings; landslides and slumps caused total damage to many buildings; docks in several ports destroyed by submarine slides and tsunami; sea-wave damage on U.S. coast and elsewhere; 131 lives lost; shorelines rose 10 meters in places and settled 2 meters elsewhere
1965	April 29	Washington, Puget Sound	VII–VIII	Property loss $12.5 million, mostly in Seattle; felt over an area of 350,000 square kilometers; 3 persons killed and 3 died apparently of heart attacks

1969	October 1	California, Santa Rosa	VII–VIII	Property loss of $6 million; felt over an area of 30,000 square kilometers
1971	February 9	California, San Fernando	VIII–XI	$500 million direct physical loss, 65 killed, more than 1000 persons injured; felt over an area of 230,000 square kilometers
1975	August 1	California, Oroville	IX	3.8 kilometer-long fracture zone; felt over 120,000 square kilometers; $2 to $3 million property damage
1975	November 29	Hawaii	VIII	Largest earthquake to strike Hawaii since 1868; tsunami reached heights of 14.6 meters above sea level 25 kilometers west of epicenter; dislocations of faults along a zone 25 kilometers long; subsidence as much as 3.5 meters; $4.1 million property damage
1979	February 28	Southeastern Alaska	VII	First major earthquake since 1899 to occur between Yakutat Bay and Prince William Sound; M_s 7.1; ground acceleration at 73 kilometers distance was 0.16g
1979	August 6	California, Coyote Lake	VII	Ground displacement along Calaveras fault; maximum horizontal displacement was 5 to 6 millimeters; 16 injuries and $500,000 property damage

IMPORTANT EARTHQUAKES OF THE UNITED STATES AND CANADA *(continued)*

Year	Date (local time)	Place	MM Intensity	Remarks
1979	October 15	California, Imperial Valley	IX	91 injured; 1565 homes and 450 businesses damaged; estimated $30 million damage including heavy agricultural damage; felt over 128,000 square kilometers; maximum acceleration recorded 27 kilometers from epicenter was 1.74g; 25 kilometers of fault displacement along the Imperial Valley fault; 55 centimeters maximum lateral displacement and 19 centimeters of vertical displacement
1980	January 24 January 26	California, Livermore	VII	50 injuries; estimated $11.5 million damage; felt over 75,000 square kilometers; 1500 meters of discontinuous surface rupture showing a maximum of 5 to 10 millimeters of right lateral displacement
1980	May 18	Washington, Mount Saint Helens	IV	This earthquake occurred seconds before the explosion that began the eruption of Mount Saint Helens volcano; the eruption killed 31, left 33 missing, and caused between $500 million and $2 billion damage

1980	May 25 May 27	California-Nevada border Mammoth Lakes	VII	Four earthquakes occurred with M_L 6.1, M_L 6.0, M_L 6.1, and M_L 6.2; total of 13 injured; felt over approximately 250,000 square kilometers; $2 million estimated damages; 17-kilometer-long zone of discontinuous surface rupture associated with the Hilton Creek fault, with maximum vertical displacement of 50 millimeters and 200 millimeters of slip on a single fracture; many landslides and rocks fall in snow-covered unpopulated area
1980	November 8	California, off coast of northern California	VII	6 injured; $1.75 million property damage; highway overpass collapsed; small landslides and some liquefaction occurred; although the source was under the Pacific Ocean off the coast, it was felt over 97,000 square kilometers
1983	May 2	California, Coalinga	VIII	45 injured with no loss of life; heavy property damage in Coalinga and broken oil pipelines in area; felt over all of central California; thousands of aftershocks, estimated $31 million in damage
1986	January 31	Ohio, Painesville	VI	17 injured, some damage, felt throughout Ohio and in 14 other states
1986	October 1	California, Whittier	VIII	8 killed, many injured, 22,000 homeless, $213 million damage
1986	November 24	California, El Centro	VI	94 injured, $2.6 million damage

IMPORTANT EARTHQUAKES OF THE UNITED STATES AND CANADA *(continued)*

Year	Date (local time)	Place	MM Intensity	Remarks
1988	November 25	Chicoutimi, Quebec	VI	Largest earthquake in New England area since 1755
1989	October 17	California, Santa Cruz Mountains	X	63 killed, 3757 injured, $5.6 billion damage; most of the deaths occurred in the collapse of the second level of an elevated freeway in Oakland; the Bay Bridge was disabled for over one month due to collapse of one section of the span; severe damage occurred in portions of the Marina District in San Francisco, the downtown mall in Santa Cruz, and in scattered properties from Watsonville and Santa Cruz to San Francisco and Oakland
1992	April 25	California, Petrolia, Cape Mendocino	IX	Thrust of subduction plate; damage generally moderate; high peak acceleration recorded; no deaths.
1992	June 28	Landers, California	VIII	Largest magnitude (M_s 7.5) in U.S. in 40 years; over 70 kilometers of surface faulting in Mojave Desert; horizontal offset over 3 meters; one death occurred
1994	January 17	Northridge, California	VIII	Slip of blind-thrust fault. 56 deaths and over $15 billion damage. Collapsed freeways. Cracked welds in some steel-frame structures.

Source: U.S. National Oceanic and Atmospheric Administration and U.S. Geological Survey.

Appendix C

Abridged Modified Mercalli Intensity Scale

Note: The mean maximum acceleration and velocity values for the wave motion are for firm ground but vary greatly depending on the type of earthquake source.

Average peak velocity (centimeters per second)	Intensity value and description	Average peak acceleration (g is gravity = 9.80 meters per second squared)
	I. Not felt except by a very few under especially favorable circumstances. (I Rossi-Forel scale)	
	II. Felt only by a few persons at rest, especially on upper floors of buildings. Delicately suspended objects may swing. (I to II Rossi-Forel scale).	
	III. Felt quite noticeably indoors, especially on upper floors of buildings, but many people do not recognize it as an earthquake. Standing automobiles may rock slightly. Vibration like passing of truck. Duration estimated. (III Rossi-Forel scale)	

Average peak velocity (centimeters per second)	Intensity value and description	Average peak acceleration (g is gravity = 9.80 meters per second squared)
1–2	IV. During the day felt indoors by many, outdoors by few. At night some awakened. Dishes, windows, doors disturbed; walls make creaking sound. Sensation like heavy truck striking building. Standing automobiles rocked noticeably. (IV to V Rossi-Forel scale)	0.015g–0.02g
2–5	V. Felt by nearly everyone, many awakened. Some dishes, windows, and so on broken; cracked plaster in a few places; unstable objects overturned. Disturbances of trees, poles, and other tall objects sometimes noticed. Pendulum clocks may stop. (V to VI Rossi-Forel scale)	0.03g–0.04g
5–8	VI. Felt by all, many frightened and run outdoors. Some heavy furniture moved; a few instances of fallen plaster and damaged chimneys. Damage slight. (VI to VII Rossi-Forel scale)	0.06g–0.07g
8–12	VII. Everybody runs outdoors. Damage negligible in buildings of good design and construction; slight to moderate in well-built ordinary structures; considerable in poorly built or badly designed structures; some chimneys broken. Noticed by persons driving cars. (VIII Rossi-Forel scale)	0.10g–0.15g

Average peak velocity (centimeters per second)	Intensity value and description	Average peak acceleration (g is gravity = 9.80 meters per second squared)
20–30	VIII. Damage slight in specially designed structures; considerable in ordinary substantial buildings with partial collapse; great in poorly built structures. Panel walls thrown out of frame structures. Fall of chimneys, factory stacks, columns, monuments, walls. Heavy furniture overturned. Sand and mud ejected in small amounts. Changes in well water. Persons driving cars disturbed. (VIII + to IX Rossi-Forel scale)	0.25g–0.30g
45–55	IX. Damage considerable in specially designed structures; well-designed frame structures thrown out of plumb; great in substantial buildings, with partial collapse. Buildings shifted off foundations. Ground cracked conspicuously. Underground pipes broken. (IX + Rossi-Forel scale)	0.50g–0.55g
More than 60	X. Some well-built wooden structures destroyed; most masonry and frame structures destroyed with foundations; ground badly cracked. Rails bent. Landslides considerable from river banks and steep slopes. Shifted sand and mud. Water splashed, slopped over banks (X Rossi-Forel scale)	More than 0.60g

Average peak velocity (centimeters per second)	Intensity value and description	Average peak acceleration (g is gravity = 9.80 meters per second squared)
	XI. Few, if any, (masonry) structures remain standing. Bridges destroyed. Broad fissures in ground. Underground pipelines completely out of service. Earth slumps and land slips in soft ground. Rails bent greatly.	
	XII. Damage total. Waves seen on ground surface. Lines of sight and level distorted. Objects thrown into the air.	

Appendix D

Geologic Time Scale

Relative duration of major geologic intervals	Era	Period	Epoch	Approximate duration of millions of years	Millions of years ago
Cenozoic	Cenozoic	Quaternary	Holocene	Approx. the last 10,000 years	
			Pleistocene	2.5	2.5
		Tertiary	Pliocene	4.5	7
			Miocene	19.0	26
			Oligocene	12.0	38
			Eocene	16.0	54
			Paleocene	11.0	65
Mesozoic	Mesozoic	Cretaceous		71	136
		Jurassic		54	190
		Triassic		35	225
Paleozoic	Paleozoic	Permian		55	280
		Carboni-ferous	Pennsylvanian	45	325
			Mississippian	20	345
		Devonian		50	395
		Silurian		35	430
		Ordovician		70	500
		Cambrian		70	570
	Precambrian			4030	

Scale (millions of years): 0, 50, 100, 150, 200, 250, 300, 350, 400, 450, 500, 550, 4600

Appendix E

Conversion Tables

Metric-English

Length	
1 millimeter (mm) [0.1 centimeter]	= 0.0394 inch (in)
1 centimeter (cm) [10 mm]	= 0.3937 in
1 meter (m) [100 cm]	= 39.37 in
	= 3.28 feet (ft)
1 kilometer (km) [1000 m]	= 0.621 mile (mi)
Area	
1 square centimeter (cm^2)	= 0.155 square inch (in^2)
1 square meter (m^2)	= 10.76 square feet (ft^2)
	= 1.196 square yards (yd^2)
1 hectare (ha)	= 2.4710 acres (a)
1 square kilometer (km^2)	= 0.386 square mile (mi^2)
Volume	
1 cubic centimeter (cm^3)	= 0.0610 cubic inch (in^3)
1 cubic meter (m^3)	= 35.314 cubic feet (ft^3)
	= 1.31 cubic yards (yd^3)
1 cubic kilometer (km^3)	= 0.240 cubic mile (mi^3)
1 liter (l)	= 1.06 quarts (qt)
	= 0.264 gallon (gal)
1 cubic meter	= 8.11×10^{-4} acre feet
Mass	
1 kilogram (kg) [1000 grams (g)]	= 2.20 pounds (lb)
	= 0.0011 ton (tn)
1 metric ton (MT) [1000 kg]	= 1.10 tn

Pressure

1 kilogram per square centimeter (kg/cm^2)	= 14.20 pounds per square inch (lb/in^2)
1 pascal (10 dynes/cm^2)	= 47.9 pounds per square foot (lb/ft^2)
1 bar	= 100 kilopascals

Velocity

1 meter per second (m/s)	= 3.281 feet per second (ft/s)
1 kilometer per hour (km/h)	= 0.9113 ft/s
	= 0.621 mile per hour (mi/h)

Metric Unit Conversion

Physical Quantity	**cgs Unit**	**Equivalent in SI**
Mass	g	10^{-3} kg
Length	cm	10^{-2} m
Time	s	1 s
Force	dyne	10^{-5} N (newton)
Pressure and	dyne/cm^2	10^{-1} Pa (pascal)
elastic moduli	bar	10^{5} Pa
Energy	erg	10^{-7} J (joule)
Density	g/cm^3	10^{3} kg/m^3
Moment	dyne-cm	10^{-7} J
Viscosity	poise	10^{-1} Ns/m^2

Appendix F

How to Determine Fault Planes from First Motions

Given: A set of first motion measurements for an earthquake (i.e., the P wave polarity) as a compression (•) or a dilatation (°) at a set of seismographic stations. The azimuth (direction along the great circle) from the epicenter to each station is known, as well as the angle at which the P wave ray is incident to the vertical at the ground surface.

Aim: To determine the strike and dip of the δ slipped fault-plane at the focus (hypocenter) and the direction of the horizontal component of slip.

Equipment: An equal angle projection net (such as the Wulff net on p. 321 and used in the figure on page 319, on which lines of longitude map great circles) and tracing paper. Glue the net to a piece of stiff cardboard and push a pin or thumbtack through the cardboard, from the back, at the center of the net. Push the tracing page onto the net and trace onto it: the circumference of the net; orthogonal diameters marked North, East, South, and West; and the rotation pinpoint O.

Construction I: How to plot a station (S) polarity correctly on the net (say Azimuth N20°E and ray incidence angle 40°). We follow the steps shown in the figure on the facing page.
Steps
(1) Mark the point X on the circumference of the net at N20°E as in part (a).
(2) Rotate the paper until X coincides with N on the net. Mark S, 40° along the longitude line from O to X (part (b)).
(3) Rotate the tracing paper until N on the paper again coincides with N on the net (as in part (c)). Then S is the position of the station on the projection.

Construction II: How to draw the great circle between two stations, R and S, plotted on the projection. We follow the steps in the second set of rotations given in the figure on page 319.
Steps
(1) Plot R and S as in Construction I (part (d)).

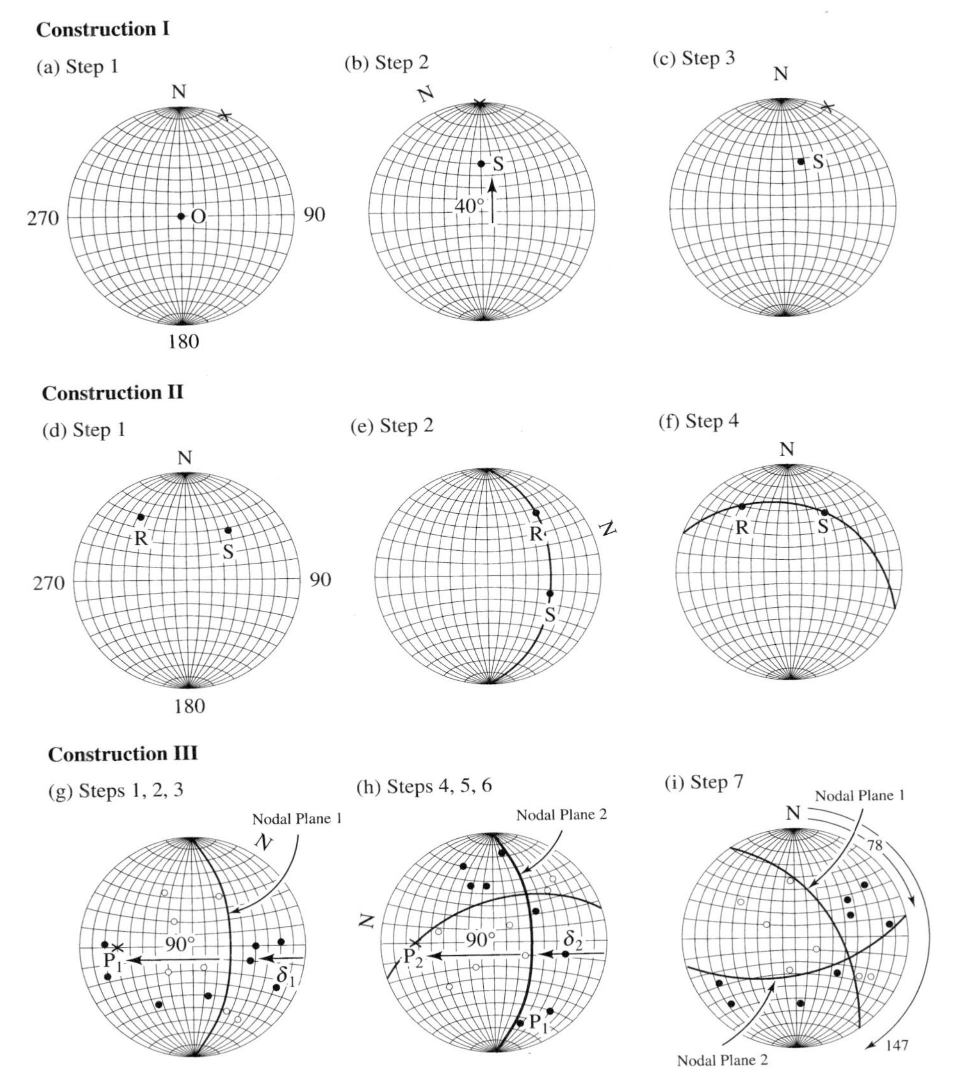

Equal-area (Schmidt) projection net.

(2) Rotate the tracing paper until R and S lie on the same longitude line. Trace this great circle (part (e)).
(3) Rotate the tracing paper until N on the paper coincides with N on the net. The great circle joining R and S is shown on the projection (part (f)).

Construction III: How to locate the fault plane and the auxiliary plane for the plotted station data by rotations of the tracing paper around the pin. See the third sequence of projection diagrams.

Steps
(1) Plot all the station polarities on the tracing paper as described in Construction I, indicating a compression (•) or a dilatation (°).
(2) Rotate the paper until a longitude line separates compressions and dilatations. Trace the longitude line (part (g)). This great circle represents nodal plane 1. The dip δ_1 of nodal plane 1 is measured from the point of intersection of the nodal plane with the equator along the equator to the circumference.
(3) Measure 90° along the equator from its intersection with nodal plane 1; mark P_1, (see part (g)). The ray OP_1 is normal to nodal plane 1.
(4) Rotate the tracing paper to find a second longitude line, now passing through P_1, which separates the remaining compressions and dilatations.
(5) Trace this great circle which defines fault plane 2, on the net (part (h)). As before, the dip δ_2 is the distance from its intersection with the equator along the equator to the circumference.
(6) Measure P_2 along the equator 90° from its intersection with nodal plane 2. P_2 should lie on nodal plane 1 (part (h)).
(7) Rotate the tracing paper until N on the paper coincides again with N on the projection net.
The strike of the nodal plane is measured clockwise around the circumference from N to the great circle of the nodal plane (part (i)).

Construction IV: How to measure the horizontal direction of slip of the causative fault at the focus, from the projection great circles.
If nodal plane 1 is the fault plane, then nodal plane 2 is called the *auxiliary plane* (and vice versa). The slip vector is normal to the auxiliary plane, and is therefore OP_2. The strike of the horizontal component of the slip vector is measured clockwise around the circumference of the projection from N to OP_2.
Note: One fault plane above corresponds to the actual geological fault, the auxiliary plane is at right angles to it. Independent information (such as surface faulting) is needed to tell them apart.

Answers to the Demonstration Case
Dip and strike of nodel plane 1 = 60°, 147°.
Dip and strike of nodal plane 2 = 60°, 78°.
Horizontal direction of slip on nodal plane 1 = 348°.

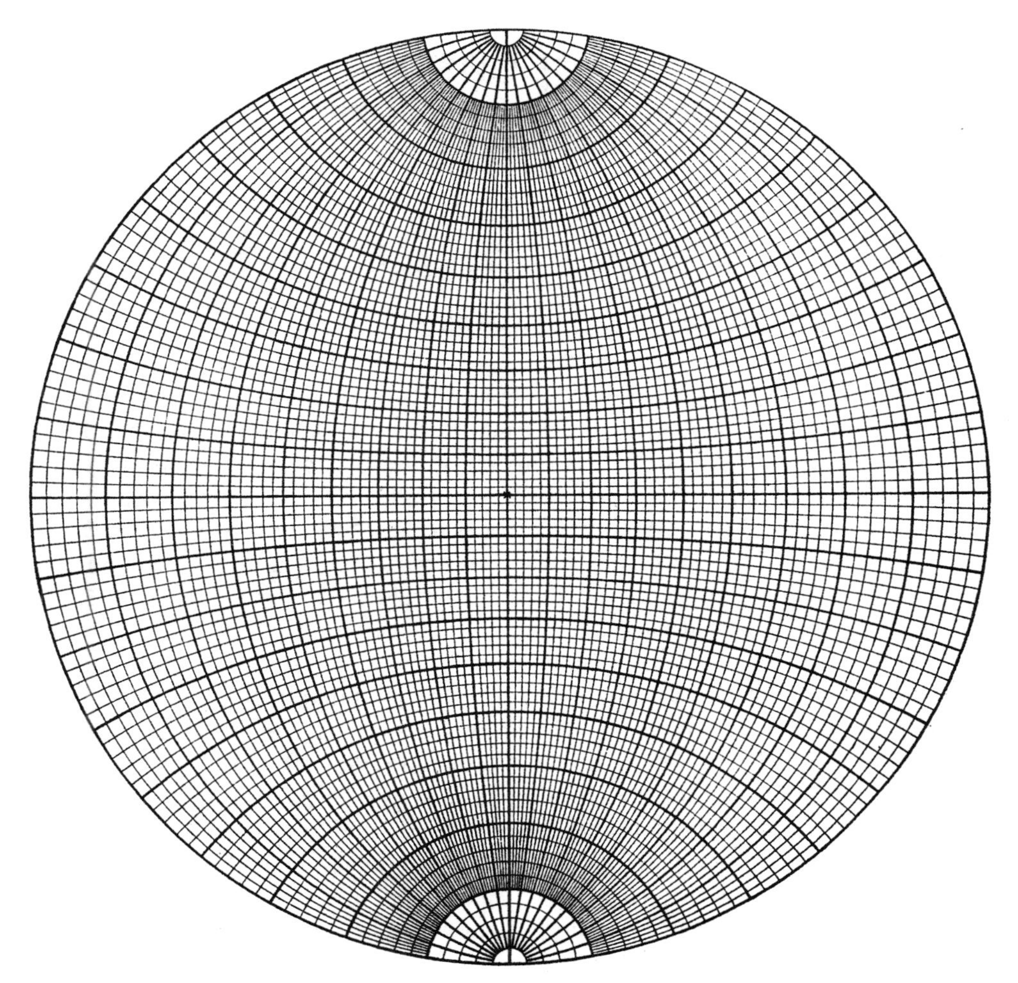

Equal-angle stereographic net.

Appendix G

Sample Calculations of Magnitudes, Moment, and Energy of an Earthquake

The following calculations are for an Alaskan earthquake recorded at Oroville, California. The energy factor (equation 4) gives us an idea of the scale of energy release that is possible for earthquakes of different magnitude. For instance, 30 earthquakes of magnitude 6 are needed to release the equivalent amount of energy in the Earth's crust that is released by just one magnitude 7 earthquake; and 900 earthquakes of magnitude 5 are needed to produce the same energy. It follows, therefore, that even if small earthquakes occur in swarms in a particular area, they do very little to reduce the reservoir of strain energy needed for a major earthquake. But tectonic energy is drained away into heat and seismic waves in a truly gigantic way by a major earthquake like that of 1906 along the San Andreas fault with a magnitude M_s of 8 1/4. This earthquake released about 10^{17} Joules of strain energy within 60 seconds! (Only a fraction went into ground shaking.)

It is well known that, as the threshold of earthquake size being considered in a seismic region is lowered, the number of earthquakes above that magnitude rapidly increases (see Appendix A, p. 298). The rate of occurrence n of shocks above a given magnitude is again logarithmic and is measured by a parameter b (see equation 6). The smaller b is, the more numerous are the earthquakes in a given time span. When b is determined for a seismically active region, the total seismic energy released over a period can be calculated by using the energy factor.

Magnitude is also sometimes roughly estimated from the length of surface fault rupture L (in kilometers — see equation 7).

These calculations all follow from substitutions of measurements made directly on a seismogram into empirical formulas.

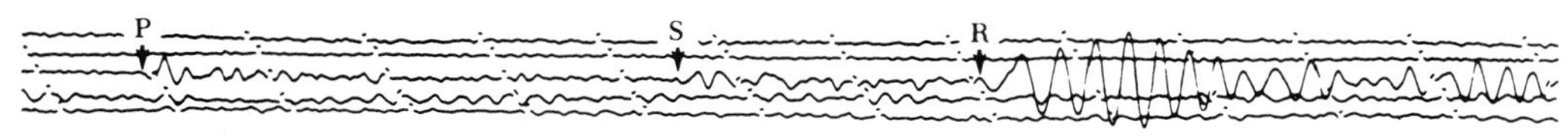

Typical P, S, and Rayleigh waves on seismogram. One minute between gaps.

Let A be the amplitude, and T the period of a wave measured at a distance Δ from the source.

Measured values from above seismogram (reduced to ground motion)

$$\text{P wave, } A = 1.4 \text{ microns } (10^{-6} \text{ meters}),\ T = 12 \text{ seconds}$$

$$\text{Rayleigh wave, } A = 4.3 \text{ microns},\ T = 20 \text{ seconds}$$

$$\Delta = 28^\circ$$

Body-wave magnitude m_b ($25^\circ < \Delta < 90^\circ$)

$$\begin{aligned} m_b &= \log A - \log T + 0.01\Delta + 5.9 \\ &= 0.15 - 1.08 + 0.28 + 5.9 \\ &\approx 5.3 \end{aligned} \tag{1}$$

Surface-wave magnitude M_s ($25^\circ < \Delta < 90^\circ$)

$$\begin{aligned} M_s &= \log A + 1.66 \log \Delta + 2.0 \\ &= 0.63 + 2.40 + 2.0 \\ &\approx 5.0 \end{aligned} \tag{2}$$

Moment M_o (Newton meters)

$$\log M_o = 10.92 + 1.11\, M_s \tag{3}$$

$$\text{Therefore } M_o \approx 3 \times 10^{16} \text{ Newton meters}$$

(N.B. Formula (3) is not applicable to the largest earthquakes.)

Seismic energy E (Joules)

$$\begin{aligned} \log E &= 4.8 + 1.5\, M_s \\ &= 12.3 \\ E &= 2.0 \times 10^{12} \text{ Joules} \end{aligned} \tag{4}$$

Relation between moment magnitude M_w and seismic moment M_o in Newton-meters (see Chapter 7)

$$M_w = \tfrac{2}{3} \log M_o - 6.0 \approx 4.9 \tag{5}$$

Relation between n and M_s

$$\log n = a - bM_s \tag{6}$$

Relation between M_s and fault rupture length in kilometers L (worldwide data)

$$M_s = 6.10 + 0.70 \log L \tag{7}$$

Appendix H

The Elements of Wave Motion

All readers will be familiar with the motions on the surface of water that occur when wind blows across a lake or a stone is dropped into a pond. Sound and light are other common examples of wave motion. The physical nature of all such waves can best be understood by thinking of waves traveling along an elastic string. The string might be started to vibrate by being plucked, as with a violin, or struck, as with a hammer blow in a piano.

The figure on page 325 shows part of the side-to-side vibration of the string as the wave travels down the string from left to right. The distance between the crests (or troughs) of the wave measures the length of the wave, and the maximum excursion of the wave measures the wave amplitude.

For a traveling simple harmonic wave, like that shown, the displacement y at a given time t is a function of position x and time t:

$$y = \mathrm{A} \sin \frac{2\pi}{\lambda} (x - vt)$$

Here A is the *amplitude* as shown; the *wave length* λ (Greek lambda) is the distance between crests B and D, and v is the *wave velocity*.

In time T, called the *period*, the wave travels a distance λ, for example, from B to D. The period T is thus the time of a complete vibration.

The velocity of travel of the wave is the wavelength divided by the period:

$$v = \frac{\lambda}{T} = f\lambda \qquad (1)$$

Here f is the *frequency*, the number of vibrations per second (units: cycles per second or Hertz). The angular frequency ω, in radians per second, is $\omega = 2\pi f$ (ω is the Greek letter omega.)

Thus, from (1), the sideways motion of the point O ($x = 0$) can be written:

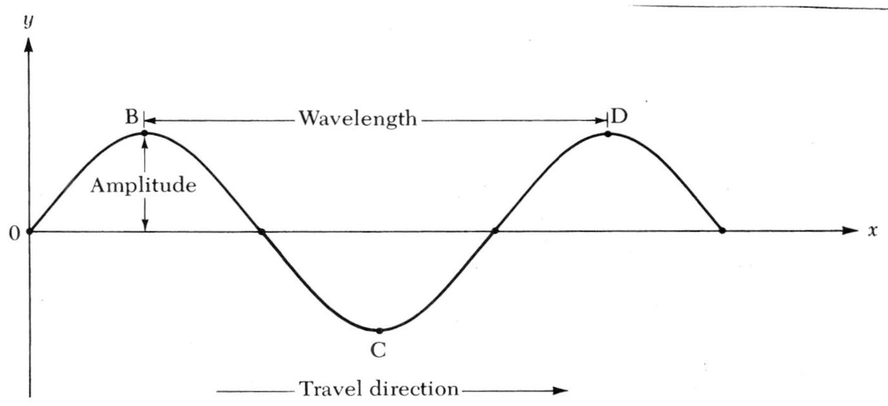

Transverse wave at time t.

$$\text{Displacement} = y = A \sin \frac{2\pi}{T} t = A \sin \omega t \tag{2}$$

$$\text{Velocity} = \mathring{y} = \frac{2\pi A}{T} \cos \frac{2\pi}{T} t = \omega A \cos \omega t \tag{3}$$

$$\text{Acceleration} = \ddot{y} = -\frac{4\pi^2 A}{T^2} \sin \frac{2\pi}{T} t = -\omega^2 A \sin \omega t \tag{4}$$

$$= -\omega^2 y \tag{5}$$

In earthquakes, the values of ground acceleration, velocity, and displacement vary a great deal, depending on the frequency of the wave motion. High-frequency waves (higher frequencies than 10 hertz) tend to have high amplitudes of accelerations but small amplitudes of displacement, compared with long-period waves, which have small accelerations and relatively large velocities and displacements. (These relations follow from equations 1 to 5.)

We can now think of the wave represented in the figure as an earthquake wave traveling through the rock to the ground surface (as a P or S seismic wave; see Figure 1.10).

The following table gives a handy comparison between ground accelerations of different wave frequencies and ground motion amplitudes.

Frequency (Hz)	Acceleration (fraction of gravity g)		
	Displacement		
	0.1 cm	1.0 cm	2.0 cm
10	0.04g	0.4g	0.8g
5	0.02g	0.2g	0.4g
1	0.004g	0.04g	0.08g
0.5	0.002g	0.02g	0.04g

Note that the peak horizontal ground acceleration in the 1989 Loma Prieta earthquake measured on rock near the San Francisco Bay Bridge was 0.2g at a frequency of about 5 Hz. The table indicates that the corresponding backward and forward movement of the ground with the same frequency had an amplitude of about 1 centimeter. The actual measured size of the seismic displacement was greater but had a much lower frequency.

Appendix I

Propagation of Elastic Waves

The elasticity of a homogeneous, isotropic solid can be defined by two constants, k and μ.

k is the modulus of incompressibility*, or bulk modulus
for granite, k is about 27×10^{10} dynes per square centimeter;
for water, k is about 2.0×10^{10} dynes per square centimeter.

μ is the modulus of rigidity
for granite, μ is about 2.6×10^{11} dynes per square centimeter;
for water, $\mu = 0$.

Within the body of an elastic solid with density ρ, two elastic waves can propagate:

P waves Velocity $\alpha = \sqrt{\left(k + \frac{4}{3}\mu\right)/\rho}$

for granite, $\alpha = 4.8$ kilometers per second;
for water, $\alpha = 1.4$ kilometers per second.

S waves Velocity $\beta = \sqrt{\mu/\rho}$
for granite, $\beta = 3.0$ kilometers per second;
for water, $\beta = 0$ kilometers per second.

Along the free surface of an elastic solid, two surface elastic waves can propagate:

Rayleigh waves Velocity $c_R < 0.92\beta$, approximately
where β is the S-wave velocity in the rock.

Love waves (for a layered solid) Velocity $\beta_1 < c_L < \beta_2$
where β_1 and β_2 are S-wave velocities in the surface and deeper layers, respectively.

The dimensions of a harmonic wave are measured in terms of period T and wavelength λ (see Appendix H).

**For unit conversion see Appendix E.*

✍ An Earthquake Quiz

(Answers on p. 334)

1. What mechanical idea did the Chinese scholar Chang Heng probably use in constructing the first known seismoscope? (See Chapter 3, Figure 3.1.)

2. What is the main difference between a seismograph and a seismoscope?

3. Why were the earliest seismographic stations in the United States established in California? Why at astronomical observatories?

4. With a ruler, measure the time interval between the phases marked P and S on the seismogram of Chapter 3, Figure 3.5. Then, using the values 8.0 kilometers per second for the velocity of P waves, and 4.4 kilometers per second for that of S waves, check the distance between the Berkeley station and the earthquake focus.

5. Measure with a ruler the largest wave amplitude shown in Chapter 3, Figure 3.5. If you are told that the seismograph that gave the record of earthquake waves in that figure amplifies the actual displacement of the ground approximately 3000 times, and that the figure has been reproduced at the original scale, calculate the greatest amplitude (as a fraction of a millimeter) of the actual ground motion. Could you feel such a small motion?

6. By using your ruler on Chapter 8, Figure 8.4, determine the bracketed duration, in seconds, of the strong ground acceleration (amplitude above 0.05g) recorded on the east component at the Hollywood Storage Parking Lot in the 1971 San Fernando earthquake.

7. If the mean speed of P waves is 10 kilometers per second through the Earth, what is the approximate time for a P wave to travel from one side of the Earth to the other?

8. Why is the Earth's mantle considered to be solid?

9. Could a seismograph on a ship at sea detect an earthquake?

10. Why do seismologists object to the common usage of "tidal" wave in reference to a "tsunami"?

11. Some earthquakes in Japan are felt in Tokyo only by people at the top of a tall building. Why?

12. Why are many earthquakes not accompanied by observations of fault rupture?

13. Read Zechariah 14:4 in the Bible. What is your interpretation of this geological event?

14. Why do you expect no great quake on the moon like that near San Francisco in 1906?

15. Based on the observed relative motion of the great tectonic plates of the Earth, what would be the direction of strike-slip faulting that you would expect to accompany large shallow earthquakes on the Digdig fault in the Philippines? The Anatolian fault in Turkey? The San Andreas fault in California? The Pyrenees?

16. Why are the sites around lakes, bays, and estuaries likely to lead to higher Modified Mercalli intensities than sites on high ground?

17. The 1964 Alaska Good Friday earthquake was rated by measuring seismograms as body-wave magnitude $m_b = 6.5$, but surface wave magnitude $M_s = 8.6$. Explain the large difference. Why does the M_s value best describe the size of this earthquake?

18. What is the point of operating sensitive seismographs around large reservoirs? What is the lowest number of seismographs you would install to locate local earthquakes?

19. Why are small shallow earthquakes often recorded during a volcanic eruption?

20. What percentage of the heat energy reaching the Earth's surface from the interior is released by earthquakes each year?

21. Is there a difference between weather prediction and earthquake prediction? In what sense are these problems different?

22. Calculate how many hours of advance warning in California may be possible before the arrival of a tsunami generated by an earthquake in Alaska.

23. Determine the Richter magnitude (M_L) of an earthquake whose largest wave recorded by a standard seismograph is 10 millimeters at a distance of 100 kilometers from the epicenter.

24. What is the energy in Joules of this earthquake?

25. Why are at least three seismographic stations required to locate earthquake epicenters on the basis of P-wave arrival times?

26. Show from the left-hand scale in Box 8.1 in Chapter 8 that a handy approximation to remember between the S-minus-P-wave travel times and

distance d kilometers to the earthquake focus is $S - P \simeq d/8$ seconds (for short distances).

27. From the S-minus-P interval on the seismograms of Chapter 8, Figure 8.4, estimate how far the earthquake source is from the strong-motion instrument.

28. At the present rate of slip (say 3 centimeters per year) on the San Andreas fault, how long will it be before Los Angeles is a suburb of San Francisco? Will it matter?

29. A large dam impounds an extensive lake about 200 meters deep. What is the extra pressure (in bars, that is, 10^6 dynes per square centimeter or 10^5 Pascals) placed on the Earth's crust by the water load? How does this quantity compare with the stress released along a rupturing shallow fault (5 to 50 bars, approximately)? Does this result necessarily imply that reservoir loads can produce earthquakes?

30. It has often been observed that some farm animals become disturbed just before a substantial earthquake in the area. Can you make some suggestions that might explain this?

31. Many earthquakes are accompanied by audible noises. Can you explain them in terms of the P waves of seismology?

32. After the great India (Assam) earthquake of June 12, 1897, farmers found piles of sand in their fields that hindered cultivation. Explain this circumstance.

33. The earthquake harvester: In the Kwanto (Japan) earthquake of 1923, potatoes emerged from soft farm ground. Why?

34. Fish are reported stunned or killed by strong earthquakes originating from under the sea. Explain.

35. Can you explain why the flow of natural springs is often affected by large earthquakes? Why would some cease to flow and others have an increased flow?

36. Compute the energy available to go into seismic waves when the great rock avalanche fell down the face of Mount Huascarán in the Peru earthquake of May 31, 1970. (Assume 50 million cubic meters of rock fell 1 kilometer.) What is the equivalent earthquake magnitude?

37. You are lying in bed when you notice a light fixture, 1 meter long, hanging from the ceiling start to swing. Having read this book, you make some measurements. You count swings and find that 10 occur in 20 sec-

onds before stopping. The free end of the fixture swings through a maximum of 1 centimeter. You then calculate that an earthquake of a certain magnitude has occurred, centered a certain distance away. What is the rough magnitude and distance?

38. The jelly earthquake machine: Make a model of an elastic crust of the Earth by pouring a stiff jelly mixture into a wide shallow square or rectangular pan. Make a vertical cut down the center of the jelly with a sharp knife. Then slide, by pulling horizontally in opposite directions, opposite sides of the jelly, parallel to the cut. Describe what happens when the sides of the cut slip.

39. How long does it take a stone to fall 60 meters? Now estimate how long it will take rubble to fall from the top of a high building in an earthquake.

40. In an earthquake, a fault ruptures through your property. Would you gain or lose land in strike-slip faulting? Normal faulting? Thrust faulting? Discuss the legal implications for such property in a built-up area.

41. You have just taken off in an airplane. Will you know if a great earthquake then shakes the airport?

42. Why did Japanese peasants associate catfish with the cause of earthquakes?

43. In the Mexico earthquake of September 19, 1985, most collapsed buildings in Mexico City were 10- to 12-story concrete-frame apartment houses. Why would such buildings be more likely to collapse in this earthquake than shorter buildings?

44. The State Office of Emergency Preparedness has just issued a warning that there is a 50 percent chance of a damaging earthquake occurring in your area next week. What would be your program of preparedness?

45. Suppose that five damaging earthquakes have occurred in your area in the last 100 years. During that time about 500 earthquakes (most small) have been recorded by the local seismographic station. What are the simple odds that the next earthquake will be damaging?

46. The dice earthquake game: You and your opponent take turns in rolling two dice. The sum of the numbers in any throw will lie between 2 and 12. Suppose these numbers refer to intensities (II to XII) on the Modified Mercalli scale. An intensity of X or above scores 0; a lower intensity scores 10. The player to accumulate a score of 100 wins. Do large intensities always follow small ones?

47. For the mathematically minded: What is the probability in the game above of getting an intensity of X or XI in one throw of the two dice?

48. The biggest earthquakes, such as the 1960 Chile earthquake, shake the whole Earth, like a hammer ringing a bell. The deepest free vibrations, or tones, of the Earth are 53 minutes a period. How long would a simple pendulum be to have such a period?

49. Gravity on the moon is only one-sixth of that on Earth. Explain why, if quakes of equal magnitude occurred on the moon and the Earth, the seismic hazard on the lunar surface would be greater than on Earth.

50. The figure at the bottom of the page shows waves due to kangaroos jumping near a seismograph at Woomera, southern Australia. The jumps are recorded as a series of sharp peaks. Measure the average frequency at which the hops occur. Explain the increase and decrease of amplitude and the duration of the kangaroo signal. [Seismogram obtained by I. C. F. Stewart and P. J. Setchell, *Search* 5, (1974): 107.]

51. In 1982 a small aircraft collided with Soufrière volcano on St. Vincent, West Indies, killing all aboard. A nearby seismograph recorded the impact with seismic magnitude $m_b = 1.0$. For airspeed 250 kilometers/hour and plane weight 2650 kilograms, determine the kinetic energy lost on impact and the maximum coupling efficiency for seismic energy (see Appendix G).

52. From the discussion in Chapter 5, identify the type of fault that corresponds to the earthquake beach-ball diagrams at the top of page 334. Dark areas are ground compressions and white areas are dilatations (rarefactions). The boundary FF′ denotes the fault plane and north is at the top.

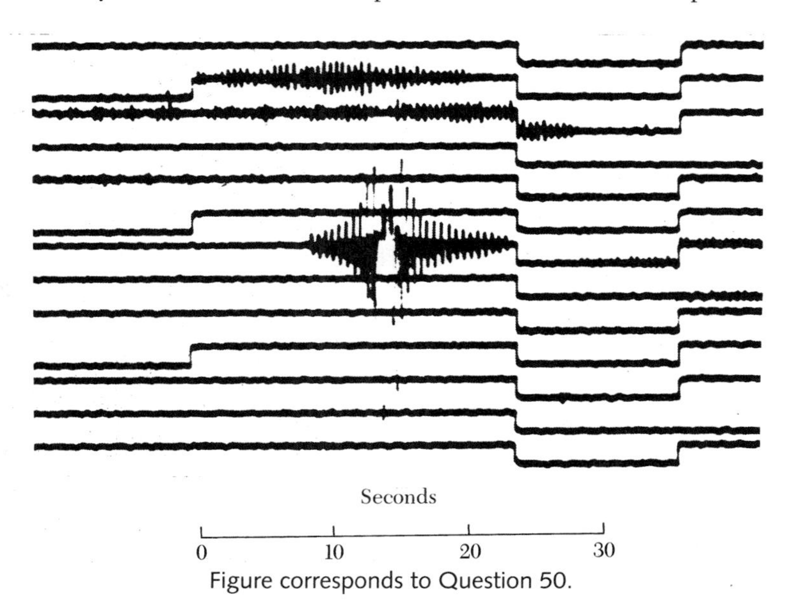

Figure corresponds to Question 50.

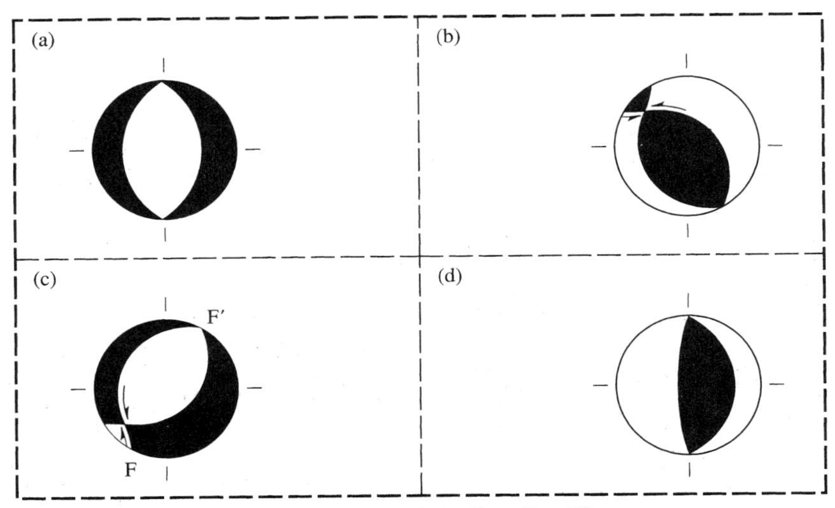

Figure corresponds to Question 52.

Answers to Quiz

1. An inverted pendulum.
2. Seismoscopes register no time marks.
3. California is a seismic region. Astronomers kept accurate time.
4. 110 kilometers.
5. 0.003 millimeter. No.
6. About 6 seconds.
7. 21 minutes.
8. S waves pass through.
9. Yes, from the P wave.
10. There is no connection with the sun or moon.
11. Long-period seismic waves are amplified by building response there.
12. Many faults are deep underground or submarine.
13. Dip-slip faulting with strike EW.
14. There is no plate-tectonic activity.
15. Left-lateral. Right-lateral. Right-lateral. Left-lateral.
16. Soil fracture including liquefaction.
17. Surface waves correspond better to fault rupture length.
18. For the detection of increased seismicity. Three.

19. Movements of liquid magma affect rock strength.
20. Less than 0.001 percent.
21. Yes. Weather observations occur within the atmosphere, and faster variations occur in the atmosphere.
22. 22 hours.
23. Four.
24. 10^{17}.
25. One is required for each of the three unknowns (latitude, longitude, origin time).
27. About 15 kilometers.
28. 20 million years.
29. 20 bars. Similar. No.
30. Perhaps small foreshocks.
31. High-frequency P waves refracted into the air.
32. Liquefaction of sandy soils.
34. Shock of intense P wave.
36. 1.2×10^{15} Joules. About 6.9.
37. Greater than 7, over 100 kilometers away.
39. 3.5 seconds.
40. No change; gain; lose.
41. Probably not. Seismic air waves are too small.
43. Corner columns unconnected on two sides.
45. 1/100.
47. 5/36.
48. About 2300 kilometers.
49. Ground accelerations would much exceed lunar gravity.
50. 2 hops per second.
51. 6×10^{6} Joules, 10^{-1}.
52. (a) Normal dip-slip on north-striking fault dipping 50° W or 40° E. (b) Thrust left-oblique displacement. (c) Normal right-oblique displacement on fault which strikes N 20° E and dips 30° NW. (d) Pure reverse slip on north-striking fault that dips 60° W or 30° E.

Glossary

accelerometer: a seismograph for measuring ground acceleration as a function of time.

active fault: a fault along which slip has occurred, variously in historical or Holocene or Quaternary time, or earthquake foci are located.

active margin: continental margin characterized by volcanic activity and earthquakes (i.e., location of transform fault or subduction zone).

aftershocks: smaller earthquakes following the largest earthquake of a series concentrated in a restricted crustal volume.

amplitude (wave): the maximum height of a wave crest or depth of a trough.

aseismic region: one that is almost free of earthquakes.

asperities (fault): roughness on the fault surface subject to slip.

asthenosphere: the layer below the lithosphere that is marked by low seismic-wave velocities and high seismic-wave attenuation. It is a soft layer, probably partially molten.

attenuation: the reduction in amplitude of a wave with time or distance traveled.

auxiliary fault-plane: a plane orthogonal to the fault plane.

basin depression: depression in which sediments collect.

barrier (fault): an area of fault surface resistant to slip because of geometrical or structural changes.

Benioff zone: a narrow zone, defined by earthquake foci, that is tens of kilometers thick dipping from the surface under the Earth's crust to depths of up to 700 kilometers. (Also **Wadati-Benioff zone.**)

blind thrust: a thrust-fault deep in the crust with no or only indirect surface expression such as a fold structure.

body wave: a seismic wave that travels through the interior of an elastic material; P and S seismic waves.

body-wave magnitude: magnitude M_b of an earthquake as estimated from the amplitude of body waves.

bore: a single water wave with an almost vertical front.

bulk modulus: See **incompressibility.**

characteristic earthquake: an earthquake with a size and generating mechanism typical for a particular fault source.

coda: the concluding train of seismic waves that follows the principal part of an earthquake.

continental shelf: part of the continental margin between the coast and the continental slope; slopes about 0.1°.

core (of Earth): the central part of the Earth below a depth of 2900 kilometers. It is thought to be composed mainly of iron and silicates and to be molten on the outside with a solid central part.

creep (slow fault slip): slow slip occurring along a fault without producing earthquakes.

crust (of Earth): the outermost rocky shell of the Earth.

damping: loss of energy in wave motion due to transfer into heat by frictional forces.

density: the mass per unit volume of a substance, commonly expressed in grams per cubic centimeter.

digital recording: a series of discrete digits.

dilatancy (of rocks): the increase in the volume of rocks mainly due to pervasive microcracking.

dip: the angle by which a rock layer or fault plane deviates from the horizontal. The angle is measured in a plane perpendicular to the strike.

dip-slip fault: a fault in which the relative displacement is along the direction of dip of the fault plane; the offset is either normal or reverse.

dispersion (wave): the spreading out of a wave train due to each wave length traveling with its own velocity.

duration (of strong shaking): the time interval between the first and last peaks of strong ground motion above a specified amplitude.

earthquake: the vibrations of the Earth caused by the passage of seismic waves radiating from some source of elastic energy.

earthquake occurrence (recurrence) interval: the average interval of time between the occurrence of earthquakes in a particular region.

elastic rebound theory: the theory of earthquake generation proposing that faults remain locked while strain energy slowly accumulates in the surrounding rock and then suddenly slip, releasing this energy in the form of heat and seismic waves.

epicenter: the point on the Earth's surface directly above the focus (or hypocenter) of an earthquake.

fault: a fracture or zone of fractures in rock along which the two sides have been displaced relative to each other parallel to the fracture. The total fault offset may range from centimeters to kilometers.

fault plane: the plane that most closely coincides with the rupture surface of a fault.

first motion: on a seismogram, the direction of motion at the beginning of the arrival of a P wave. Conventionally, upward motion indicates a compression of the ground; downward motion, a dilatation.

flower structure: more-or-less symmetrical splays into sub-faults near the intersection of the main fault with the ground surface.
focal depth (of earthquakes): the depth of the focus below the surface of the Earth.
focus (hypocenter): the place at which rupture commences.
foreshocks: smaller earthquakes preceding the largest earthquake of a series concentrated in a restricted crustal volume.
frequency: number of oscillations per unit time; unit is Hertz (Hz), which equals 1 cycle per second.
geodimeter: a surveying instrument to measure the distance between two points on the Earth's surface.
gouge: crushed, sheared, and powdered rock altered to clay.
graben: a crustal block of rock, generally long and narrow, that has dropped down along boundary faults relative to the adjacent rocks.
Gutenberg discontinuity: discontinuity in seismic velocity that marks the boundary between the core and the mantle; named after seismologist Beno Guttenberg.
hazard (seismic): dangerous physical effects of earthquakes, such as landslides, ground shaking, tsunamis.
hertz: the unit of frequency equal to 1 cycle per second, or 2π radians per second.
Holocene: about 10,000 years before the present.
hypocenter: same point as the focus
incompressibility: an index of the resistance of an elastic body, such as a rock, to volume change.
inner core (of Earth): central solid region of the Earth's core, probably mostly iron; radius about 1221 kilometers.
intensity (of earthquakes): a measure of ground shaking obtained from the damage done to structures built by humans, changes in the Earth's surface, and felt reports.
interplate earthquake: earthquake with its focus on a plate boundary.
intraplate earthquake: earthquake with its focus within a plate.
island arc: chain of islands above a subduction zone (e.g., Japan, Aleutians).
isoseismal: contour lines drawn to separate one level of seismic intensity from another.
isostasy: the way in which the lithosphere "floats" on the asthenosphere.
lava: magma, or molten rock, that has reached the surface.
left-lateral fault: a strike-slip fault on which the displacement of the far block is to the left when viewed from either side.
Lehmann discontinuity: boundary between outer and inner core of the Earth.
liquefaction (of soil): process of soil and sand behaving like a dense fluid rather than a wet solid mass during an earthquake.

lithology: physical character of rocks.
lithosphere (of Earth): the outer, rigid shell of the Earth above the asthenosphere. It contains the crust, continents, and plates.
Love waves: seismic surface waves with only horizontal shear motion transverse to the direction of propagation.
lurching of ground: disruption of soil by lateral spreading under gravity.
magma: molten rock material that forms igneous rocks upon cooling.
magnitude (of earthquakes): a measure of earthquake size, determined by taking the common logarithm (base 10) of the largest ground motion recorded during the arrival of a seismic wave type and applying a standard correction for distance to the epicenter. Three common types of magnitude are Richter (or local) (M_L), moment (M_w), and surface wave (M_s).
mantle (of Earth): the main bulk of the Earth, between the crust and core, ranging from depths of about 40 to 3470 kilometers. It is composed of dense silicate rocks and divided into a number of concentric shells.
mare: a dark, low-lying lunar plain, filled to an undetermined depth with volcanic rocks. (Plural: maria).
meizoseismal region: the area of strong shaking and significant damage in an earthquake.
microseism: weak, almost continuous background seismic waves or Earth "noise" that can be detected only by seismographs often caused by surf, ocean waves, wind, or human activity.
microzonation: the division of a town or county into smaller areas according to the variation in seismic hazard.
Mohorovičoć discontinuity (M-discontinuity): the boundary between crust and mantle, marked by a rapid increase in seismic P-wave velocity to more than 8 kilometers per second. Depth: 5 kilometers (under oceans) to 45 kilometers (under mountains).
moment (of earthquakes): a measure of earthquake size related to the leverage of the forces (couples) across the area of the fault slip. The rigidity of the rock times the area of faulting times the amount of slip. Dimensions are dyne-cm (or Newton-meters).
moment magnitude: magnitude M_w of an earthquake estimated by using the seismic moment.
normal fault: a dip-slip fault in which the rock above the fault plane has moved downward to the rock below.
oblique faulting: the slip on the fault has components both along the dip and along the strike of the fault.
origin time: the time of initiation of the seismic waves at an earthquake source (usually given in Universal Time, UT).
outer core (of Earth): outer liquid shell of the Earth's core, probably iron with some oxygen; inner radius, 1221 kilometers, outer radius, 3480 kilometers.

paleoseismology: that part of earthquake studies that deals with evidence for earthquakes before instrumental recording of seismic waves or damage from felt reports.
passive margin: continental margin formed during initial rifting apart of continents to form an ocean; frequently has thick sedimentary deposits.
period (wave): the time interval between successive crests in a sinusoidal wave train; the period is the inverse of the frequency of a cyclic event.
plate (tectonic): a large, relatively rigid segment of the Earth's lithosphere that moves in relation to other plates over the deeper interior. Plates meet in convergence zones and separate at divergence zones.
plate tectonics: a geological model in which the Earth's crust and uppermost mantle (the lithosphere) are divided into a number of more-or-less rigid segments (plates).
precursor: a change in the geological conditions that is a forerunner to earthquake generation on a fault.
prediction (of earthquakes): the forecasting in time, place, and magnitude of an earthquake; the forecasting of strong ground motions.
Quaternary: about 2 million years before the present.
P wave: the primary or fastest wave traveling away from a seismic event through the rock and consisting of a train of compressions and dilatations of the material.
probability: the number of cases that actually occur divided by the total number of cases possible.
probability of exceedence of a given earthquake size: the odds that the size of a future earthquake will exceed some specified value.
Rayleigh waves: seismic surface waves with ground motion only in a vertical plane containing the direction of propagation of the waves.
recurrence interval: the average time interval between earthquakes in a seismic region.
resonance: the largest vibration of a mechanical system (such as a soil layer) due to enhancement of the energy at a frequency special to that system.
reverse faulting: the rock above the fault plane (the "hanging" wall) moves up and over the rock below ("foot" wall).
ridge (midoceanic): a major linear elevated landform of the ocean floor, many hundreds of kilometers in extent. It resembles a mountain range with a central rift valley.
rift: region where the crust has split apart, usually marked by a rift valley (e.g., East African Rift, Rhine Graben).
right-lateral fault: a strike-slip fault on which the displacement of the far block is to the right when viewed from either side.
rigidity: an index of the resistance of an elastic body to shear. The ratio of the shearing stress to the amount of angular rotation it produces in a rock sample.

risk (seismic): The probability of life and property loss from an earthquake hazard within a given time interval and region.
runup height: the elevation of the water level above the immediate tide level when a tsunami runs up onto the coastal land.
sag fault: a narrow geological depression found in strike-slip fault zones. Those that contain water are called sag ponds.
scarp (fault): a cliff or steep slope formed by displacement of the ground surface.
seafloor spreading: the process by which adjacent plates along midoceanic ridges move apart to make room for new seafloor crust. This process may continue at 0.5 to 10 centimeters per year through many geological periods.
seiche: oscillations (standing waves) of the water in a bay or lake.
seismic discontinuity: a surface or thin layer within the Earth across which P-wave and/or S-wave velocities change rapidly.
seismic gap: an area in an earthquake-prone region where there is a below-average release of seismic energy.
seismic moment: See **moment (of earthquakes).**
seismic wave: an elastic wave in the Earth, usually generated by an earthquake source or explosion.
seismicity: the occurrence of earthquakes in space and time.
seismograph: an instrument for recording as a function of time the motions of the Earth's surface that are caused by seismic waves.
seismology: the study of earthquakes, seismic sources, and wave propagation through the Earth.
seismometer: the sensor part of the seismograph, usually a suspended pendulum.
seismoscope: a simple seismograph recording on a plate without time marks.
shadow zone: the area on the Earth's surface protected from seismic wave shaking by some blocking object in the Earth.
slip (fault): the relative motion of one face of a fault relative to the other.
soil amplification: growth in the amplitude of earthquakes when seismic waves pass from rock into less rigid material such as soil.
strain (elastic): the geometrical deformation or change in shape of a body. The change in an angle, length, area, or volume divided by the original value.
stress (elastic): a measure of the forces acting on a body in units of force per unit area.
stress drop: the sudden reduction of stress across the fault plane during rupture.
strike of fault: the line of intersection between the fault plane and the surface of the Earth. Its orientation is expressed as the angle west or east of true north.

strike-slip fault: a fault whose relative displacement is purely horizontal.
strong ground motion: the shaking of the ground near an earthquake source made up of large amplitude seismic waves of various types.
subduction zone: a dipping ocean plate descending into the Earth away from an ocean trench. It is usually the locus of intermediate and deep earthquakes defining the Wadati-Benioff zone.
surface-wave magnitude: magnitude of an earthquake estimated from measurements of the amplitude of surface waves.
surface waves (of earthquakes): seismic waves that follow the Earth's surface only, with a speed less than that of S waves. There are two types of surface waves — Rayleigh waves and Love waves.
swarm (of earthquakes): a series of earthquakes in the same locality, no one earthquake being of outstanding size.
S wave: the secondary seismic wave, traveling more slowly than the P wave and consisting of elastic vibrations transverse to the direction of travel. It cannot propagate in a liquid.
tectonic earthquakes: earthquakes resulting from sudden release of energy stored by major deformation of the Earth.
tectonics: large-scale deformation of the outer part of the Earth resulting from forces in the Earth.
teleseism: an earthquake that occurs at a distant place — usually overseas.
thrust fault: a reverse fault in which the upper rocks above the fault plane move up and over the lower rocks at an angle of 30° or less so that older strata are placed over younger.
tomographic: construction of the image of an internal object or structure from measurements of seismic waves at the surface.
transform fault: a strike-slip fault connecting the ends of an offset in a midoceanic ridge, an island arc, or an arc-ridge chain. Pairs of plates slide past along transform faults.
travel-time curve: a graph of travel time versus distance for the arrival of seismic waves from distant events. Each type of seismic wave has its own curve.
trench: long, narrow arcuate depression in the seabed which results from the bending of the lithospheric plate as it descends into the mantle at a subduction zone.
triple junction: point where three plates meet.
tsunami: a long ocean wave usually caused by seafloor displacement in an earthquake.
viscoelastic material: a material which can behave as an elastic solid on a short-time scale and as a viscous fluid on a long one.
volcano: an opening in the crust that has allowed magma to reach the surface.
volcanic earthquakes: earthquakes associated with volcanic activity.

volcanic rock: igneous rock which involves the eruption of molten rock.
volcanic tremor: the more-or-less continuous vibration of the ground near an active volcano.
volcanism: geological process which involved the eruption of molten rock.
Wadati-Benioff zone: see **Benioff zone.**
wavefront: imaginary surface or line that joins points at which the waves from a source are in phase (e.g., all at a maximum or all at a minimum).
wavelength: the distance between two successive crests or troughs.

Further Reading

Titles preceded by an asterisk are recommended elementary discussions on earthquakes, and most parts are suitable for the general reader and first-year college students.

Adams, W. M., ed. *Tsunamis in the Pacific Ocean.* Honolulu: East-West Center Press, 1970.

Ambraseys, N. N., C. P. Melville, and R. D. Adams. *The Seismicity of Egypt, Arabia and the Red Sea, A Historical Review.* Cambridge, Eng.: Cambridge University Press, 1994.

*Anderson, C. J. "Animals, Earthquakes, and Eruptions," *Field Museum of Natural History Bulletin.* Chicago: vol. 44, no. 5, 1973, pp. 9–11.

*Bolt, B. A. *Earthquakes and Geological Discovery.* Scientific American Library. New York: W. H. Freeman and Company, 1993.

*Bolt, B. A. *Inside the Earth.* New York: W. H. Freeman and Company, 1982. Reprinted TechBooks Virginia, 1992.

*Bolt, B. A. *Nuclear Explosions and Earthquakes: The Parted Veil.* San Francisco: W. H. Freeman and Company, 1976.

*Bolt, B. A., W. L. Horn, G. A. Macdonald, and R. F. Scott, *Geological Hazards.* Berlin: Springer-Verlag, 1975.

Cox, A., ed. *Plate Tectonics and Geomagnetic Reversals.* San Francisco: W. H. Freeman and Company, 1972.

*Davison, C. *The Founders of Seismology.* Cambridge, Eng.: Cambridge University Press, 1927.

Decker, R. W., and B. Decker. *Volcanoes,* 3d ed. New York: W. H. Freeman, 1998.

Fowler, C. M. R. *The Solid Earth–An Introduction to Global Geophysics.* Cambridge, Eng.: Cambridge University Press, 1990.

Freeman, J. R. *Earthquake Damage and Earthquake Insurance.* New York: McGraw-Hill, 1932.

*Gere, J. M., and H. C. Shah. *Terra Non Firma.* New York: W. H. Freeman and Company, 1984.

Gutenberg, B., and C. F. Richter. *Seismicity of the Earth and Associated Phenomena.* Princeton, N.J.: Princeton University Press, 1954.

Hass, J. E., and D. S. Mileti. *Socioeconomic Impact of Earthquake Prediction on Government, Business and Community.* Boulder: Institute of Behavioral Sciences, University of Colorado, 1976.

*Herbert-Gustar, A. L., and P. A. Mott. *John Milne: Father of Modern Seismology.* Tenterden, Eng.: Paul Norburg Pub. Ltd., 1980.
*Iacopi, R. *Earthquake Country.* San Francisco: Lane Book Company, 1998.
Kearey, P., and F. J. Vine. *Global Tectonics.* Oxford: Blackwell Scientific Publications, 1990.
*Keller, E. A., and N. Pinter. *Active Tectonics. Earthquakes, Uplift and Landscape.* New Jersey: Prentice Hall, 1996.
Kulhanek, O. *Anatomy of Seismograms.* Amsterdam: Elsevier, 1990.
Lawson, A. C. *The California Earthquake of April 18, 1906. Report of the State Earthquake Investigation Commission.* Washington, D.C.: Carnegie Institution, 1908.
McCalpin, J. P., ed. *Paleoseismology.* San Diego: Academic Press, 1996.
*McPhee, J. A. *Assembling California.* New York: Farrar, Straus and Giroux, 1993.
Murck, B. W., B. J. Skinner, and S. C. Porter. *Environmental Geology.* New York: John Wiley, 1996.
*Oliver, J. *Shocks and Rocks. Seismology and the Plate Tectonics Revolution.* Washington, D.C.: American Geophysical Union, 1996.
Palm, R. *Earthquake Insurance.* Boulder: Westview Press, 1995.
Panel on Earthquake Prediction of the Committee of Seismology, *Predicting Earthquakes.* Washington, D.C.: National Academy of Sciences, 1976.
Plafker G. and J. P. Galloway, eds. *Lessons Learned from the Loma Prieta, California, Earthquake of October 17, 1989.* U.S.G.S. Circular 1045, 1989.
*Press, F., and R. Siever. *Earth,* 2d ed. New York: W. H. Freeman and Company, 1998.
Reiter, L. *Earthquake Hazard Analysis — Issues and Insights.* New York: Columbia University Press, 1990.
Richter, C. F. *Elementary Seismology.* San Francisco: W. H. Freeman and Company, 1958.
Rikitake, T. *Earthquake Prediction.* Amsterdam: Elsevier, 1976.
Rothé, J. P. *The Seismicity of the Earth, 1953–1965.* Paris: UNESCO, 1969.
Scholz, C. H. *The Mechanics of Earthquakes and Faulting.* Cambridge University Press, 1990.
Steinbrugge, K. V. *Earthquakes, Volcanoes, and Tsunamis.* New York: Scandia America Group, 1982.
Simkin, T., L. Siebert, L. McClelland, D. Bridge, C. Newhall, and J. H. Latter. *Volcanoes of the World.* New York: Academic Press, 1981.
van Rose, S., and R. Musson. *Earthquakes, Our Trembling Planet.* Nottingham: British Geological Survey, 1997.
*Walker, B. *Earthquake.* Alexandria, Va.: Time-Life Books, 1982.
Wallace, R. E., ed. The San Andreas Fault System, California. U.S. Geological Survey, Professional Paper 1515, 1990.

*Yanev, P. *Peace of Mind in Earthquake Country.* San Francisco: Chronicle Books, 1991.

Yeats, R. S., K. Sieh, and C. R. Allen. *The Geology of Earthquakes.* Oxford: Oxford University Press, 1997.

In addition, valuable sources of current information on earthquakes, available to the public and to schools by subscription, include the following:

Earthquakes and Volcanoes (bimonthly). U.S. Geological Survey, U.S. Government Printing Office, Washington, D.C. 20402.

California Geology (monthly). California Division of Mines and Geology, Sacramento, CA 95812.

Earthquake Sounds. A tape cassette containing sounds recorded in various earthquakes is available with catalog (by K. V. Steinbrugge) from the Seismological Society of America, 201 Plaza Professional Building, El Cerrito, CA 94530.

Earthquake Slides. Photographs of earthquake effects, copies of seismograms, and seismicity maps can be obtained from the National Geophysical and Solar Terrestrial Data Center, Code D62, NOAA/EDS, Boulder, CO 80302. An extensive collection of slides illustrating seismic damage is available at the Earthquake Engineering Research Center library, University of California, Berkeley.

The Homeowner's Guide to Earthquake Safety. An instructive simple description of ways to mitigate seismic risk for homes. Sold through the California Seismic Safety Commission, 1755 Creekside Oaks, Sacramento, CA 95833.

A valuable field guide for studying earthquakes, *Learning from Earthquakes,* is available from the Earthquake Engineering Research Institute, 499 14th St., Suite 320, Oakland, CA 94612.

World Wide Web (Internet) Resources

Much interesting information on earthquakes is now available on the Internet. There are home pages that provide access to information about worldwide seismicity, historical earthquakes, hazard reduction, the latest destructive earthquakes, photographs, video clips, and even profiles of seismologists.

The following Web sites are listed for convenience under the chapters of the book. Many of these sites, however, provide links to other fascinating displays at additional sites. The creativity and enthusiasm of people who study earthquakes is high, so log on and explore with the Internet.

Chapter 1 *What We Feel in an Earthquake*
http://www.iris.edu (click on the image to initiate display)

The IRIS Seismic Monitor is an interactive display of global seismicity that allows users to monitor earthquakes in near real-time, view records of ground motion, learn about earthquakes, and visit seismic stations around the world.

Chapter 2 *Where Earthquakes Occur*
Surfing the Internet for Earthquake Data:
http://seismo.ethz.ch/seismosurf/seismobi.html

Edinburgh University World-Wide Earthquake Locator:
http://geovax.ed.ac.uk/quakes/quakes.html

Caltech Seismological Laboratory:
http://www.gps.caltech.edu/seismo/seismo.page.html

The "Seismo Lab" at Caltech features the earthquake "Record of the Day" (and previous "Records of the Day"), data on southern California earthquake mechanisms, current research projects at the laboratory, and links to other seismology-related Web servers.

Chapter 3 *Measuring Earthquakes*
U.S. Geological Survey, Pasadena Office:
http://www.socal.wr.usgs.gov/USGS/

The USGS office in Pasadena, California, is the source of earthquake information for southern California. Data on the "Last Significant Earthquake" in southern California, as well as "Past," "Present," and "Future" seismic activity, are features of this site.

National Earthquake Information Center:
http://www.neic.cr.usgs.gov/

The National Earthquake Information Center, located in Golden, Colorado, is the gateway for worldwide seismic information.

Build Your Own Seismograph:
http://cea-ftp.cea.berkeley.edu/~edsci/lessons/indiv/davis/hs/seismograph.html

With a few materials and some time, you can build you own seismograph.

Chapter 4 *Exploring Inside the Earth*
Welcome to the Planets:
http://pds.jpl.nasa.gov/planets/

This NASA site provides data and images for each of the planets, a glossary, and information on the spacecraft that produced the images.

Earth's Interior:
http://www.seismo.unr.edu/ftp/pub/louie/class/100/interior.html

Here is basic detailed information on a range of topics pertinent to the Earth's interior.

Chapter 5 *Faults in the Earth*
Ask-A-Geologist:
http://walrus.wr.usgs.gov/docs/ask-a-ge.html

Questions addressed to this site will be answered by a geologist with the U.S. Geological Survey. The site also includes a list of Frequently Asked Questions (FAQ).

California Division of Mines and Geology, Seismic Hazards Mapping Program:
http://www.consrv.ca.gov/dmg/shezp/ftlindex.html

This site gives an index of 182 faults in California.

Chapter 6 *The Causes of Earthquakes*
Asteroid and Comet Impact Hazards:
http://ccf.arc.nasa.gov/sst/main.html

This site at the NASA Ames Space Science Division is a gateway to a wealth of information about near-Earth objects, Earth-crossing asteroids, and efforts to identify them. Links to a variety of related sites include the Cretaceous-Tertiary asteroid Event (page 185), Collisions with Earth, and The Collision of Comet Shoemaker-Levy 9 with Jupiter.

Chapter 7 *Earthquakes and Plate Tectonics*
Plate Motion Calculator:
http://manbow.ori.u-tokyo.ac.jp/tamaki-html/plate_motion.html

The Plate Motion Calculator at the Ocean Research Institute of the University of Tokyo models present-day relative motion or absolute motion across any selected plate boundary.

Plate Tectonics:
http://volcano.und.nodak.edu/vwdocs/msh/ov/ovpt.html

The University of North Dakota's Volcano World is the home for this general overview of plate tectonics.

Earth's Interior and Plate Tectonics:
http://bang.lanl.gov/solarsys/earthint.htm

This is a starting point for exploring the above topics. The site has many illustrations and a glossary.

Chapter 8 *The Size of an Earthquake*
U.C. Berkeley Seismological Laboratory:
http://www.seismo.berkeley.edu

Included here is recent earthquake information and information on how to make your own seismogram.

Chapter 9 *Volcanoes, Tsunamis, and Earthquakes*
Electronic Volcano (Dartmouth College):
http://www.dartmouth.edu/~volcano/

This site, for serious students, is encyclopedic in scope and is frequently updated.

Smithsonian Global Volcanism Program:
http://www.volcano.si.edu.gvp/

User-friendly, this site presents a global view and has an excellent list of annotated links.

Scott's Hotlist of Volcanoes and Earthquakes:
http://www.rcch.com/hotlist/0.58htm

Hawaii's Center for Volcanology:
http://www.soest.hawaii.edu/GG/hvc.html

Maintained by the University of Hawaii (Manoa), this site includes images and data for past and current eruptions in the Hawaiian Islands.

Cascades Volcano Observatory:
http://vulcan.wr.usgs.gov/home.html

This USGS laboratory gives information about the Mount Saint Helens eruption and monitoring of the Cascade volcanoes. The site provides links to images and information on how to prepare for a Cascade volcano eruption.

University of North Dakota:
http://volcano.und.nodak.edu/vw.html

This site, at the University of North Dakota, is a "front door" for almost everything available on volcanoes. Features include Today in Volcano History, What's Erupting Now?, Ask a Volcanologist, Volcanic Peaks and Monuments, and Volcanoes of the World.

Japan's Volcano Research Center:
http://hakone.eri.u-tokyo.ac.jp/vrc/VRC.html

Current volcanic activity in Japan is the focus of this site at the University of Tokyo. Reports are current on the major Japanese volcanoes, including Fuji and Unzen.

Tsunamis:
http://www/geophys.washington.edu/tsunami/intro.html

The University of Washington Geophysics Program provides "Tsunami!: The WWW Tsunami Information Resource," a gateway for general tsunami information and research. It gives information on recent tsunamis and computer simulations.

National Geophysical Data Center:
http://www.ngdc.noaa.gov/hazard/hazards.html

Historic tsunami data consisting of about 6600 records from 49 B.C. to present on wave height, damage, travel times, arrival times, and number of deaths, plus earthquake epicenters, magnitudes, and depths. The center also maintains tsunami maps, photograph and slide sets, tsunami catalogs, tide gauge records, and bathymeric data.

Chapter 10 *Events That Precede an Earthquake:*
U.S. Geological Survey, Frequently Asked Questions about Earthquakes:
http://quake.wr.usgs.gov/more/eqfaq.html

Chapter 11 *Dangers from an Earthquake*
Federal Emergency Management Agency:
http://www.fema.gov (home page)

This site includes links to other FEMA publications in earthquake engineering and seismic safety. FEMA publications, telephone (800) 480-2520 or (303) 555-0123.

Chapter 12 *Reducing Earthquake Risk*
U.S. Census Bureau:
http://www.census.gov/

This is the gateway to the Census Bureau's information resources. The Population Clocks provide up-to-the-minute estimates of the U.S. and world populations and data on population growth. Links provide easy access to basic population data for every state and county.

Pacific Earthquake Engineering Center:
http://peer.berkeley.edu/

This is a doorway to an excellent library on earthquake engineering.

U.S. Geological Survey:
http://quake.usgs.gov/hazprep/index.html

This site presents information on earthquake hazards and preparedness.

ABAG (Association for Bay Area Governments):
http://www.abag.ca.gov/bayarea/eqmaps/fixit/fixit.html

Given here are techniques for mitigating earthquake hazards. It shows easy and low-cost ways to make your home safer in earthquakes. The site in the San Francisco Bay area includes damage photos and fix-it diagrams.

Index